BINA

PROCEEDINGS OF A WORKSHOP ON
OPTICAL SURVEYS FOR QUASARS

Sponsored by the National Optical
Astronomy Observatories

Held at the University of Arizona

Tucson

January 6-8, 1988

Organizing Committee

Patrick S. Osmer, Craig B. Foltz, Richard F. Green
Paul C. Hewett, Maarten Schmidt

Editors of the Proceedings

Patrick S. Osmer, Alain C. Porter, Richard F.
Green, Craig B. Foltz

A SERIES OF BOOKS ON RECENT DEVELOPMENTS IN ASTRONOMY AND ASTROPHYSICS

©Copyright 1988 Astronomical Society of the Pacific
390 Ashton Avenue, San Francisco, California 94112

All rights reserved.

Printed by Brigham Young University Print Services, Provo, Utah 84602

First published 1988

Library of Congress Catalog Card Number: 88-71919
ISBN 0-937707-19-8

D. Harold McNamara, Managing Editor of Conference Series

ASTRONOMICAL SOCIETY OF THE PACIFIC
CONFERENCE SERIES

Volume 2

PROCEEDINGS OF A WORKSHOP ON OPTICAL SURVEYS FOR QUASARS

Edited by

Patrick S. Osmer, Alain C. Porter, Richard F. Green, Craig B. Foltz

ASTR
seplae

Sd
8|6|90
JJ

TABLE OF CONTENTS

PREFACE

The last two years have seen great advances in and
discoveries from optical surveys for quasars. In particular,
the efforts to build high speed plate scanning machines have
come to fruition, as has parallel work to develop software to
process and detect automatically quasars in slitless spectra
and multicolor data. Multi-fiber feeds enable simultaneous
spectroscopy of 50 or more objects so that an order of
magnitude increase in the size of feasible surveys,
especially of quasars as faint as 21st magnitude, is now
practical. Digital detectors with high quantum efficiency
are now in use for both surveys and followup spectroscopy.

As a result, significant progress has occurred in the
subjects of quasar evolution, the spatial distribution of
quasars and their association with galaxies, the discovery of
quasars at $z > 4$, and the occurrence of gravitational lensing
in quasars. There have also been important developments in
the related fields of radio and X-ray surveys and the
theoretical implications of the new observations.

Consequently, the organizers realized that it would be timely
to hold a workshop on optical surveys for quasars to discuss
the recent developments, compare the results of different
programs, and consider future directions and problems for the
field. Although the workshop was organized and announced
rather informally and on short notice, nearly every effort or
group working in the field was able to send at least one
representative to the meeting that was held in Tucson from
January 6 to 8, 1988, thus confirming the value of having the
meeting. There were 66 registered participants plus numerous
drop-ins, with the attendance reaching nearly 100 at some
sessions. The format for the workshop consisted of oral and
poster presentations with considerable time for discussion,
which in fact was well taken up with lively interchanges on
all the topics discussed.

This volume presents the oral and poster papers together with
discussion edited from tape recordings of the sessions. The
editors are grateful to the contributors for submitting
articles of such high quality. Many new results are
presented in preliminary form, ahead of publication in
archival journals. Thus, we believe this volume will serve
as a useful summary of the state of the field in early 1988.

It is a great pleasure to acknowledge the support of John
Jefferies, the Director of NOAO who originally conceived of
the initiatives program that led to the funding of the

workshop, and of Sidney Wolff, the present Director of NOAO, who maintained the funding despite the financial pressures on the budget. Pat Cochrane and Peggy Stephens provided crucial logistical and secretarial support without which the meeting would not have occurred nor this volume published. We thank the University of Arizona for making available the meeting room in which the workshop was held.

Patrick S. Osmer

LIST OF PARTICIPANTS REGISTERED FOR
QUASAR WORKSHOP JANUARY 1988

S. Anderson	L. Miller
J. Baldwin	K. Mitchell
E. Borra	P. Osmer
B. Boyle	J. Peacock
L. Bryant	B. Peterson
A. Cavaliere	A. Porter
F. Chaffee	W. Romanishin
R. Clowes	A. Sandage
A. Cowley	N. Sanduleak
D. Crampton	A. Savage
R. Davies	M. Schmidt
S. Demers	D. Schneider
D. Engels	P. Seitzer
C. Foltz	T. Shanks
E. Gosset	P. Shaver
R. Green	M. Smith
R. Griffiths	J. Stocke
D. Groote	P. Strittmatter
F.D.A. Hartwick	J. Surdej
J. Heisler	D. Trevese
P. Hewett	J. P. Swings
A. Hoag	E. Turner
A. Iovino	D. Turnshek
K. Kellerman	F. Vagnetti
T. D. Kinman	P. Veron
R. G. Kron	S. Warren
G. MacAlpine	R. Webster
T. Maccacaro	D. Weedman
B. Marano	R. Weymann
H. L. Marshall	P. Wehinger
P. McCarthy	R. Windhorst
J. McGraw	J. Wright
R. McMahon	G. Zamorani

THE LIGHTER SIDE OF THE WORKSHOP

Selected Quotes from the Tapes of the Discussion, as compiled
by Alain Porter

Maccacaro : What do you mean by QSO?

Cavaliere : We are trying to collect pieces of ignorance.

Turner : It's important to understand what a dull tool the
covariance function is.

Weymann : I was chagrined, disturbed, and puzzled by the
remark that Bruce Peterson made.

Kellerman : I'm depressed.

Schmidt : You need a Richard Green with an objective prism.

Marshall : Maybe I should finish the controversy.
Schmidt : Why don't you start a new one?

Schmidt : What is known about the Veron sample?
Boyle : Absolutely nothing.

Peacock : We're going to have to put our money where our
mouth is.

Marshall : I don't remember being worried by anything.
Hewett : Perhaps we can solve that at coffee.

Kron : It's always very pleasing when someone takes your work
so seriously that they bother to prove you're wrong.

Weedman : (attributed to Smith) How did Kodak know where the
quasar cutoff occurred?

Stocke : I realize that (BL Lacs) seem boring to some of you
who love contrasty spectra, but one BL Lac object is now
worth 100 QSOs.

Boyle : I hate having this terrible weight on my shoulders.

MacAlpine : Ed, if you were given unlimited access to any
facilities you wanted, and 10 postdocs, what would you do
first?
Turner : Go on vacation.

Green : I'm drawing a poor response!

Marshall : I agree with your statement. It's a motherhood type of statement.

Schmidt : Ah, look here. This makes no sense.

Schmidt : It's a very simple calculation.
Shanks : That's why I got Professor Rees to work it out!

Marshall : I'm sweating.
Schmidt : Good!

Marshall : Okay. I'll be quiet.

Weedman : Does it take 100 hours to analyze a UK Schmidt plate?
Savage : No, it takes 100 hours to TAKE the bloody thing!

INTRODUCTION

The organization of the articles in this volume follows that of the workshop. Within each part the oral papers are presented in the order they were given, followed by the poster papers in alphabetical order.

Part 1 covers the evolution of the luminosity function for z < 2.2, in particular the topic of luminosity evolution. As this work is based primarily on surveys done with the ultraviolet excess technique, all papers using the technique will be found here. In addition, the section contains papers on selection effects and on present and future deep surveys for faint quasars.

Part 2 covers the evolution of the luminosity function for z > 2.2 and the search for the highest redshift quasars. It contains the papers using the slitless spectrum and multicolor techniques. The articles discuss the methods and their efficiency. The recent results on quasars at z > 4 and their implications for evolution are in this section.

Part 3 covers the topics of gravitational lenses and of radio surveys and radio properties of quasars. The possibility that undetected microlensing of quasars can distort the interpretation of data from magnitude limited samples and new observational results on searches for lenses are included in this section. Radio surveys have completely different selection properties from optical surveys, and the results of major efforts on radio quasars and galaxies are presented.

Part 4 covers the relation of quasars to galaxies and galaxy clustering and the evidence for quasar clustering at large and small scales. For z < 0.6 it is possible to observe directly how the association of quasars and galaxies evolves, and recent results are reviewed. Evidence for quasar clustering at small scales has been accumulating; this topic is extensively discussed and debated here.

Part 5 covers the results of some of the X-ray surveys for quasars and BL Lac objects. In addition there are theoretical papers on the implications of quasar evolution for their physical nature and on the possibility that dust in the intergalactic medium could account for the scarcity of high redshift quasars. There is also a presentation of luminosity dependent effects in quasar spectra.

Part 6 is a result of an effort made on the occasion of the workshop to collect information on the different optical surveys for quasars. It presents a compilation of recently completed and ongoing surveys from material supplied by the researchers themselves. It is intended to serve as a guide to all workers in the field on what is being done.

THE EVOLUTION OF OPTICALLY-SELECTED QSOs

B.J. BOYLE
University of Edinburgh, Blackford Hill, Edinburgh, EH9 3HJ, U.K.
and
Anglo-Australian Observatory, PO Box 296, Epping, NSW 2121, Australia.

T. SHANKS
University of Durham, South Road, Durham DH1 3LE, U.K.

B.A. PETERSON
Mount Stromlo and Siding Spring Observatories,
Woden, ACT 2606, Australia.

ABSTRACT The recently completed AAT QSO survey is used to obtain a new determination of the QSO luminosity function and its evolution. The survey contains spectroscopic identifications for 420 faint ($B < 20.9$ mag) ultra-violet excess (UVX) selected QSOs. Using maximum likelihood techniques we find that the evolution of bright ($M_B < -23$) low redshift ($z < 2.2$) QSOs is well represented by *luminosity* evolution models of the form $L(z) = L_0(1 + z)^{3.2 \pm 0.1}$. There is marginal evidence for additional *density* evolution amongst the faintest ($M_B > -23$) QSOs in the survey, but selection effects in the original catalogues may be responsible for this effect.

INTRODUCTION

The need for a large sample of faint ($B > 19.5$ mag) QSOs to discriminate between different forms of QSO evolution has long been recognised (see e.g. Koo 1986). With the introduction of fibre-optic systems such as FOCAP at the Anglo-Australian Telescope (AAT), such samples can now be obtained without requiring prohibitively large amounts of telescope time. We have used the FOCAP system to obtain spectra for over 1400 QSO candidates selected by the UVX method from machine measurements of UK Schmidt photographic plates. We present here a brief description of the results obtained from the completed survey, which will be published in full in the Monthly Notices of the Royal Astronomical Society, (Boyle *et al.* 1988).

THE AAT SURVEY

COSMOS machine measurements (MacGillivray and Stobie 1985) of UK Schmidt telescope U and J plates were used to obtain positions and magnitudes for stellar images in
7 ($5° \times 5°$) high Galactic latitude fields. Stellar magnitudes were calibrated using deep
($B < 21$ mag, $U < 20.7$ mag) photoelectric, electronographic or CCD sequences. Faint
($B < 20.9$ mag) UVX ($U - B < -0.35$) stellar objects in randomly chosen circular (0.35
deg^2) areas on each Schmidt field were then selected for spectroscopic observation with
the FOCAP fibre-optic system at the AAT (Gray 1984).

In total, 34 FOCAP fields have been observed and we have obtained spectra for
more than 1400 UVX objects, of which 420 have been identified as QSOs. In addition,
56 narrow emission line galaxies have been found. The remaining objects are Galactic
stars, including 41 white dwarfs. Integration times of 9000 seconds per FOCAP field
were sufficient to classify 90–95% of the objects in the survey and to obtain unambiguous
redshifts for $\sim$ 90% of the QSOs. A detailed analysis of the selection effects present in
the survey (Boyle *et al.* 1987) reveals that, for $z < 2.2$, the survey is unlikely to be more
than 10-15% incomplete with only a slight dependence on magnitude or redshift. The
weak redshift dependence is well illustrated by the smooth number-redshift relation for
the AAT survey plotted in figure 1, with no apparent 'holes' or 'peaks' due to selection
effects, particularly in the redshift region $0.6 < z < 0.9$.

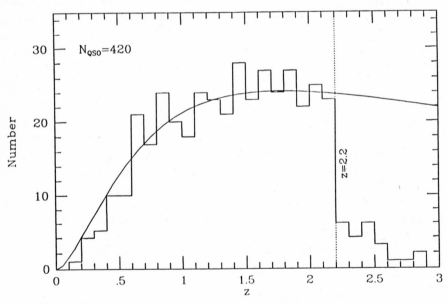

Fig.1 The number-redshift diagram for QSOs identified in the AAT survey. The
solid line indicates the prediction of the luminosity evolution model.

THE N(m) RELATION

The QSO number-magnitude, n(m), relation derived from the AAT survey is presented in figure 2, together with QSO surface densities obtained from similar spectroscopic surveys. All surface densities have been corrected to exclude $z > 2.2$ QSOs. From figure 2 we can clearly see that the QSO n(m) relation exhibits a 'break' from a steep (dlogn/dB = 0.86) power-law slope at bright magnitudes to a flatter slope (dlogn/dB = 0.31) beyond $B \simeq 19.5$ mag, with the surface density of $B < 20.9$ mag QSOs in the AAT survey (37 ± 3 deg^{-2}) being considerably lower than that predicted by an extrapolation of the 0.86 power law slope (160 deg^{-2}). At faint magnitudes the AAT counts also agree well with those derived by other authors: Crampton *et al.* (1987) obtain a QSO surface density at $B < 20.5$ mag of 27 deg^{-2} (cf. 30 ± 3 deg^{-2} obtained here) and Marano *et al.* (1987) find 31 QSOs deg^{-2} at $B < 20.84$ mag (cf. 34 ± 3 deg^{-2} from the AAT sample). The 'break' observed in the n(m) relation is in marked constrast to the pure power law slope predicted by density evolution models, This 'break' is caused by a similar feature in the QSO LF which, as we shall now see, will help us to determine the form of the QSO evolution.

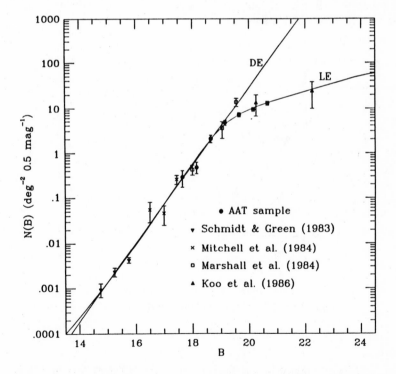

Fig.2   The number-magnitude relation for QSOs with $z < 2.2$ identified from UVX surveys with complete spectroscopic identification. DE: Density evolution prediction, LE: Luminosity evolution prediction.

THE QSO LUMINOSITY FUNCTION

We construct first a binned estimate of the QSO LF based on the V_a statistic (see Schmidt and Green 1983). To increase our coverage of the QSO LF at bright magnitudes, QSOs from the spectroscopic surveys of Schmidt and Green (1983), Mitchell *et al.* (1984), Marshall *et al.* (1984) and Crampton *et al.* (1987) have been included in this and the maximum likelihood analysis of the LF. The apparent magnitude, M_B, of each QSO was calculated following the procedure of Schmidt and Green (1983), assuming a power law spectrum with spectral index $\nu = 0.5$. Estimates for the QSO LF obtained in this manner are presented in figure 3 for both an open ($q_0 = 0$) and closed ($q_0 = 0.5$) universe. The QSO LF was calculated at 1 magnitude intervals in four separate redshift bins over the range $0.3 < z < 2.2$. The $z > 0.3$ limit was applied to the data in order to avoid low redshifts where incompleteness in the selection of QSOs may occur as a result of their 'extended' nature on the original photographic plates (see Schmidt and Green 1983). Errors on the estimates in each bin are equal to $[\sum(1/V_a)^2]^{1/2}$, see Marshall (1985). In both cases we can see that the LF exhibits a 'break' from a steep power law slope at bright magnitudes to a much flatter slope at faint magnitudes. The redshift dependence of this feature appears to be predominantly in the direction of increasing luminosity toward higher redshift.

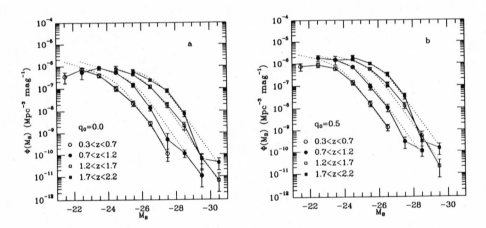

Fig.3   The QSO luminosity function derived for a) $q_0 = 0.0$ and b) $q_0 = 0.5$ universes. Dotted lines represent luminsoity evolution fits to bright ($M_B < -23$) QSOs.

A more quantitative decription of the QSO LF may now be obtained through the use of likelihood techniques (Marshall *et al.* 1983) to generate 'best-fit' models for a variety of parametric forms chosen to represent the LF and its redshift dependence. These models can then be tested for goodness-of-fit using 2D Kolmogorov-Smirnoff (KS) statistics (Peacock 1983). Guided by the estimate of the QSO LF from the $1/V_a$

analysis, we choose to represent the QSO LF as a smooth two power law function thus:

$$\Phi(M_B, z) = \frac{\Phi^*}{[10^{0.4(M_B - M_B(z))(\alpha+1)} + 10^{0.4(M_B - M_B(z))(\beta+1)}]}$$

where the redshift dependence, $M_B(z)$, is expressed as a $(1 + z)$ power law luminosity evolution (PLE), corresponding to:

$$M_B(z) = M_B^* - 2.5k_L log(1 + z)$$

It will also be interesting to establish how much, if any, additional *density* evolution is required to fit the data. We use a luminosity and density evolution (LDE) model (see Koo 1986) whose parametric form may be expressed thus:

$$\Phi(M_B, z) = \frac{\Phi^* \rho_D(z)}{[10^{0.4(M_B - M_B(z))(\alpha+1)} + 10^{0.4(M_B - M_B(z))(\beta+1)}]}$$

where the density evolution function, $\rho_D(z)$, is expressed as an exponential in look-back time, τ:

$$\rho_D(z) = exp(k_D \tau)$$

The 'best-fit' values for α, β, M_B^*, k_L and, where applicable, k_D were obtained by minimising the likelihood function, S, given by Marshall *et al.* (1983). Errors quoted on the 'best-fit' parameters were obtained from ΔS (Lampton *et al.* 1976) and correspond to the 68% confidence level. In table 1 we present the 'best-fit' values for a number of models applied to the data, together with the 2D KS probability, P_z, for each model.

TABLE 1   'Best-fit' parameters for evolution models.

Model	q_0	$M_{B_{max}}$	z_{min}	N_{QSO}	α	β	M_B^*	k_L	k_D	P_z
A (PLE)	0.0	−23	0.3	581	−3.84	−1.61	−23.10	3.34		0.42
B (PLE)	0.5	−23	0.3	567	−3.79	−1.44	−22.42	3.15		0.55
C (PLE)	0.0	−21	0.3	610	−3.87	−1.50	−23.07	3.36		0.19
D (PLE)	0.5	−21	0.3	610	−3.76	−1.21	−22.26	3.20		0.01
E (PLE)	0.5	−23	0.0	624	−3.66	−1.36	−22.03	3.38		0.06
F (LDE)	0.0	−23	0.3	581	−3.84	−1.61	−23.10	3.34	0.00	0.42
G (LDE)	0.5	−23	0.3	567	−3.84	−1.56	−22.58	3.07	0.18	0.61
H (LDE)	0.5	−21	0.3	610	−3.78	−1.34	−22.74	2.77	1.71	0.26
68% confidence regions:					±0.15	±0.20	±0.25	±0.10	±0.10	

We find that, at the 5% level, the power law luminosity evolution models provide a perfectly acceptable fit to evolution of bright $(M_B < -23)$ QSOs for both $q_0 = 0$ (model A) and $q_0 = 0.5$ (model B). The predicted number-magnitude and number-redshift relations for this model in the $q_0 = 0.5$ universe are illustrated in figures 1 and 2. Very little additional density evolution is required to obtain an equally satisfactory fit to the data at these magnitudes (models F and G). Indeed, for $q_0 = 0$, the maximum likelihood value for k_D in the luminosity and density evolution model is $k_D = 0$, i.e. simply luminosity evolution. The 3σ upper limit for k_D in the $q_0 = 0.5$ case is $k_D = 0.5$, equivalent to $k_L = 1$ (see Marshall et al. 1983). This may be compared with $k_L = 6$ obtained in pure exponential luminosity evolution models (Boyle et al. 1988). Thus, over the redshift range $0.0 < z < 2.2$, any evolution in density for bright ($M_B < -23$) QSOs is more than 50 times slower than the corresponding evolution in luminosity.

Although the inclusion of low luminosity ($-21 < M_B < -23$) QSOs does not alter the acceptability of the luminosity evolution model in the $q_0 = 0$ universe (model C) it does reduce the acceptability of the same model (model D) in the $q_0 = 0.5$ universe to the 1% level. This poor fit can be seen from figure 3, where the predicted LFs are plotted as dotted lines. The discrepancy between the observed and predicted LFs at low luminosities and low redshifts is immediately apparent. As can be seen from table 1, in this case (model H), additional density evolution is required to fit the data. However, we note that incompleteness in the original data-sets may be responsible for the observed effect. In particular, the loss of low luminosity QSOs from UVX-selected catalogues through reddening by the increased contribution of the host galaxy (see Marshall 1985) could prove a major factor. It is clear that further observations of faint QSOs will be required before any density evolution is confirmed at these magnitudes. Finally, the inclusion of low redshift ($z < 0.3$) QSOs does not reduce the acceptability of the luminosity evolution modles to below 5%, even for $q_0 = 0.5$ (model E).

DISCUSSION

There are two principle different physical interpretations of the 'luminosity evolution' observed above. We are either witnessing

    (a) the evolution of individual long-lived ($\sim 10^{10}$ yr) QSOs or

    (b) the evolution in the statistical properties of an ensemble of short-lived ($\sim 10^8$ yr) objects.

As is well known, alternative a) predicts that massive black holes ($10^9 - 10^{10} M_\odot$) will reside in the centres of QSOs and Seyferts at low redshift, whereas b) implies that less massive black holes ($\sim 10^8 M_\odot$) must exist in a much greater fraction of current 'inactive' galaxies at the present day. At present there is, however, no compelling observational evidence for favouring one alternative over the other, although we will now discuss the results presented here in the light of these two alternatives. Firstly, the limits placed on the amount of density evolution permissable for bright ($M_B < -23$) QSOs will more seriously constrain short-lived models for QSO evolution at these magnitudes, although theoretical models for both the short- and the long-lived interpretations will also be constrained by the narrow range for k_L allowed by the models.

In contrast, the possible departure from luminosity evolution revealed by the faint ($-21 < M_B < -23$) QSOs, at least for a $q_0 = 0.5$ Universe would appear to argue against a long-lifetime for these QSOs. We note, however, that this density evolution goes the 'wrong' way for most theoretical models (see e.g Cavaliere et al. 1985) which predict an *increase* in the numbers of faint QSOs at low redshifts. By comparing the extrapolated $z = 0$ QSO LF to estimates of the Seyfert LF (see figure 4) we can also learn much about the evolutionary status of QSOs. The observed agreement at low luminosities in both the slope and normalisation of the QSO LF with the Seyfert LF provides a strong indication that luminosity evolution continues on smoothly to $z = 0$; an observation which would favour more strongly the long-lived hypothesis. This observation is reinforced by the success of the extrapolation of the QSO LF to lower luminosities (dotted line in figure 4) in predicting the surface densities of QSOs at $B > 21$ mag. However, an accurate estimate of the QSO LF at these magnitudes will require futher spectroscopic surveys at $B < 22.5$ mag.

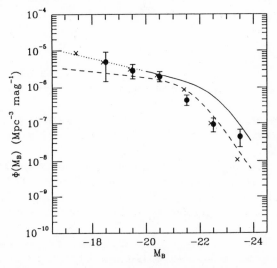

Fig.4 Comparison of the Seyfert and QSO luminosity functions. The solid line denotes the extrapolated $z = 0$ QSO LF for $q_0 = 0.5$. The filled circles represent the LF derived by Cheng *et al.* (1985) for Seyfert 1 and 1.5 nuclei. Crosses represent the composite Seyfert LF derived by Weedman (1986) and the dashed line represents the parametric form for the Seyfert LF obtained by Marshall (1986).

CONCLUSIONS

A large data-set of spectroscopically confirmed QSOs, including the recently completed AAT sample has been used to establish that the evolution of bright ($M_B < -23$) QSOs at low ($z < 2.2$) redshift is described by a simple power law increase in luminosity with redshift: $L \propto (1 + z)^{3.2 \pm 0.1}$. Any additional density evolution required amounts to less than 2% of the equivalent evolution in luminosity over this redshift range. However, at fainter magnitudes some density evolution may be needed to fit the data, although this observation is complicated by many selection effects. Whilst the choice between long- and short-lived interpretations of the observed luminosity evolution is still not resolved, future observations, particulary those concerned with better estimating the faint end of the QSO LF, will prove useful in this respect.

Acknowledgements

We are greatly indebted to the staff of the AAO for making the FOCAP facility available, without which this project would not have been possible. We would also like to thank the staff of the UK Schmidt Unit and the COSMOS measuring machine at the Royal Observatory, Edinburgh for the provision and measurement of the plate material involved in this project. We also acknowledge useful discussions with Herman Marshall and John Peacock. Finally, BJB acknowledges the support of an SERC research fellowship during the course of this work and TS thanks the Royal Society for the receipt of a fellowship.

References

Boyle, B.J., Fong, R., Shanks, T. and Peterson, B.A. 1987, Mon. Not R. astr. Soc., 227, 717.

Cavaliere, A., Giallongo, E. and Vagnetti, F. 1985, Astrophys. J., 296, 402.

Cheng, F.Z., Danese, J., De Zotti, G. and Franceschini, A. 1985, Mon. Not. R. astr. Soc., 212, 857.

Crampton, D., Cowley, A.P. and Hartwick, F.D.A. 1987, Astrophys. J., 314, 129.

Gray, P.M. 1984, In *Instrumentation in Astronomy* SPIE proc., 455, 57.

Koo, D.C. 1986, In *The Structure and Evolution of Active Galactic Nuclei*, p317, eds Giurcin,G.,Mardirossian,F.,Mezzetti,M. and Ramella,M., Reidel, Dordrecht, Holland.

Koo, D.C., Kron, R.G. and Cudworth, K.M. 1986, Publs astr. Soc. Pacif., 98, 397

Lampton, M., Margon,B. and Bowyer, S. 1976, Astrophys. J. 208, 177.

MacGillivray, H.T. and Stobie, R.S. 1985, Vistas in Astronomy, 27, 433.

Marano, B., Zamorani,G. and Zitelli,V. 1987, Mon. Not. R. astr. Soc., submitted.

Marshall, H.L. 1985, Astrophys. J., 299, 109.

Marshall,H.L. 1986, In *The Structure and Evolution of Active Galactic Nuclei*, p627, eds Giurcin,G.,Mardirossian,F.,Mezzetti,M. and Ramella,M., Reidel, Dordrecht, Holland.

Marshall, H.L., Avni, Y., Braccesi, A., Huchra, J.P., Tanenbaum, H., Zamorani, G. and Zitelli, V. 1984, Astrophys. J., 283, 50.

Marshall, H.L., Tanenbaum, H., Huchra, J.P., Braccesi, A. and Zitelli, V. 1983, Astrophys. J., 269, 42.

Mitchell, K.J., Warnock,A. and Usher,P.D. 1984, Astrophys. J. 287, L3.

Peacock, J.A. 1983, Mon. Not. R. astr. Soc, 202, 615.

Schmidt, M. and Green, R.F. 1983, Astrophys. J., 269, 352.

Weedman, D.W. 1986, In *The Structure and Evolution of Active Galactic Nuclei*, p215, eds Giurcin,G.,Mardirossian,F.,Mezzetti,M. and Ramella,M., Reidel, Dordrecht, Holland.

DISCUSSION

Surdej : How can you discriminate between luminosity evolution and a conspiracy of gravitational lensing to get the same effect, namely, more luminous quasars at higher redshifts?

Boyle : I haven't investigated this personally, but I take it from people like John Peacock that gravitational lensing will not amount to anything like the amount of evolution required in this case.

Peacock : It's very hard to see how the selection effects could conspire to move the thing a lot. The selection effects for just about any lensing situation you can imagine are much stronger at higher luminosities.

Turner : I will talk about this tomorrow in some detail. I think it possible that at least some of the effects might be due to undetected gravitational lensing.

Clowes : Have you made any allowance for the K correction?

Boyle : The only allowance I made is to assume a power law spectral index of 0.5. I have not allowed for the effect of emission lines, either. At present I am having a look at their effect plus the Baldwin effect, and my initial impressions are that it doesn't make an awful lot of difference.

Warren : The Baldwin effect could be important where you have a lack of faint objects at low redshift and where the UV excess is reduced because of emission lines in the B band. Also, could you say what you did about the data at faint magnitudes?

Boyle : Well, I didn't include the Koo and Kron data because I wasn't sure about its completeness or selection effects. I am aware that the Baldwin effect could scatter faint objects out of the sample. All I can say in defense is that the median U-B for the sample is -0.7, while our limit for selection is -0.35 or even -0.3 in later samples, when more fibers were available.

Warren : But, don't the reddest U-B's occur at just where you've got a low density of objects?

Boyle : Yes, around 0.3 to 0.7 in redshift. I see what you're driving at, but I hesitate to say if it is due to the Baldwin effect.

Peterson : I'll comment that there seems to be no color selection or increase in objects that are missed when you divide the sample into the apparently bright and apparently faint objects.

EVOLUTION OF LOW LUMINOSITY QUASARS

HERMAN L. MARSHALL
Space Sciences Laboratory, University of California, Berkeley,
CA 94720

Abstract   New data samples from faint quasar surveys are added to data from brighter surveys. As shown in previous papers by other authors, there is a significant feature in the optical luminosity function. Here it is shown that this break is comparable to a feature in the luminosity function derived for Seyfert galaxies. The quasar luminosity function is determined in several redshift intervals and is tested against several evolution models. A pure luminosity evolution model is derived that fits all the data adequately.

Background

As background for this talk, I would first like to start with some background material concerning the evolution of quasars and low luminosity active galactic nuclei. In 1985 (Marshall 1985a), I pointed out that the luminosity function of quasars with $z < 2.2$, $M_B < -23$, and $B < 20$ appeared to be well fit by a power law functional form independent of redshift, and that the function giving the evolution of that luminosity function was well fit with a power law in $(1+z)$. Furthermore, the model of Schmidt and Green (1983, SG) could be fit to the data only by using different values of the parameters than those they suggested. The basic reason for our differences was that SG chose to attempt to fit one model to all the available data in high redshift samples as well as data from faint ($B > 20$) samples. While it is well known that the pure luminosity evolution model suggested by my results could not be extrapolated to faint magnitudes or high redshifts without predicting far too many objects (cf. Marshall 1985a, Koo and Kron 1988), it was certainly allowed that modifications to the model would potentially alleviate these difficulties (since the data being considered are outside the original fit region).

I suggested one such modification in 1985 (Marshall 1985b,

1987). Examining a complete sample of Seyfert galaxies from the CfA redshift survey (Huchra and Burg 1988), I found that the luminosity function of quasars could be extended to low luminosities (comparable to those of Seyfert galaxies) by continuing the luminosity function with a different power law slope. Using a maximum likelihood method similar to that described by Marshall *et al.* (1983), I determined the values of the break luminosity and the slope below the break. The extended luminosity function also fit the (albeit sparse) data for quasars that was available at the time. I now show that the newly published data for quasars at low luminosities and modest redshifts can be fit by a similar extension of the basic pure luminosity evolution model and that the values of the fitted parameters are nearly identical.

Data

The new data are taken from Boyle (1986) and Koo and Kron (1988). The first sample is described in Boyle *et al.* (1987; see also Boyle, these proceedings) and is completely identified to B = 20.9 while the second is completely identified for B < 21.1 and partially identified to B = 22.6 (assuming B = J + 0.1, as suggested by Koo and Kron). The data from Boyle consist of about 150 quasars and Koo and Kron's sample contributes another 28 objects. Koo and Kron state that quasar candidates were chosen for spectroscopy on a random basis, not related to apparent magnitude for instance, so I will assume that the area of their survey is reduced to 0.156 sq. deg. (32 of 59 sources are spectroscopically identified at "quality" 3 or greater). The BF sample of quasars (Marshall *et al.* 1984) are added to those of Boyle *et al.* and Koo and Kron to bolster the total number of quasars with B < 20, in case the break luminosity is in the region of interest at any redshift. Figure 1 shows the Hubble diagram for the 3 samples used here.

For the following analysis, only data with 0.4 < z < 2.2 and L > 0.2 are considered. As in prior papers, luminosities and luminosity parameters are given at 2500 Å in the quasar rest frame in units of 10^{30} erg^{-1} s^{-1} Hz^{-1} for H_0 = 50 km s^{-1} Mpc^{-1}. The luminosity limit corresponds to M_B = -21.8, higher than the value sometimes assumed to distinguish quasars from Seyfert galaxies. I use a fainter limit because the Boyle *et al.* sample does not appear to show incompleteness until this luminosity. This is not an objective criterion because the data have never been taken that can dispute this claim. This is not true in the case of the Bright Quasar Sample (SG). The CfA redshift survey provides a test of its incompleteness due to the rejection of non-stellar images (Marshall 1985a), indicating that a limit at M_B = -23 was appropriate. Similarly, no data exist to test the completeness of the Koo and Kron sample. Although they only find narrow line objects for $z < 0.9$, I shall assume that this sample is also complete in this region. Objects identified as having narrow lines are excluded from the analyses, pending a better understanding of their relationship to the quasars under investigation.

Fit Results

A pure luminosity evolution model is assumed in what follows; luminosity functions take the form

$$
\phi(L) = \begin{cases} \dfrac{\phi_0}{L^*}\left(\dfrac{L}{L^*}\right)^{-\beta} & L < L^* \\[3ex] \dfrac{\phi_0}{L^*}\left(\dfrac{L}{L^*}\right)^{-\alpha} & L \geq L^* \end{cases} \tag{1a}
$$

where

$$
L^* = L_Q^*(1+z)^k. \tag{1b}
$$

Eq. 1b incorporates all evolution effects. There are five parameters of the model: ϕ_0, α, β, L_Q^*, and k. Several of these parameters are taken or calculated from the model fits of Marshall (1985a, model G) to higher luminosity data, for which evolution is much more apparent, so that parameters α and k are "known" and $\phi_0 = \rho_0(L_Q^*)^{-\beta+1}$ (Marshall 1987). The fit method is taken from Marshall et al. (1983) and is similar (except for the assumption of a cosmological model) to that of Marshall (1985b, 1987). This notation differs somewhat from that of Marshall et al. (1983) so that the notion of a "luminosity at zero redshift" need not be introduced.

The likelihood fit for model G (where $\alpha = 3.6$, k = 3.2, and $\rho_0 = 25$ Gpc^{-1} for $q_0 = 0.5$) gives $L_Q^* = 0.23 \pm 0.035$ and $\beta = 1.35 \pm 0.30$. The uncertainties are highly correlated (see Figure 2). The model passes the Kolmogorov-Smirnov (K-S) tests defined by Marshall et al. (1983) (Figure 3). These results are not yet directly comparable to those of Marshall (1987) for Seyfert galaxies because a smooth luminosity function was assumed there. These values are consistent with the results of Marshall (1987) at the 1σ level, indicating that the luminosity function of Seyfert galaxies is quite similar to that of quasars, once evolution effects are accounted using a pure luminosity evolution model. By use of the likelihood ratio test, I find that $\beta = 2$ is rejected but $\beta = 1$ is not. These facts will be discussed below.

I have also tested the fit of the model proposed by Koo and Kron (1988). The primary difference between their model and that of pure luminosity evolution is in the shape of the high luminosity end of the luminosity function and the dependence of the normalization of the low end on redshift. In their model, the normalization of the low end does not depend on cosmic time whereas there is a slight dependence on cosmic time in the pure luminosity evolution fitted here. The K-S tests show that the model of Koo and Kron can be rejected by the data with better than 99.99% confidence. In order to investigate this result further, I now examine the dependence of the quasar luminosity function on redshift.

The Luminosity Function in Intervals of Redshift

I now add the high luminosity quasar data from Schmidt and Green (1983), Mitchell, Warnock, and Usher (1984) and the brighter sample from Marshall *et al.* (1984) so that the shape of the luminosity function can be readily compared in different redshift intervals. For the lack of any better, more physical reasoning, I chose intervals of equal temporal length in the assumed cosmology ($q_0 = 0.5$, Friedmann universe). The intervals must be chosen in such a way that they are small enough that evolution is not detectable over any given interval (as defined by the K-S test proposed by Avni and Bahcall 1980) but large enough that there are sufficient numbers of quasars in each interval (of order 20). Intervals chosen to be of equal redshift length (Marshall 1988) were found to show evolution effects and were not used here. I have divided up the interval from $z = 0.4$ to $z = 2.2$ into 6 intervals, giving $\Delta t = 0.072 t_0$, where t_0 is the age of the universe in the cosmology chosen. The luminosity function is calculated by summing $1/V_a$, where V_a is defined by Avni and Bahcall as the "available" volume in which a quasar is detectable. The K-S tests show that the chosen intervals are acceptably small.

Figure 4 shows the cumulative luminosity functions as well as the predictions of the extension of the pure luminosity evolution (PLE) model fitted here and the model of Koo and Kron. The fit of the models to the data can be tested using the second of the K-S tests defined by Marshall *et al.* (1983), the one based on C(L|z). For a relatively narrow redshift interval, this test reduces to a test of the luminosity function alone. The results are that *the PLE model fits the observed quasar data in every interval.* There was no adjustment of the normalization in each redshift interval to get the best fit. The Koo and Kron model, however, fits poorly in all redshift intervals examined here. Not only is the shape at high luminosity incorrect (a power law is a better description) but the low end is shown to have a problem in the sense that their model overpredicts the number s of quasars. One may argue that one may be misled by the use of cumulative luminosity functions but I point out that, because the distributions are rather steep, the high luminosity data actually have very little effect on the fainter data.

<u>Discussion</u>

There are many implications of these results. As mentioned earlier, the slope of the faint end is found to be between 1 and 2. However, β = 1 is not ruled out at the 90% confidence level. Had this been the case, then zeroth moment of the distribution, the total number of quasars at any given redshift, would diverge if the functional form is extended to zero luminosity. Thus, on physical grounds, another break in the distribution would be required below the currently observed range of data. This would imply that there could be large numbers of extremely low luminosity active nuclei in galaxies that could be found by large redshift surveys of galaxies such as that by Koo and Kron and Filipenko and Sargent (1985). It is important to know if further data can be obtained to reject β = 1. On the other hand, β > 2 is rejected, so the first moment of the distribution converges near the break luminosity. This means that one may calculate the background light due to quasars or the ambient energy density due to quasar light at any given redshift less than 2.2 (cf. Donahue and Shull 1987).

I have also examined the behavior of the PLE model at high redshifts. By combining the sample of Crampton, Cowley and Hartwick (1987, CCH) with that of Koo and Kron, I find that the PLE model predicts that 16.1 quasars would be found in their surveys in the interval 2.2 < z < 2.4, whereas 5 are actually found. This observation is inconsistent with model prediction at the 99.95% confidence level. I point out that such a discrepancy is not found between the model prediction and numbers of objects found in the combined UV excess surveys (as used for Fig. 4) if the interval 2.0 < z < 2.2 is considered. There are (at least) two possible explanations. The first is that the evolution of quasars suddenly changes character between z = 2.1 and z = 2.3, or that (one or both of) the two quasar surveys are more than 50% incomplete. I prefer that latter explanation because it seems unlikely that physical changes in quasar evolution are occurring exactly at the point where the UV excess method breaks down. I have also found a similar effect if I examine the interval 2.2 < z < 2.7; 11 quasars are observed in this interval but 39 are predicted. Interestingly, however, the observed z and L|z distributions are well fit by the PLE model (after renormalizing).

There are several issues that may affect these results adversely. One is that a full 5-parameter fit was not performed on the complete data set. I have chosen to fit two regions of the z-L plane independently and fit two parameters using data in one region (B < 20)and the other three in the other region (B > 20). This is probably not a major problem because of the extremely good fits that are obtained. Furthermore, because of the way that Eq. 1b follows the L(z) line at B = 20, these two regions are fairly independent (in the context of the model). Another effect is incompleteness in the UV excess surveys. There is very little effect if the total numbers are affected (e.g., the slope of the bright end of the luminosity function changes if the SG sample is incomplete) but variations of completeness with redshift affect the form of the evolution function. Incompleteness at very low luminosities ($L < 0.2$ or $M_B > -21.8$) is hard to quantify and certainly affects the low end of the luminosity function. Similarly, some of the narrow line objects may be actually Seyfert 1.8-2.0, meaning that they may be related in some way to the quasars we hope to understand. On the other hand, it is possible that they constitute a separate population whose evolution can be measured.

Desired Future Surveys

I now suggest several surveys that should significantly improve our understanding of quasar evolution and luminosity functions. The first and most important is one that shall be debated over the course of this conference. We need a survey for high redshift quasars that is well-defined, complete (within its definition) and fully identified spectroscopically. Because the current models rely on broad-band magnitudes for luminosity estimation, broad-band magnitude limits are practically necessary. Surveys limited by line equivalent widths are very difficult to incorporate in the current models in a quantitative, rigorous fashion. Surveys limited by line flux may be useful, if the relation between line and broad-band luminosities can be well determined. The need for such samples is obvious: one may probe the epoch of quasar formation and the early behavior of the quasar luminosity function.

A complete, spectroscopic survey of high redshift galaxies would be invaluable for examining the question of the total number of quasars and how this number depends on cosmic time. As pointed out earlier, the total numbers of faint quasars at modest redshifts does not appear to be converging yet, so one must examine large numbers of galaxies spectroscopically to find the faintest active nuclei. Fortunately, the contrast between the nucleus and the stellar component increases with redshift because many galaxies can be resolved at z ~ 0.5. Fortunately, we needn't go much deeper than B = 22 for now, because it is quite apparent that the current samples are missing objects due to the contribution of the galaxy light even at z ~ 0.7 (cf. Koo and Kron 1988).

Finally, many of these low luminosity quasar candidates may show narrow lines, so it is necessary to determine the fraction of these that have broad lines at some predefined level. Thus, high resolution (1Å) spectra with moderate signal are needed to classify these faint quasar candidates. It is very important to know if they belong with the rest of the *bona fide* quasars and, if not, whether they evolve as well.

Acknowledgements

I thank Dave Koo and Richard Kron for discussions of their model. This work has been supported under NASA contract NAS5-30180.

References

Avni, Y. and Bahcall, J.N. 1980, Ap. J., 235, 694.
Boyle, B. 1986, Ph.D. thesis, Durham University.
Boyle, B., Fong, R., Shanks, T., and Peterson, B. A. 1987, M.N.R.A.S., 227, 717.
Crampton, D., Cowley, A. P., and Hartwick, F. D. A. 1987, Ap. J., 314, 129 (CCH).
Donahue, M. and Shull, J. M. 1987, Ap. J., submitted.
Filipenko, A. V. and Sargent, W. L. W. 1985, Ap. J. (Supplements), 57, 503.
Koo, D. C. and Kron, R. G. 1988, Ap. J., 325, 92.
Marshall, H. L. 1985a, Ap. J., 283, 109.
Marshall, H. L. 1985b, in Structure and Evolution of Active Galactic Nuclei, ed. G. Giuricin, F. Mandirossian, M. Mezzetti, and M. Ramella, (Dordrecht: Reidel), p. 627.
Marshall, H. L. 1987, Astron. J., 94, 628.
Marshall, H. L. 1988, Active Galactic Nuclei: proceedings of the Georgia State University Conference, ed. H. R. Miller and P. J. Wiita, (New York: Springer-Verlag), in press.
Marshall, H. L., Avni, Y., Braccesi, A., Huchra, J. P., Tananbaum, H., Zamorani, G., and Zitelli, V. 1984, Ap. J., 269, 42.

18 HERMAN L. MARSHALL

Marshall, H. L., Avni, Y., Tananbaum, H., and Zamorani, G. 1983,
Ap. J., 269, 35.
Mitchell, K. J., Warnock, A., III, and Usher, P. D. 1984, Ap. J. (Letters),
287, L3.
Schmidt, M. and Green, R.F. 1983, Ap. J., 269, 357 (SG).

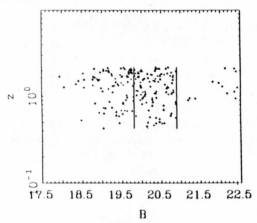

B

1. Hubble diagram for the data employed in the fit for parameters
of the faint end of the quasar luminosity function. *Filled circles*:
quasars from the BF sample of Marshall *et al.* (1984); *triangles*:
quasars from Boyle (1986); *stars*: quasars from Koo and Kron
(1988).

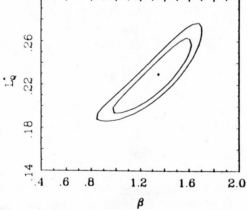

β

2. Confidence regions (1σ and 90%) for L_Q^* and β, the
parameters of the faint end of the quasar luminosity function in the
PLE model (see text). The parameter L_Q^* is in units of 10^{30} erg
s^{-1} Hz^{-1} at 2500 Å. Note that β = 1 is allowed at 90% confidence
but that β = 2 is easily rejected.

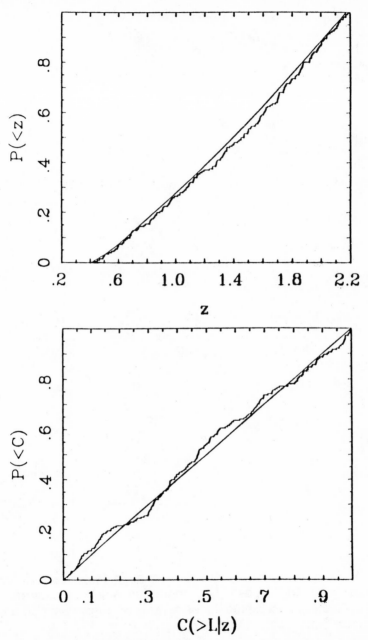

3. Predicted and observed distributions compared for the Kolmogorov-Smirnov tests as described by Marshall *et al.* (1983). The quantity C(L|z) represents the probability that a quasar would be found with luminosity L or greater given the observed redshift.

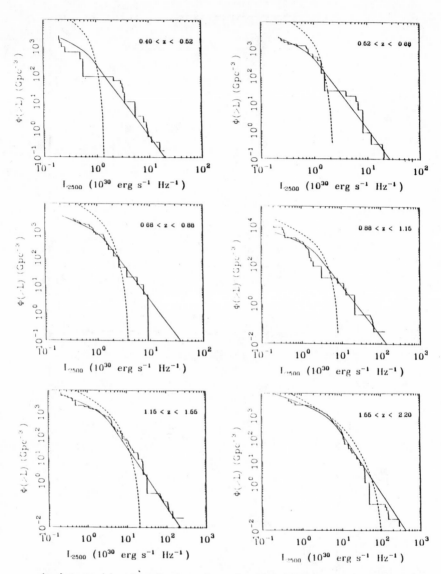

4. Integral luminosity functions for six different redshift intervals.
Jagged curves: observed integral luminosity function (see text);
solid curves: PLE model (see text); *dashed curves*: Koo and Kron
(1988) model.

DISCUSSION

Shanks : What was the reason for the bad fits in two of your cumulative distributions that you looked at in the very beginning?

Marshall : You're asking, are the low redshift intervals not fitting in the low luminosity intervals?

Shanks : Maybe I've misunderstood, but you said that two cumulative distributions at the very beginning of your talk had you worried, but you never seemed to say why.

Marshall : I don't remember being worried by anything!

Hewett : Perhaps we can solve that at coffee.

Kron : It's always very pleasing when someone takes your work so seriously they bother to prove you're wrong. But I'd like to make a couple of comments. You're quite right that our model doesn't fit at all to low redshift quasars nor to high luminosity quasars at any given redshift. But, the model never was intended to be universally optimal. It was invented to look at the characteristics of the bend of the luminosity function and how the change in slope of the luminosity function changes the density and luminosity with increasing redshift. The bulk of our data are between redshifts 1 and 3 with luminosities around the breakpoint, which is what we were trying to model.

Marshall : Yes, I can imagine that a restatement of the same scheme that you have in that paper might well fit the data but would look exactly like a pure luminosity evolution model. It might be like a synchrotron nebula, where the bright end objects decay out of the sample and have to be replenished. Note that in pure luminosity evolution the faint end of the luminosity function does actually shift from one redshift bin to another, and your model had it not shifting at all. However, I can't necessarily tell the difference, because the shift is very small.

Boyle : I'd like to come to the defense of Crampton et al. You mentioned that their survey didn't agree with your model. While I have seen the same problem, I find when I do the KS test that the survey is fine.

Cowley : What Herman is saying is that if the model doesn't fit the observations, it must be the observations are wrong!

Marshall : The only reason I think I can say this is because I'm trying to make an argument based on consistency with previous samples, and I think there is a problem.

Crampton : Quasars of redshift 2 are so easy to detect in our survey that we worry about them, in fact. If you take our numbers straight without any correction for effects of emission lines, you're probably already getting too many. They're so easy to find, it's embarrassing.

Marshall : I didn't look only at the 2.2-2.4 region in your survey. Basically, the normalization just seemed too low for the whole high redshift interval. Otherwise the density would have to drop dramatically just in the narrow range, and intuitively I don't feel comfortable with that.

Cowley : Why should they be out there? The technique is capable of detecting them out to 3.4; there's certainly no problem at 2.2.

Turner : A statistical technicality, do I understand that you use the KS test as a way of fitting these luminosity functions?

Marshall : No, I use the maximum likelihood method for fitting them, and I use KS test to see if the fits are acceptable.

Turner : Good. The KS test is not very good for fitting parameters. While I am a great fan of it in many ways, it is very poor for testing deviations in tails of distributions that have areas of high and low populations. It will almost entirely test your function around the area in luminosity where you have a great number of observed quasars. It would be possible to do other tests that would be more sensitive to the bright end.

Marshall : That's certainly true. I applied the KS test because many people understand it and it was the least objectionable a priori test for a first test of the data.

SELECTION EFFECTS IN QSO SURVEYS

Bruce A. Peterson
Mt. Stromlo and Siding Spring Observatories
The Australian National University
Woden A. C. T. 2606, Australia

ABSTRACT. Selection effects in QSO surveys must be taken into account when the surveys are used to derive the QSO luminosity function and its evolution. The influence of selection effects on number-redshift diagrams from several QSO surveys is discussed in the context of the luminosity function derived by Boyle *et al.* (1987) from UVX surveys.

1. INTRODUCTION

All surveys have an apparent flux limit. This is usually taken as understood. Problems arise from additional effects which are not taken into account when the luminosity function is derived from a magnitude limited survey. These selection effects have been discussed by Clowes (1981) for surveys based on slitless spectra, and by Savage and Peterson (1982) for surveys of flat spectrum radio sources. In addition, Wampler and Ponz (1985) have raised a number issues relating to the limiting magnitude of a survey and to the effects of emission lines and variability.

If the the Universe is homogeneous, the number-redshift diagram of two surveys with the same magnitude limit should be the same. If this is not the case, and different methods were used to identify the QSOs, then we may expect that there are additional selection effects that must be understood before we can interpret the luminosity functions derived from these surveys.

Before comparing number-redshift diagrams from actual surveys, let us first examine some model diagrams. For objects with a constant co-moving density, Figure 1 shows the relative number of objects in a

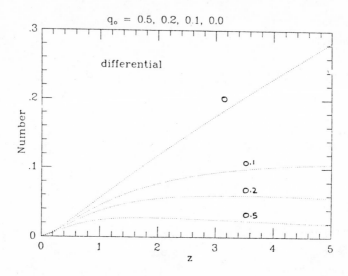

Figure 1. The number of objects per redshift interval for $q_o = 0, 0.1,$ 0.2, and 0.5, assuming a constant co-moving density of objects.

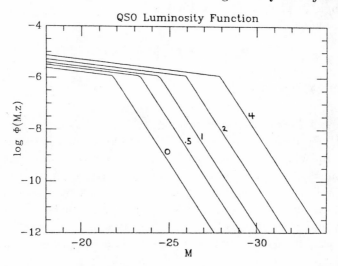

Figure 2. The QSO luminosity function derived from UVX QSOs by Boyle *et al.* (1987) at $z = 0, 0.5, 1, 2,$ and 4.

redshift interval $\Delta z = 0.2$ centered on z for various values of q_o. For $q_o > 0$, the number per redshift interval rises steeply to $z \approx 1$ and then levels off. This illustrates the amount of volume in a redshift interval for various models.

In order to consider the effects of an apparent magnitude cutoff on the number-redshift diagram, a luminosity function must be introduced. Boyle *et al.* (1987) have shown that QSOs found in UVX surveys can represented by a luminosity function that undergoes pure luminosity evolution and consists quite simply of a steep exponential for the most luminous QSOs and a flat exponential for the dim QSOs. This luminosity function is shown in Figure 2 at various redshifts. Keep in mind that it is derived from data in the interval $0.3 < z < 2.2$, but for purposes of discussion its evolution will be extrapolated outside this interval using the redshift dependence derived from data within this interval.

$H_o=50$ $q_o=0.5$ $B_S=21$ $z_c=1.0,1.6,2.2$

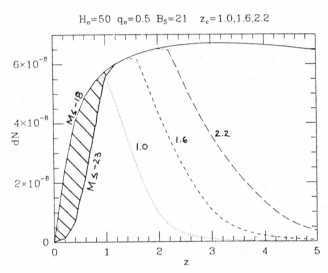

Figure 3.   The expected number of QSOs per redshift interval in a magnitude limited survey assuming the luminosity function in Figure 2 and a survey limit of $B_S = 21$. The broken lines represent the case where the lumnosity function stops evolving at $z_c = 1.0$, 1.6, and 2.2. The hatched area shows the effect of requiring that the QSOs must have absolute magnitudes brighter than -23.

The solid line in Figure 3 is the number-redshift diagram for a $B \leq 21$ magnitude limited survey of objects that have the co-moving luminosity function shown in Figure 2, but cut off at $M > -30$. The broken lines represent the number-redshift diagram for a magnitude limited survey of objects that have the same co-moving luminosity function except that the evolution stops at $z = z_c$, and the co-moving luminosity function is unchanged for $z > z_c$. In the example where the evolution stops at some z_c, the number of survey objects reaches a maximum near the redshift where the evolution stops, and then drops following the decline of the bright end of the luminosity function. If the absolute luminosity is restricted to $M < -23$, then the low redshift end of the number-redshift diagram is slightly truncated because the low luminosity QSOs with $M > -23$ are seen only at low redshift, where they are a major contributor to the observed number of QSOs. If the absolute luminosity cutoff is increased, from $M > -30$ to $M > -34$, there is no perceptible change in the number-redshift diagram because the additional high luminosity objects are too rare.

2. RADIO SELECTED QSOS

At this point it is useful to compare the model number-redshift diagram in Figure 3 to that shown in Figure 4 for the sample of flat spectrum radio QSOs described by Jauncey *et al.* (1984) and by Savage (1988). Except for a slight excess of QSOs with $z > 3$, the radio number-redshift diagram is consistent with the model luminosity function that evolves to $z \approx 1.4$ and then has a constant co-moving density for all $z > 1.4$.

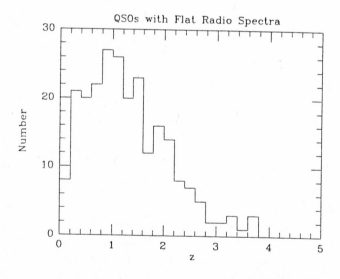

Figure 4. The number-redshift distribution for QSOs with flat radio sepctra (Jauncey *et al.* 1984, Savage 1988).

3. UVX QSOS

Surveys of UVX QSOs are subject to redshift dependent color changes which cause the QSOs to lose their ultra-violet excess and drop out of the survey. The main changes occur at $z \approx 0.75$ when Mg II and Fe II are in the B-band and C IV is not yet in the U-band, at $z \approx 1.65$ when C IV is in the B-band and Lα is not yet in the U-band, and for $z \geq 2.2$ when Lα enters the B-band. Figure 5 shows these changes in color as a function of redshift for the UVX QSOs in the survey of Boyle *et al.* (1987). The solid line represents the redshift-color variation for the QSOs discussed by Veron (1983).

The number-redshift diagram is shown in Figure 6 for the Boyle *et al.* (1987) UVX survey and in Figure 7 for the Schmidt and Green (1983) UVX survey. Both surveys have a deficiency in numbers at $z \approx 0.75$, while only the Schmidt and Green survey appears affected at $z \approx 1.65$.

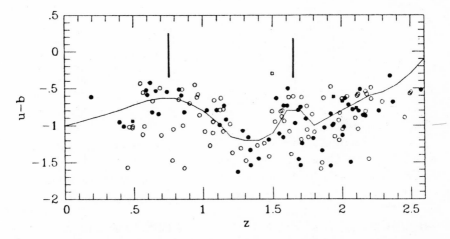

Figure 5. $u - b$ verses z for QSOs in the UVX survey of Boyle *et al.* (1987). Open circles denote QSOs with $b > 20$, filled circles those with $b < 20$. The solid line represents the QSOs discussed by Veron (1983). Vertical lines mark the redshifts $z = 0.75$ and 1.65 were the increase in $u - b$ may cause QSOs to be missed in a UVX survey.

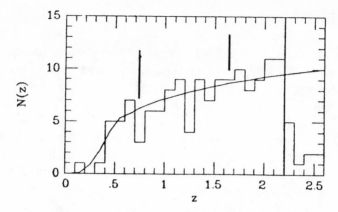

Figure 6. The number-redshift diagram for QSOs in the UVX survey of Boyle *et al.* (1987). Vertical lines mark the redshifts $z = 0.75$ and 1.65 were the increase in $u - b$ may cause QSOs to be missed in a UVX survey.

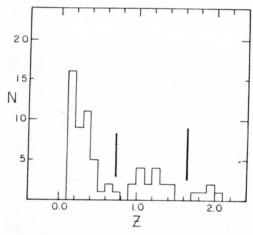

Figure 7. The number-redshift diagram for QSOs in the UVX survey of Schmidt and Green (1983) as given by Wampler and Ponz (1985). Vertical lines mark the redshifts $z = 0.75$ and 1.65 were the increase in $u - b$ may cause QSOs to be missed in a UVX survey.

4. QSOS SELECTED FROM SLITLESS SPECTRA

Because the identifications in slitless spectra surveys depend upon detecting emission lines, QSOs with weak lines may be systematically missed in surveys of this type. Table 1 compares the rest frame equivalent widths of the Lα emission line in the grens survey, (Crampton *et al.*

1987) the APM survey (Foltz *et al.*, 1987), the radio survey (Jauncey *et al.* 1984, Savage 1988), and a survey based upon variability (Hawkins 1986, Peterson and Hawkins 1988). The values for the APM survey are

Table 1.
Lα Rest Frame Equivalent Widths

Sample	$\overline{W}$	σ_W	n
Radio	55.8	36.5	37
Variable	60.5	40.7	19
Grens	84.9	45.7	59
APM	95.2	34.0	69

my estimates from the spectra published in the *Astronomical Journal* and should be regarded with caution. With this in mind, we see that the two surveys based upon slitless spectra, the grens and APM, do have systematically stronger line objects than the radio and variable surveys which should not be affected so much by line strength. This is an indication that weak line objects may be being missed in the surveys based upon slitless spectra. However, the high redshift QSOs found by Hazard *et al.* (1986) have rest frame Lα equivalent widths similar to those of the radio and variable samples.

The grens survey of Crampton *et al.* (1987) is similar in limiting magnitude to the $B \leq 21$ UVX survey of Boyle *et al.* (1987). Comparing the grens number-redshift diagram in Figure 8 with the UVX number-redshift diagram in Figure 6, we see that the grens QSOs peak around $z = 1$ and then decline, in contrast to the UVX QSOs which increase right out to the UVX survey limit at $z = 2.2$. Here we have an example of two surveys with similar limiting magnitudes, but different selection criteria, that produce different number-redshift diagrams!

Figure 9 shows the QSO surface density in four redshift intervals for QSOs in the grens and UVX surveys. The four redshift intervals correspond to four different combinations of emission lines seen in the grens spectra; Mg II only, C III only, C III and C IV, Lα and C IV. The surface density of UVX QSOs is higher except in the redshift interval where the grens QSOs are identified by C III only. Because the only redshift interval where the grens survey has the higher surface density is the interval where the grens identifications are based upon the relatively weak C III line, it must be that there are too many UVX QSOs, rather than too few grens QSOs, in the other redshift intervals where the grens identifications are based upon stronger emission lines.

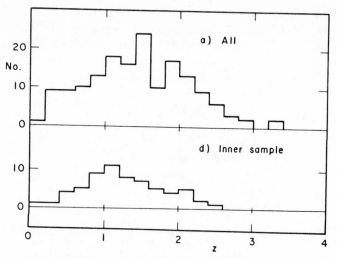

Figure 8.   The number-redshift diagram for **a)** all confirmed QSOs detected with the blue grens survey by Crampton *et al.* (1987) and **d** their "inner" complete sample with $m < 20.5$.

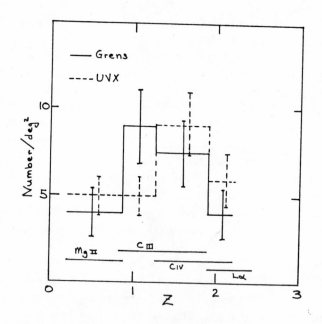

Figure 9. The surface densities of QSOs found in the grens (Crampton *et al.* 1987) and UVX (Boyle *et al.* 1987) surveys. The redshift range of each strong emission line used to identify grens QSOs is also shown.

Figure 10 shows the K-corrections from Koo and Kron (1988) with the four redshift intervals of Figure 9 marked. Except in the single redshift interval where the surface brightness of grens QSOs is greater than that of the UVX QSOs, the K-correction is negative. This means that the continuum magnitude limit of the UVX survey is actually fainter than the broad band $B = 21$, resulting in too many UVX QSOs being included in these redshift bins. Correcting the UVX survey magnitudes for the presence of emission lines should bring the number-redshift diagram into better agreement with that of the grens survey. This discrepancy between the UVX and grens surveys is also reduced by the addition of more grens data that has been recently obtained (Crampton 1988).

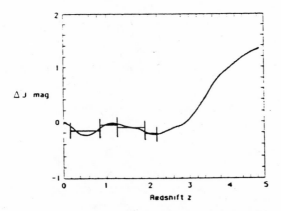

Figure 10. The K-corrections derived by Koo and Kron (1988) with the four redshift intervals of Figure 9 indicated.

5. THE $< V/V_M >$ STATISTIC

In order to take into account both a limiting optical magnitude and a limiting radio flux-density Schmidt (1968) used the $< V/V_M >$ statistic.

In a subsequent series of papers by Longair and Scheuer (1970), Rees and Schmidt (1971), and Carswell and Weymann (1972) the relative insensitivity of the $< V/V_M >$ statistic to the actual redshifts of the QSOs in the survey was discussed.

This relative insensitivity of V/V_M to redshift is illustrated in Figure 11. Note that QSOs near the survey magnitude limit have $V/V_M \approx 1$ regardless of their redshift. Figure 11 shows how V/V_M is closely related to S_0/S, the ratio of the flux density of the survey limit, S_0, to the flux density of the individual object S.

All surveys have a limited dynamic range. Many photographic surveys extend only a few magnitudes brighter than the survey limit. Note that the effect of a bright cutoff is to exclude objects with small S_o/S and small V/V_M that would fall in the lower left portion Figure 11. If no correction is made for a bright survey limit, $< V/V_M >$ is biased to larger values and will indicate evolution even when none is present.

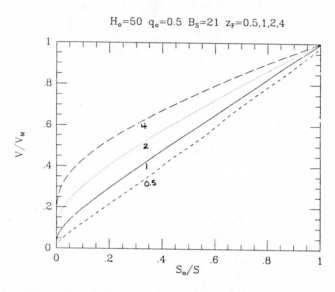

Figure 11. The relation between V/V_M and S_o/S. S_o is the flux-density limit of the survey. S is the flux-density of the QSO. Each curve is a locus of QSOs with the same redshift but different absolute luminosities.

REFERENCES

Boyle, B. J., Fong, R., Shanks, T. and Peterson, B. A. 1987, *Mon. Not. R. astr. Soc.*, **227**, 717.

Carswell, R. and Weymann, R. 1972, *Mon. Not. R. astr. Soc.*, **156**, 19p.

Clowes, R. G. 1981, *Mon. Not. R. astr. Soc.*, **197**, 731.

Crampton, D. 1988, "The CFHT/MMT Blue Grens Survey", this conference.

Crampton, D., Cowley, A. P. and Hartwick, F. D. A. 1987, *Astrophys. J.*, **314**, 129.

Foltz, C. B., Chaffee, F. H. Jr., Hewett, P. C., MacAlpine, G. M., Turnshek, D. A., Weymann, R. J. and Anderson, S. F. 1987, *Astron. J.*, **94**, 1423.

Hawkins, M. R. S. 1986, *Mon. Not. R. astr. Soc.*, **219**, 417.

Hazard, C., McMahon, R. G. and Sargent, W. L. W. 1986, *Nature*, **322**, 38.

Jauncey, D. L., Battey, M. J., Wright, A. E., Peterson, B. A. and Savage, A. 1984, *Astrophys. J.*, **286**, 498.

Koo, D. C. and Kron, R. G. 1988, *Astrophys. J.*, **325**, xxx.

Longair, M. S. and Scheuer, P. A. G. 1970, *Mon. Not. R. astr. Soc.*, **151**, 45.

Peterson, B. A. and Hawkins, M. R. S. 1988, (in preparation).

Rees, M. J. and Schmidt, M. 1971, *Mon. Not. R. astr. Soc.*, **154**, 1.

Savage, A. 1988, "The Parkes Complete Sample of Flat Spectrum Radio Sources, $S \geq 0.5$ Jy", this conference.

Savage, A. and Peterson, B. A. 1982, "Unbaised Searches for Quasars beyond a Redshift of 3.5", In I. A. U. Symposium No. 104, *Early Evolution of the Universe and its Present Structure*, eds. G. O. Abell and G. Chincarini, (Dordrecht: Reidel), pp. 57–64.

Schmidt, M. and Green, R. F. 1983, *Astrophys. J.*, **269**, 352.

Schmidt, M. 1968, *Astrophys. J.*, **151**, 393.

Veron, P. 1983, "Quasar Surveys and Cosmic Evolution", In 24th Liège Astrophysical Colloquium, Institut d'Astrophysique, *Quasars and Gravitational Lenses*, pp. 210–235

Wampler, E. J. and Ponz, D. 1985, *Astrophys. J.*, **298**, 448.

DISCUSSION

Osmer : I think it's hard to compare the results of the grens or grism approach and UVX results because of the selection effects. It's especially hard for visually selected grism samples because the detection limit depends on both line strength and continuum magnitude. Gratton and I showed last year in a paper in the PASP how important the effects can be. Schmidt, Schneider, and Gunn have allowed for selection effects in their work, and the APM surveys can do so as well. However, I would guess there are similar effects in UVX surveys, and I am not sure they have been corrected for.

Peterson : Yes, the point I was trying to get across is that you have to worry about the K correction for your sample. You really want to use continuum magnitudes in limiting the sample. You also want to worry about any evolutionary effects that may be present in the lines. I would like people to address these issues when they analyze their data.

Cowley : There's another effect that may be real and is caused by looking at only tiny regions on the sky. We now have 2 areas we have surveyed, with probably double the number of objects observed in the first paper. The redshift histograms of the two areas are quite different, so it matters whether your line of sight intersects a cluster or group.

Peterson : Yes, and then again, as Brian pointed out, the actual surface densities that the UVX surveys yield from different areas is quite consistent, so I don't know what all this is telling us.

Boyle : Did you say that in the grens surveys the $n(z)$ relation provided evidence for a turnaround about redshift 1.5, similar to that seen in the radio surveys?

Peterson : I said that the redshift distribution is consistent with a luminosity function that stops evolving at redshifts around 1.2 or thereabouts.

Boyle : I find that unusual in that the grens quasars are the ones with the high equivalent widths whereas the radio selected ones have lower equivalent widths. Yet, they seem to show the same evolution.

Peterson : The redshift distributions are very similar.

Chaffee : I think one of the things we hope to find out from the APM survey, and that's why we are taking high S/N data, is information on the line strengths as function of epoch or luminosity. It should be very useful for sorting out the selection effects coming from the lines.

Peterson : That is the type of information needed to understand what's going on.

QSO SURVEYS TO B = 24

RICHARD G. KRON
Yerkes Observatory, The University of Chicago
Box 0258, Williams Bay, WI 53191

ABSTRACT   This contribution presents results of a very faint spectroscopic survey for QSOs, and proposes strategies for new surveys reaching to B = 24.   Section 1 summarizes the completed survey to B = 22.6 (Koo and Kron 1988).   Section 2 shows that our picture for the evolution of the luminosity function based on this survey is generally consistent with results for QSOs at z > 4.   Section 3 comments on the application of color selection at longer wavelengths to existing 4-m plates. Finally, section 4 considers various possibilities and problems for probing to still fainter limits.

1. A SURVEY TO B = 22.6

Koo and Kron (1988, KK) derived the luminosity function and its dependence on redshift from the deep survey of Koo, Kron, and Cudworth (1986, KKC), plus 18 new redshifts obtained since KKC. The candidates were selected on IIIa-J plates, the criteria being a stellar image brighter than B = 22.6 (J = 22.5), and U-J, J-F colors unlike those of normal Galactic stars.   Spectroscopy is complete brighter than J = 21.0, and the spectroscopic sampling is representative for the fainter objects, $21.0 < J \leq 22.5$.

There are a total of 30 spectroscopically-confirmed QSOs in 0.29 deg^2.   Contaminants in the original list are mostly white dwarfs and peculiar stars at the brighter magnitudes, and narrow emission-line objects at the fainter magnitudes.   The ratio of *bona fide* QSOs (broad lines) to the total in the original candidate list is about 0.55.   It turns out that the narrow emission-line objects could have been mostly isolated by their colors and lack of variability, and the white dwarfs could have been mostly isolated by their proper motions and lack of variability.   Thus if required it appears to be possible to construct candidate lists with even higher reliability, before spectroscopy is attempted.

The KK data yield an estimate of the faint end of the luminosity function in the range $1 < z < 3$, which can be described fairly accurately by

$$\log \rho \; = \; 0.38 \; M_J \; + \; 12.6 \quad (H_o = 50 \text{ km sec}^{-1} \text{ Mpc}^{-1}),$$

approximately independent of z and q_o. Here ρ is the number per comoving Gpc^3 per mag, and M_J is the restframe absolute J magnitude. I will refer to this result in the following as the "fiducial power law". The KK survey, taken by itself, shows no evidence for evolution. The picture becomes clearer when these faint data are combined with brighter data from other surveys. The agreement between the various surveys where they overlap is good. The result is that the luminosity function breaks away from the relatively shallow slope at the faint end to a steeper slope at brighter luminosities. The break luminosity occurs at higher luminosity at higher z, and moreover this high-luminosity part becomes flatter at higher z. Thus we recover the result known for a long time (Petrosian 1973) that the high-luminosity QSOs evolve much more strongly than the low-luminosity QSOs.

Our interpretation of the character of the evolution differs from previous work in the sense that we see no decrease in comoving density as z increases at *any* sampled luminosity. Figure 1 shows a schematic representation of the variation of the luminosity function with redshift that we claim is consistent with all of the surveys. In particular, we propose that no cutoff at high redshifts has yet been detected, or to put this more strongly, there is no characteristic feature at any redshift. Apparently the emissivity due to QSOs is continuing to increase with redshift out to the highest redshifts yet sampled.

For our redshift interval $3.16 < z \leq 3.98$ $(0.5 < \log z \leq 0.6)$ we have an upper limit from our survey that nominally falls slightly below the fiducial power law. Clearly the fiducial power law cannot be rejected as a hypothesis on the basis of expecting to see one or two objects and actually seeing none. Recently Steve Majewski used the same selection as KKC in SA 28 and found a QSO with $z = 3.19$ (F = 19.79, U-J = 2: , J-F = 0.02, F-N = 0.93). This admittedly puts it just over the $\log z = 0.5$ bin boundary, yet the effort expended to find this object was minimal. Later (section 3) I will comment on how to get a better limit to the number of high-z, faint QSOs from a similar methodology.

2. THE FAINT END OF THE Z > 4 LUMINOSITY FUNCTION

The recent discoveries of a number of QSOs at z > 4 (Shaver 1987) allows a direct test of the fiducial power law in a regime outside of its original formulation. The best result for this purpose is from the South Galactic Pole field (Warren *et al.* 1987). Within 30 deg², three QSOs have so far been found with z > 4. Specifically, the interval of log z is 0.60 < log z < 0.65, and the interval of R magnitude is 19 ≤ R ≤ 20. The question is whether this number of QSOs is consistent with the evolution picture and luminosity function of KK. Some degree of incompleteness can be compensated for by choosing the Δm and Δz bins just to accommodate the three QSOs. The R band is at $\lambda \sim 6500$, or $\lambda_{rest} \sim 1250$ at the typical redshift. The correction to go from emitted $\lambda 1250$ to emitted $\lambda 4650$ was obtained from spectro-photometry of QSOs (Malkan 1983). I estimate that the observed-frame <R> = 19.5 is approximately equivalent to a K-corrected <J> = 19.1. From the usual Friedmann relations it is easy to derive the distance modulus and the differential volume element, and thereby the comoving density. I obtain from the Warren *et al.* (1987) survey, for $H_o = 50$ km sec⁻¹ Mpc⁻¹, log ρ = 0.41 Gpc⁻³ mag⁻¹ at M_J = -30.4 (q_o = 0), and log ρ = 1.47 Gpc⁻³ mag⁻¹ at M_J = -28.6 (q_o = 0.5), which is shown in Figure 1.

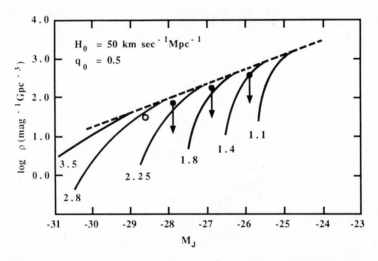

Fig. 1   Evolution picture from Koo and Kron (1988). Upper limits at z = 3.5 are from the same reference, and the circle is the Warren *et al..* (1987) result at z = 4.2.

The Warren *et al.* point is consistent with the fiducial power law, especially given the likelihood of some incompleteness in the survey and some error in the extrapolation from $M_J \sim -25$ to $M_J \sim -30$. However, the serendipitous discovery of a QSO at $z = 4.4$ by McCarthy and Dickinson cannot be reconciled with this picture unless considered as a fluke.

3. A NEW FAINT HIGH-REDSHIFT SURVEY

I return to the question of the KK upper limit at $\log z = 0.55$ that is below the fiducial power law (Figure 1). If this result is right and there are truly fewer low-luminosity QSOs at large redshift, then we would have found another feature in the luminosity function, and direct evidence that lower-luminosity systems developed later. Therefore it is important to obtain a better limit.

A check on this can already be done with data in hand. To get a better limit at $z > 3$ we need to move to redder bands since, with the Lyman continuum in the U band, high-redshift QSOs are very faint in U (also in J because of the Lyman a forest). For the past many years Koo and I, and now S. Majewski, have been obtaining IV-N + RG695 Mayall plates of our fields, some of which are of outstanding quality from the point of view of image size and sensitivity. An early run (late August 1979) was especially good, yielding such plates in two fields in Hercules and two in SA 68. In SA 68 we have the excellent-seeing plate MPF 1286, plus very good 127-02 plates (these are the archaic version of IIIa-F), so this field is an obvious starting place for this project. We intend (Bershady, Munn, Majewski, Koo, Zhan) to repeat the same procedures as in Koo and Kron (1982), except that objects would be independently identified on the IV-N plates as well, and we will be dealing with the J-F, F-N diagram (Koo and Kron 1982) instead of the U-J, J-F diagram. What I have just described is of course the extension to 4-m depth of the red color-selected survey of Warren *et al.* (1987).

If candidate QSOs are selected according to their F magnitude, then the redshift beyond which the K-correction becomes large and positive is about 4.5, as opposed to $z = 3.2$ for the IIIa-J selection. At $z = 4$ there would be some advantage of IIIa-F over IIIa-J for detection of QSOs. With this approach we expect to see three to six QSOs to F = 21.5 in a 0.3 $\deg^2$ field with $z > 3.5$. If this is still not adequate as a limit, we have enough additional fields with good plate material that we could expand the area surveyed by a factor of four.

Figure 2 shows the J-F, F-N plot for stellar objects in another field, SA 28. This is for stellar objects with 19 < J < 21, and presumably the same diagram selected by F magnitude would appear generally similar. The locus gives the predicted redshift trajectory in this color-color plane assuming the same energy distribution described in KK to get the $K_J(z)$ correction. Obviously there will be a range of energy distributions and therefore a range in the colors at each redshift, but the locus ought to be at least vaguely representative. Note that the number of good candidates is not large, but that is what is expected.

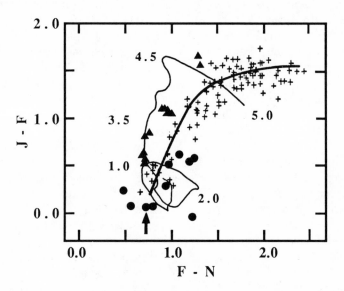

Fig. 2 J-F *vs.* F-N diagram for stars with 19 < J < 21 in SA 28. The light line is the QSO redshift trajectory with indicated redshifts, and the dark curve is the color-color relation for Population I dwarf stars. The dots are objects that would have been selected as QSO candidates based only on the U-J, J-F colors; the triangles are QSO candidates that would not have been selected by U-J, J-F alone. The arrow points to the QSO with z = 3.19.

4. REQUIREMENTS FOR QSO SURVEYS TO B = 24

Finally I would like to remark on prospects for a survey to B = 24. This number was chosen because it is significantly fainter than what we have achieved so far, but not so faint that the photometry will be limited by confusion from faint galaxies in the background. Also B = 24 is not so faint to preclude obtaining redshifts for a reasonable sample. The spirit of these comments is what can be accomplished with current instrumentation, rather than the potential of next-generation telescopes or detectors.

Given that spectroscopic follow-up will be required in any case, and that broad wavelength coverage is desirable, slitless spectroscopic techniques cannot compete with direct multiband imaging in terms of observing efficiency because of the better control of the latter in filtering out sky lines.

As an example the field of view of the "four-shooter" at Palomar (see poster by Scott Anderson), 1600 x 1600 pixels at 0.33 arc sec per pixel, is 8.5 arc min on a side. An extrapolation of the Koo/Kron counts would imply about ten QSOs in this field in the interval 22.5 < B ≤ 24.0, which means that one would need to survey only a few such fields in order to decide whether the extrapolation was reasonable.

4.1   The Photometric System

High photometric precision is required to discriminate Galactic stars from QSOs (e.g., Figure 2). The ratio of number of stars to QSOs is probably not going to change very much from B = 22.5 to B = 24, namely about 10 to 1. In order for the investment of telescope time in the follow-up spectroscopy to be effective, the success rate has to be high in identifying real QSOs. Of greater significance is the desire that the survey be reasonably complete, meaning that the colors of candidate QSOs be allowed to approach the colors of normal stars. The required level of precision seems achievable, based on the high-quality results obtained for faint photometry in globular star clusters (e.g. Hesser et al. 1987).

To first order, stars are thermal sources, and to second order, stars differ from thermal objects by virtue of line blanketing in the violet. A measure of the character of this blanketing would provide strong evidence that the deviation from a thermal spectrum was, or was not, characteristic of a star. I am assuming here that the hotter subdwarf stars in the halo are in general a more important source of confusion for the QSO candidates than are the disk dM stars. The hydride subdwarfs discussed by Hartwick, Cowley, and Mould (1984) are good examples of halo stars with extremely low blanketing. Such stars can be confused with QSOs, since they would lie in Figure 2 near the z = 4.5 part of the locus, or in any case away from the normal, old-disk stars. The photometric system should be designed to recognize these and other peculiar stars. Note that there is no special reason to use broad-band or other conventional filters, since the capability to tie in to a standard photometric system is of little or no interest. Table I gives an example of a photometric system based on these ideas.

TABLE I   Multiband Detection Limits, B = 24, 4-m Telescope

band (nm)	efficiency	source	sky	S/N (3000 sec)
		(counts/sec)		
325 - 360	0.05	0.8	1.9	18
365 - 400	0.10	1.7	3.8	25
415 - 430	0.15	1.1	3.5	18
595 - 625	0.20	2.9	18	21
900 - 920	0.10	1.0	9.5	9.6

The first band samples the Balmer continuum, and the second samples the metal and hydrogen blanketing. The others are continuum bands chosen to cover a wide range in wavelength and to sit between strong sky lines. The second column gives the adopted net efficiency, including the transmission of the atmosphere. The third column gives for a 4-m telescope the count rate in these bands at B = 24 for a spectrum $f_\nu \sim \nu^{-1}$. The fourth column gives the same for the sky background in 1 arc $\sec^2$, from Turnrose (1974). The last column gives the net signal-to-noise ratio for a 3000 sec exposure and a noise-equivalent area of 3 arc $\sec^2$.

At the best sites images of 0.75 arc sec full-width-at-half-maximum can occur often enough to be of practical interest. If the profile is approximated by a Gaussian, this value is equivalent to σ = 0.32 arc sec, which can be compared to Kormendy's (1987) *median* σ = 0.34 arc sec. According to King (1983), the noise-equivalent area for a seeing profile with this central width, plus realistic extended wings, would be 3.1 arc $\sec^2$. Eighty percent of the energy in the (Gaussian) image would be enclosed in a diameter of 1.1 arc sec.

4.2 Spectroscopy

At B = 24 it is desirable to sample spectroscopically a large range of the optical band for the same reason as at brighter magnitudes. For $\lambda < 6000$, most of the sky background is continuum light, and the traditional low resolution (R ~ 500) would be adequate for broad-line sources.

Adopting the same assumptions made above for telescope size, source and sky spectra, and seeing, a B = 24 QSO could be detected at S/N = 4 per R = 500 resolution element at $\lambda 5000$ in 10000 sec. This assumes a 1 arc sec entrance aperture and an overall efficiency of 0.10, which is supposed to take into account slit losses.

The numbers given above for signal-to-noise ratio, resolution, and integration time are not too far away from what Koo and I have been doing with the Cryogenic Camera at the KPNO 4-m at B = 22.5. Our performance at this brighter magnitude is comparable because we are using slits that are generally 2.5 arc sec wide. The principal difference with respect to the Cryogenic Camera for a spectrograph designed to work on faint point sources in very good seeing is appropriate scale on the detector: to get 0.33 arc sec per 15 micron pixel, the spectrograph camera would have to be f/2.3.

For $\lambda > 6000$ the sky background is dominated by intense airglow emission that is concentrated into sharp lines. These can be removed by obtaining the spectrum at high resolution (R > 10000), subtracting the sky, and then rebinning to get R = 500. With this technique the sky background becomes essentially the continuum, and the speed of observation is increased by the ratio of the total background to the continuum background. No noise penalty is paid as long as the red sky continuum is larger than the detector noise, as will be the case for read noise < 6 electrons per pixel and exposure times of the order of an hour. The penalty paid is in pixel usage. To cover the range $\lambda\lambda 6000 - 9000$ on a single 800 x 800 detector at R = 10000 would require at least 10 cross-dispersed orders, depending on the number of pixels per nominal resolution element.

Only a very few sources can be observed at one time with a multislit spectrograph because of the need for a large image scale. If a "four-shooter" array of CCDs were used in such a spectrograph, the area covered would be substantial, but the usage of pixels in the focal plane is still highly inefficient. The alternative would be a fiber-fed spectrograph that would optimally fill the detector area with spectra of sources and sky. Despite the expected poorer performance of fibers in terms of providing for excellent sky background subtraction, it may be that in this application the format advantage will substantially outweigh this loss in efficiency.

4.3   Image Structure

QSOs can be separated from stars by photometric indices because ordinary stars are restricted to a specific range in the shapes of their spectral energy distributions.   Galaxies on the other hand have composite spectra, they may have emission lines strong enough to affect even broad-band indices, and they can appear at any redshift.   There is thus a fundamental problem in separating the much greater number of galaxies from QSOs by colors, and the only alternative is to resort to image structure.

Figure 6 of Koo and Kron (1982) shows that at least in the magnitude interval $20.5 < J \leq 22$ the size distribution of galaxies is such that most galaxies are clearly resolved.   For the bluest objects, however, no such clear bimodality in image size is evident.   (Munn, Bershady, and Majewski are currently investigating spectro-scopically this region of the color-size diagram.)   At $B = 24$ it is reasonable to suppose that galaxy sizes will be systematically smaller.

To improve the image quality one could try a system like Thompson's ISIS (Thompson and Ryerson 1984).   A simpler procedure would be to shutter the exposure and integrate only during intervals of the best seeing   --   the gain in structural information may easily make up for the loss in photons, depending on the characteristics of the seeing.   Tyson suggested this some time ago, but I am unaware of an actual application to this problem.

There is a possibility of detecting underlying galaxies around QSOs at very faint limits: the KK survey has already sampled absolute magnitudes to $M_J = -23$.   Hutchings, Crampton, and Campbell (1984) found that for their QSOs at $z < 0.7$ the surrounding fuzz was comparable in luminosity to the nucleus, *i.e.*, the galaxies are quite bright.   Also, there could be a connection between the intrinsic surface brightness of the surrounding disk and the existence of an active nucleus.   Good examples are NGC 4151 and NGC 1068, both of which have exceptionally high surface-brightness disks.   It would clearly be of great interest to know whether such disks exist at high redshift.

In summary, a survey to $B = 24$ appears to be feasible with good (but realistic) seeing, and current-generation instrumentation.   The photometric system defining the survey should be designed to recognize reliably halo stars of various types.   I have proposed in this spirit a system that features relatively narrow bands and a concentration of the observing time in the violet.   An echelle spectrograph should be used for the red spectral range to realize the speed gain of the better sky subtraction.   The slit should be fed with fibers to achieve a wide field and good efficiency in use of detector area.   The seeing has to be very good in this application too, and to take advantage of it the scale delivered by the spectrograph camera has to be unconventionally large.   The fibers have to do a good job of sky-subtraction, a requirement that includes excellent positioning and stability of the fibers.   Finally, excellent image quality is needed to address the technical problem of separating QSOs from galaxies.   It could be that a detection of fuzz around z > 1 QSOs may be one of the most important achievements of such a survey.

ACKNOWLEDGMENTS

This contribution has been greatly assisted by M. Bershady, D. Koo, S. Majewski, and J. Munn.   I also thank R. Dreiser for preparing the figures.

REFERENCES

Hartwick, F.D.A., Cowley, A.P., and Mould, J.R. 1984,
    Ap. J., **286**, 269.
Hesser, J.E., Harris, W.E., VandenBerg, D.A., Allwright, J.W.B.,
    Shott, P., and Stetson, P.B. 1987, *Pub. A.S.P.*, **99**, 739.
Hutchings, J.B., Crampton, D., and Campbell, B. 1984, *Ap. J.*, **280**, 41.
King, I.R. 1983, *Pub. A.S.P.*, **95**, 163.
Koo, D.C. and Kron, R.G. 1982, *Astr. Ap.*, **105**, 107.
Koo, D.C., Kron, R.G., and Cudworth, K.M. 1986, *Pub. A.S.P.*, **98**, 285.
Koo, D.C. and Kron, R.G. 1988, *Ap. J.*, Feb 1 (in press).
Kormendy, J. 1987, in *Nearly Normal Galaxies*, ed. S.M. Faber
    (New York:Springer-Verlag), p. 163.
Malkan, M. 1983, *Ap. J.*, **268**, 582.
Petrosian, V. 1973, *Ap. J.*, **183**, 359.
Shaver, P. 1987, *Nature*, **330**, 426.
Thompson, L.A. and Ryerson, H.R. 1984, *Proc. SPIE*, **445**, 560.
Turnrose, B.E. 1974, *Pub. A.S.P.*, **86**, 545.
Warren, S.J., Hewett, P.C., Osmer, P.S., and Irwin, M.J. 1987,
    Nature, 330, 453.

EVALUATION OF THE QSO EVOLUTION RATE IN THE PRESENCE OF PHOTOMETRIC BIAS

A. CAVALIERE
Astrofisica, Dip. Fisica II Università di Roma, Italy

E. GIALLONGO
Osservatorio Astronomico di Capodimonte, Napoli, Italy

F. VAGNETTI
Astrofisica, Dip. Fisica II Università di Roma, Italy

In the literature concerning QSO evolution contrasting claims are found: (i) Simple evolutionary models implying strong evolution would fit the data in the $L - z$ plane to within overall uncertainties $\sim$ few % in all parameters, including the evolutionary time scale. (ii) Correcting for the biases introduced by the Baldwin effect relative to strong lines as C IV and Lyα and by the Bennett effect near each survey's limit, could easily wipe out the evidence of evolution. We outline here the results of a critical re-examination of the biases and of the uncertainties.

Fig. 1 and Tab. I chart the kinds of uncertainties we examine and their effects.

Fig. 2 shows the "contamination" by the lines on the samples PG (Schmidt and Green 1983), AB and BF (Braccesi et al. 1970, 1980). On these we test preliminarly the changes caused in a simple model (luminosity evolution with a constant time scale τ) by the Baldwin anticorrelation $\log W = 0.35 M_B + 11.2$. We find $\Delta \tau / \tau \simeq 1\%$, a change small because the contamination actually concerns the small percentage of objects falling in the ranges of z and L shown by Fig. 2.

Fig. 2 illustrates also the converse effect: incompleteness arising at $z \sim 0.6 \div 0.8$ when Mg II drifts out of the U into the B band, and the color is correspondingly altered. Even if still affecting a small percentage of the objects, this systematic incompleteness introduces weak spurious trends in the evolution rate, and favours evolutionary rates accelerating into the past like $L \propto (1+z)^\beta$ which means (CGV 1986) $\tau \propto t$.

Fig. 3 visualizes the Bennett bias resulting from the effect of random uncertainties made asymmetric by the convolution with a steep LF: $N_{obs}(F) = \int_{S_0}^\infty dS \, P(F|S) \, N(S)$ (Murdoch et al. 1973), with $P(F|S) = $ probability that a source of true flux S will be observed with a flux between F and $F+dF$. We take $P(F|S) \propto \exp[-(S-F)^2/2\sigma^2]/\sqrt{\sigma}$ with total variance

$$\sigma^2 = \sigma_{lines}^2 + \sigma_{phot}^2 + \sigma_{sky}^2$$

Here σ_{lines} accounts for the random residuals from line subtraction. The component $\sigma_{phot} \propto S$ represents the errors introduced by instruments like the iris photometer: this component by itself causes no change of the slope of the LF since σ_{phot} decreases

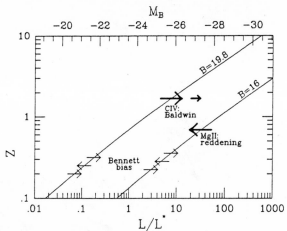

Fig. 1. An introductory illustration of the various biases discussed in the text.

for faint objects, but by the same token it actually dominates only for bright fluxes.

The uncertainty σ_{sky} = const, dominant for faint objects, is introduced by the small-scale fluctuations of the sky background falling within the small plate areas integrated to measure magnitudes (Kron 1980, Koo 1986); this implies errors increasing rapidly in magnitude when the flux F approaches σ_{sky}: $\delta m_{sky} = 2.5 \, \log\!\left(1 \pm \sigma_{sky}/F\right)$.

An important reduction of the Bennett bias occours when the LF flattens out at a survey's limit; this happens for the BF when the local LF of Cheng et al. 1985 is extrapolated toward higher z by a luminosity evolution as CGV 1985 and Weedman 1986, Boyle et al. 1987 found to constitute a remarkable fit to the data.

We base the following template analysis on the published samples AB, BF and PG. The errors stated for the samples AB+BF are $\delta m \simeq 0.1$ mag when averaged over the range $B = 17$ to 20; however, it is more realistic to take $\delta m \lesssim 0.1$ up to $B = 19.2$ where a few objects and the calibration have been actually checked with CCD frames, and to extrapolate beyond following the above equation. This implies $\delta m_{sky} \simeq 0.25$ at the survey limit $B_{lim} = 19.8$ (BF) corresponding to $F/\sigma \simeq 4$.

TABLE I   A Flow Chart for Photometric Uncertainties and their Effects

Lines	equivalent widths $W(L) + \delta W$
$\Rightarrow$	spurious trends
$\Rightarrow$	random residuals (see below)
Random uncertainties	
	residuals from line correction
	photometry errors
	sky fluctuations
$\Rightarrow$	bias when convolved with asymmetric LF
$\Rightarrow$	enhanced statistical variance

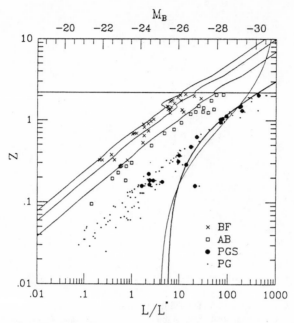

Fig. 2. The samples PG, PGS, AB, BF on the $L - z$ plane. Slanting, wiggling lines are the survey magnitude limits for the continuum, corrected for the effect of the lines: from left to right: $B_{lim} = 19.8, 19.2, 18.25$. Every PG object has its own magnitude limit, not represented here (average $B_{lim} \sim 16$) $H_o = 50$ km/s Mpc, $\Omega_o = 0.2$, $L^* = 10^{30}$ erg/s Hz at 2500 Å rest frame, spectral slope $\alpha = 0.5$. Evolutionary tracks are marked corresponding to $\tau = \tau_o(t/t_o)^n$: $n = 0$, $\tau_o = 0.15\ H_o^{-1}$ (thin line), $n = 1$, $\tau_o = 0.3\ H_o^{-1}$ (thick line). Note the   different approximations to the data in the range $z = 0.6 - 0.8$.

A subsample of the PG, originally built with a $F/\sigma > 2.5$ and with stated errors $\delta m \sim 0.3$ (Green and Morrill 1978), has been subsequently re-measured by Wampler and Ponz (1985) for $M_B \leq -24$. We have selected a reasonably complete subsample (here PGS) constituted by plates on which the objects differ by $\lesssim 0.1$ mag in the two catalogues.

The combined samples PGS+AB+BF (with full account of both the line effects and the Bennett bias) are compared through a Maximum Likelihood analysis with mathematical models based on a double-power-law shape for the true LF and on three models of luminosity evolution: (i) a uniform luminosity evolution $L \propto \exp(T/\tau)$; (iii) $L \propto (1 + z)^\beta$; and (ii) the bridging model $\tau \propto t^n$ proposed by us (CGV 1986).

We find (cf. Tab. II) for the simplest model (i) maximum shifts of the best fit values for the evolutionary parameters of $\Delta\gamma/\gamma \simeq -2\%$ and of $\Delta\tau/\tau \simeq 2\%$ with uncertainties $\delta\tau/\tau$ up to $\pm 10\%$ at the 68% confidence level. Equivalent results are found for models (iii) and (ii), considering its n, τ correlation. All models satisfy for the best fit parameters the K-S test within the 90% joint confidence level.

We stress that the evidence persists ($\tau < 0.2\ H_o^{-1}$ at 90% confidence level) of a strong evolution of the QSO population still after the most complete set of corrections performed to date, even for r.m.s. errors $\delta m \simeq 0.25$. But the final uncertainties are in

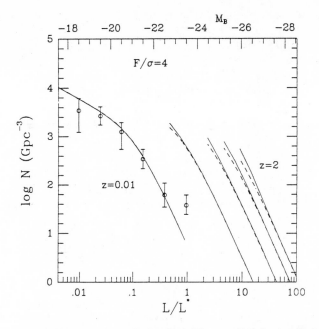

Fig. 3. To illustrate the effects expected for a realistic luminosity function at various redshifts, in the case $F/\sigma = 4$. The true LF is represented by dashed lines up to $B = 19.8$. It is assumed $\gamma_1 = 1.6$, $\gamma_2 = 3.5$, $k = 6.4$ and a r.m.s. $\delta m = 0.25$ at $B = 19.8$. Local data by Cheng et al. 1985.

fact (as expected from the above discussion) considerably larger than previous workers have found or quoted. Two points of astrophysical significance ensue.

First, the *average values* for τ as here revised are consistent with the time scale for the evolution of starburst galaxies found necessary by Danese et al. 1987 to explain the counts of mJy radiosources and the IRAS counts. This hint concurs with the morphological evidence of often disturbed host galaxies (cf. Smith et al. 1986, Yee and Green 1987) and with the clustering analysis by Iovino and Shaver 1987, to support the effectiveness of host interactions for fueling the activity, at least at $z \lesssim 1$.

Second, a realistic evaluation of the variances tends to blur second order effects concerning *trends* of $\tau(t)$, like $\tau \propto t$ [i.e. $L \propto (1 + z)^{\beta}$]. Trends of this form, potentially telling about the mechanism to fuel the central engine, are likely to be spuriously enhanced by the observational biases here discussed; in addition, their true difference from the simple case $\tau = const$ is easily swamped within the noise caused by the combination of random errors (residuals from the line correction and from sky subtraction, and measurement errors). To probe the real trend expected from interactions of the host galaxies, the data base at faint magnitudes has to be enlarged with multi-object spectroscopy (cf. Boyle et al. 1987) and the incompleteness at bright magnitudes should be bound or removed, but also the analysis should be adequately sophisticated.

Two comments are here in point. First, the noise here considered concerns mainly the high-L evolution, hence the useful statistics will concern primarily bright objects. Second, on combining various samples with their different uncertainties it may even

TABLE II   Results from a Likelihood Analysis on the Samples PGS+AB+BF

(i) Evolution: $L \propto e^{T/\tau}$ [time-scale τ = const, $k = 1/H_o\tau$]

B_{lim}	N_{obj}	δm	γ	k
19.8	61	0	3.53±0.16	6.57±0.59
19.8	61	0.25	3.46±0.15	6.45±0.64

(ii) Evolution defined by the time-scale $\tau = \tau_o(t/t_o)^n$ [$k_o = 1/H_o\tau_o$]

B_{lim}	N_{obj}	δm	γ	k_o	n
19.8	61	0	3.46±0.19	4.11±1.61	0.66±.49
19.8	61	0.25	3.37±0.19	4.38±2.19	0.55±.63

(iii) Evolution: $L \propto (1 + z)^\beta$ [time-scale $\tau \simeq \tau_o\, t/t_o$, $\beta = 1/H_o\tau_o$]

B_{lim}	N_{obj}	δm	γ	β
19.8	61	0	3.45±0.15	3.31±0.27
19.8	61	0.25	3.33±0.30	3.25±0.30

happen that opposite biases are combined. Whence the moral ensues that a true increase of information about LF shape and evolution obtains only combining rich and "homogeneous" samples, i.e., ones with sensibly matched signal-to-noise ratios: if anything, F/σ ought to be larger for the brighter samples with a magnitude limit crossing the steep branch of the LF.

REFERENCES

Boyle, B.J., Fong, R., Shanks, T., and Peterson, B.A. 1987, *M.N.R.A.S.* **227**, 717.

Braccesi, A., Formiggini, L., and Gandolfi, E. 1970, *Astr.Ap.* **5**, 264.

Braccesi, A., Zitelli, V., Bonoli, F., and Formiggini, L. 1980, *Astr. Ap.*, **85**, 80.

Cavaliere, A., Giallongo, E., and Vagnetti, F. (CGV) 1985, *Ap. J.*, **296**, 402.

———— (CGV) 1986, in *Structure and Evolution of Active Galactic Nuclei*, G. Giuricin, F. Mardirossian, M. Mezzetti, and M. Ramella Eds., Reidel, Dordrecht, p.231.

Cheng, F.Z., Danese, L., De Zotti, G., and Franceschini, A. 1985, *M.N.R.A.S.*, **212**, 857.

Danese, L., De Zotti, G., Franceschini. A., and Toffolatti, L. 1987, *Ap.J. (Letters)*, **318**, L15.

Green, R.F., and Morrill, M.E. 1978, *Pub.A.S.P.* **90**, 601.

Iovino, A., and Shaver, P. 1987, in *Evolution of Large Scale Structures in the Universe*, Proc. I.A.U. Symp. Nr. **130**, Reidel, Dordrecht, in press.

Koo, D.C. 1986, *Ap.J.* **311**, 651.

Kron, R.G. 1980, *Ap.J.Suppl.* **43**, 305.

Murdoch, H.S., Crawford, D.F., and Jauncey, D.L. 1973, *Ap.J.* **183**, 1.

Schmidt, M. and Green, R.F. 1983, *Ap.J.* **269**, 352.

Smith, E. P., Heckman, T. M., Bothun, G. D., Romanishin, W., and Balick, B. 1986, *Ap. J.*, **306**, 64.

Wampler, E.J., and Ponz, D. 1985, *Ap.J.* **298**, 448.

Weedman, D.W. 1986, in *Structure and Evolution of Active Galactic Nuclei*, G. Giuricin, F. Mardirossian, M. Mezzetti, and M. Ramella Eds., Reidel, Dordrecht, p.215.

Yee, H.K.C. and Green, R.F. 1987, preprint.

THE EXPECTED NUMBER OF QUASARS IN THE MONTREAL CAMBRIDGE SURVEY OF SUBLUMINOUS BLUE STARS.

S. DEMERS, F. WESEMAEL, G. FONTAINE, AND R. LAMONTAGNE
Département de Physique, Université de Montréal

M.J. IRWIN
Institute of Astronomy, Cambridge University

ABSTRACT   We present arguments, based on the statistical properties of known quasars and on scaling of the Palomar-Green survey, to show that we should discover nearly one hundred new bright quasars B < 16.5 in our southern hemisphere survey.

The Montreal-Cambridge Survey of subluminous blue objects is an ongoing project (Demers et al. 1987) aimed at increasing the number of known white dwarfs and subdwarfs in the southern hemisphere. Our prime objective is therefore to discover new stars, not to find quasars. The MC survey is essentially an extension of the Palomar-Green Survey (Green et al. 1987) in the south galactic pole region.

Our photographic survey is done with the CTIO Curtis Schmidt telescope; double exposure B and U plates are taken to cover the SGP area from b = -30° to b = -90°, south of a declination of -5°. The plates are measured with the APM in Cambridge; the analysis of the data produced is done at the Institute of Astronomy with the software available at the APM. Our limiting magnitude is slightly fainter than the PG survey and reach B ≃ 16.5. We intend to cover some 8600 square degrees of the southern sky, representing 360 Schmidt plates. More than half the plates have now been taken.

The plates of three observing seasons have so far been analysed and 862 UVX stars have been identified. Stellar images with U-B < -0.40 and one magnitude or more from the plate limit are selected for candidates from calibrated color magnitude diagrams. We find, on average, nine candidates among the seven to twenty thousand stars measured on each plate.

The distribution of the UVX candidate B magnitudes is compared in Fig. 1 to the B magnitudes of the 1874 objects discovered

by the PG survey. We have transformed their m_{pg} magnitudes
into B using a number of listed objects with photoelectric UBV
photometry. We note that the faint m_{pg} magnitudes require
very little correction. Fortunately the bright end of the
distribution where the correction is more important, is of no
immediate interest in the following comparison. We see from
the data on hand that the MC survey reaches 0.5 magnitude
fainter than the PG survey.

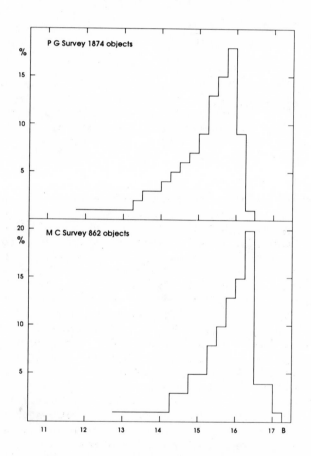

Fig. 1   Comparison of the magnitude distribution of the
          UVX objects discovered in the PG survey and dis-
          covered so far in our MC survey.

Because 92% of the quasars with z < 2.2, B < 17.0 listed by
Hewitt and Burbidge (1987), which have UBV photometry have
(U-B) < -0.40, we will assume for the following discussion
that all bright quasars with z < 2.2 are indeed UVX and would
be selected by our technique. These UVX objects should there-

fore be in our sample. We will further assume that the number
of quasars per square degree, within our magnitude range, is
the same toward the NGP or the SGP.

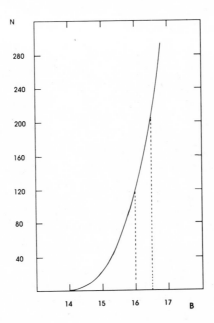

Fig. 2   Cumulative distribution of the number of quasars
          in the Hewitt and Burbidge (1987) catalog at
          galactic latitudes > 30°, brighter than a given
          magnitude.

We note from Fig. 2 that between 16.0 and 16.5 the number of
quasar nearly doubles. Because we reach fainter magnitudes,
we should, therefore, expect to detect more quasars per square
degree than the PG Survey.

When the number of known quasars, brighter than a given magnitude, in the PG survey area is compared to the number currently known in the MC survey area, one finds a rather large
shortage of quasars in the south. The difference between the
two curves of Fig. 3 in the 16.0-16.5 interval amounts to 90
objects. This difference implies that we should discover
nearly one hundred new quasars in our survey. We also note
that the results from our survey will be well matched to the
Parkes radio survey of the southern sky which will find all
the bright radio quasars (Savage 1988).

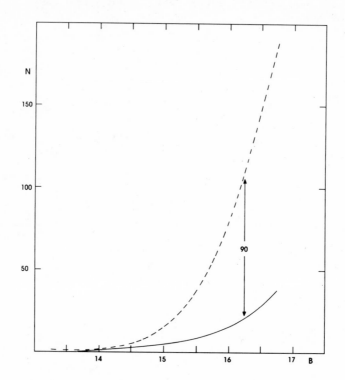

Fig. 3   Cumulative distribution of the number of quasars
brighter than a given magnitude. For the PG
survey area (dashed line) and the MC survey area
(solid line) Numbers were normalized to 8600
square degrees.

At the present time our follow-up spectrocopic observations
have been rather limited in scope due to the extremely poor
weather in Chile in 1987; mostly bright candidates have been
observed and no new quasars have been found among the 160
observed so far.

Our rediscovery rate, for the known quasars is however
excellent. Twenty three quasars, listed by Hewitt and
Burbidge (1987) with m ≤ 16.5 are located in our survey area.
Only eight are in the fields already analysed and are identified in Table I. Four satisfy our criteria and were selected
as uv bright objects. Reinspection and analysis of the plates
concerned show that the other four are much fainter than their
listed magnitudes indicate. We present in Table I details of

these eight objects; m is the magnitude listed in the Hewitt and Burbidge (1987) catalog, it is B for the object with U-B. We do not expect to be greatly troubled by many of the quasars appearing non-stellar on our plates since our analysis isophote lies between 20 and 21 in B (mag per square arcsec). Any galaxy in which the quasar is embedded would have to have an unusually high surface brightness to be detected.

TABLE I   Known bright quasars in fields analysed so far.

Name	Catalog		MC Survey		Candidate/comments
	m	U-B	B	U-B	
0050-281	16.1	-	>17.5		no barely visible
0056-363	16.5	-	15.94	-1.01	yes
0107-331	16.5	-	>17.5		no not visible
0119-286	14.7	-	15.18	-0.72	yes
0122-380	16.5	-	17.32	-0.42	no too faint
0537-441	15.96	-0.57	16.95	-0.40	no too faint
2352-342	16.4	-	16.18	-0.50	yes

ACKNOWLEDGMENTS

The MC survey is financially supported by the Natural Sciences and Engineering Research Council of Canada, by the Science and Engineering Research Council of the UK, and by the Scientific Affairs Division of NATO. G.F. wishes to acknowledge further support from a E.W.R. Steacie Memorial Fellowship. We also acknowledge the essential contribution of the Cerro Tololo Inter-American Observatory for making its facilities available to us.

REFERENCES

Demers, S., Kibblewhite, E.J., Irwin, M.J., Nithakorn, D.S., Béland, S., Fontaine, G., and Wesemael, F. 1986, A.J. 92, 878.
Green, R.F., Schmidt, M., and Liebert, J. 1986, Ap.J. Suppl., 61, 305.
Hewitt, A., and Burbidge, G. 1987, Ap.J. Suppl., 63, 1.
Savage. A. 1988, This conference.

SPECTROPHOTOMETRIC OBSERVATIONS OF THE US CATALOGS: MBQS QUASAR SURFACE DENSITIES IN THE RANGE 16.0 < B < 18.25

KENNETH J. MITCHELL
Applied Research Corporation, 8201 Corporate Drive, Suite
920, Landover, MD   20785

PETER D. USHER
Department of Astronomy, Pennsylvania State University,
University Park, PA   16802

ABSTRACT   The US survey for the selection of blue- and
ultraviolet-excess objects and the follow-up spectro-
photometric program which has produced the Medium Bright
Quasar Sample are summarized. Preliminary results in the
form of number counts and a redshift distribution for the
MBQS are presented and compared to those of other
brighter quasar samples.

THE US SURVEY FOR BLUE- AND ULTRAVIOLET-EXCESS STARLIKE OBJECTS

The US survey comprises six lists of objects selected for blue
and/or ultraviolet excess (B-UVX) which are contained within
the Palomar 1.2m Schmidt telescope fields at high galactic
latitude centered on Selected Areas SA 28, 29, 55, 57, 71 and
94 (Usher et al. 1988, and references therein). These fields
were chosen by A. Sandage for study on triple-exposure (u, v,
b) 1.2m Schmidt plates taken according to the Tonantzintla
prescription (Haro and Luyten 1962). Three bandpasses permit
two color-index dimensions, both of which enable color excess
relative to the large F and G subdwarf population to be
detected. Thus US objects with blue (b-v) and/or ultraviolet
(u-b) color excess are called B-UVX objects in order to
distinguish them from UVX objects selected by the one-
dimensional (U-B) color index only.

The primary goal of the US survey has been the selection
of B-UVX objects for followup spectroscopic classification and
photometry. The selection and classification methods of the
survey follow more or less directly from the fundamental
observation that the photometric properties of low redshift
(z<2.2) quasars and hot evolved stars overlap to some extent

in the conventional (U-B), (B-V) two-color diagram (Sandage
1965; Greenstein 1966), but are clearly distinguishable from
those of the numerous halo subdwarfs. Therefore, the essential feature of the US survey is that selection is dependent
only on color calibration relative to the halo F/G subdwarfs.
Thus the need for determining the actual Johnson (U-B), (B-V)
color indices of objects is completely bypassed, and so also
are the systematic and random errors that accumulate in the
process of calibration.

The salient features of the US survey which have contributed to its accuracy as a finding list for intrinsically B-UVX
objects are described in Usher (1981) and Mitchell (1987).
Initially, objects are selected by eye during several careful
searches of the 3-color plates. The selection includes all
objects perceived to be bluer than the numerous F/G halo
subdwarfs in either of the two color indices, as well as a
number of the probable F/G subdwarfs themselves. Uncali-
brated, differential iris photometry of the 3-color images
then produces color classes (CC) for the selected objects
based on their location in the relative (u-b, b-v) two-color
diagram, as shown in Figure 1. The CC 1-type US objects,
having a high selection completeness and colors either (U-V)<0
or "above" the blackbody line in Figure 1, can be expected to
contain nearly complete samples of quasars with Z<2.2 (Sandage
1972). Single-color Palomar Schmidt plates are used to
determine morphological classes, B-magnitudes with rms

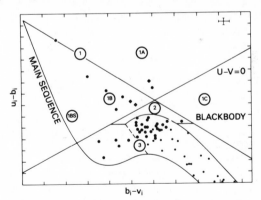

Fig. 1   A relative iris two-color diagram used for the
selection and classification of US objects in one of the
survey fields. The observed distribution of the many F/G
subdwarfs and other red stars initially selected by eye
provides a natural color calibration through which the
approximate color relationships shown in the plot can be
inferred. Blue and ultraviolet-excess (B-UVX) objects
are thereby quantitatively isolated and color classified
into one of the color class 1-type groups.

errors of ±0.1–0.2 mag which are calibrated by using the
standards of Sandage (1978) and Purgathofer (1969), and
accurate positions using reseau astrometry (Warnock and Usher
1980) or the NASA/GSFC PDS 1010A. Magnitude completeness
limits range from B ≈ 17.8 to 18.7 depending on the field.
The possibility of some incompleteness at bright magnitudes
(B<15) has been discussed by Warnock et al. (1986) and Usher
et al. (1988). Selection completeness is defined only for the
central regions of the six fields which cover a total of ~ 135
square degrees, in order to avoid edge effects on the
photographic plates (Usher 1978). Knowledge of selection
effects and completeness limits makes the US survey a useful
tool for the study of quasar population statistics.

THE MEDIUM BRIGHT QUASAR SAMPLE (MBQS)

A spectrophotometric survey of the cataloged US objects is
being carried out on the Kitt Peak 2.1 m telescope with the
Intensified Image Dissector Scanner system. A 300 l/mm
grating provides a resolution of ~ 20A over the wavelength
range $\lambda3500A–\lambda7000A$. Anywhere from 200 to 500 sky-subtracted
counts/channel are obtained in that part of the raw spectrum
where the continuum energy distribution peaks. This combination of spectral resolution, signal-to-noise, and wide
wavelength coverage well into the blue: (1) maximizes the
efficiency of observing and classifying B-UVX objects; (2)
allows accurate identifications and equivalent-width measurements to be made on quasar emission lines; and (3) provides
coverage of the zero-resdhift Balmer lines and Balmer limit
needed for the accurate classification of the major hot star
spectral types.
      The MBQS is being derived from spectroscopic observations
of magnitude-limited samples of cataloged CC1-type and CC2 US
objects, and a few CC3 objects. Spectroscopic coverage of all
CC2 objects ensures statistical completeness in the final
identified samples even if small errors exist in the original
assignment of color classes, or if quasars with only moderate
B-UVX are present in the catalogs. Intrinsically red (U-V >
0) quasars and cooler stars will not have complete spectroscopic coverage even if they exist in the incompletely
selected lists of CC3 objects. The MBQS has doubled in size
since the initial results were reported (Mitchell et al.
1984), and now includes 61 quasars with Z < 2.2, M_B < −23, and
$15.6 \leq B \leq 18.2$. Also, a complete sample of 84 hot stars of
various spectral types has been isolated (Mitchell et al.
1987). The cataloged CC1-type objects suffer only a ~ 15%
contamination by F/G subdwarfs and A stars, and only one CC2
object has been found to be an intrinsically B-UVX object (a
white dwarf), thereby providing a measure of the accuracy of

the US catalogs at these magnitudes.

Preliminary results in the form of differential number counts are presented in Figure 2a for the MBQS, and are compared to those of other bright quasar samples. The MBQS and SA29 samples are both isolated from the US survey and show, on average, a relative overabundance with respect to those quasar samples selected by the (U-B) color index only. The AB sample between 17.5 < B < 18.0 may be the exception, showing a differential surface density indistinguishable from that of the MBQS. The AB sample is apparently most incomplete at brighter magnitudes (B < 17) where no AB quasars are reported, whereas three are expected over the AB survey area based on the results of the MBQS (Mitchell et al. 1984). The apparent dearth of bright quasars near the north galactic pole has been discussed by Warnock et al. (1986), but has since been partially alleviated in the MBQS SA57 sample by extending the spectroscopic survey area in that field.

Four of the six US survey fields have complete spectro-scopic coverage of CCl-type objects for B≤17.8. Cumulative quasar surface densities to this magnitude limit in these four fields range from 0.36 to 0.51 quasars per square degree, with a mean of 0.45 and an rms dispersion of 0.07. The expected rms dispersion due to √n counting statistics is 0.14. Thus the MBQS in these four fields provides evidence that the quasar surface density for B≤17.8 is homogeneous between separate regions of sky as small as ~ 30 square degrees.

The redshift distribution of the MBQS is presented in Figure 2b. The bi-modal shape is typical of other bright quasar surveys. Preliminary results from the spectro-photometric survey of US objects indicate that this distribu-

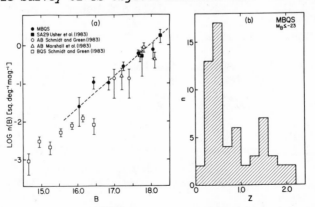

Fig. 2   (a) Differential number counts for the MBQS and other bright quasar samples (M_B<-23). The dashed line is an approximate best fit through the (filled-in) symbols corresponding to samples isolated from the US survey. (b) The MBQS redshift distribution.

tion is inherent to the bright quasar population with z<2.2 and not due to redshift-dependent color selection effects. First, the spectroscopic identifications have shown that the differential color photo- metry and the two-dimensional color classification of the US survey have combined to efficiently isolate intrinsically B-UVX objects (such as quasars) from the redder subdwarfs. In addition, because the preliminary spectroscopic coverage of the redder, CC2 US objects has so far found no MBQS quasars with Z<2.2, the number of inherently red quasars with Z<2.2 and B<18 is probably small. Therefore, incompleteness in the MBQS due to color selection effects can be expected to be small.

The MBQS when completed will provide a relatively large quasar sample at intermediate magnitudes, between the BQS and faint samples being isolated at B>19 (e.g., Koo et al., 1986).

REFERENCES

Greenstein, J. L. 1966, Ap.J., 144, 496.
Haro, G. and Luylen, W. J. 1962,Bol. Obs. Tonantzintla y Tacubaya, 3, 37.
Koo, D. C., Kron, R.G., and Cudworth, K. M. 1986, Pub. A.S.P., 98, 285.
Marshall, H. L., Tananbaum, H., Zamorani, G., Huchra, J. P., Braccesi, A., and Zitelli, V. 1983, Ap.J., 269, 42.
Mitchell, K. J. 1987, in Observational Astrophysics with High Precision Data: Proceedings of the 27th Liege International Astrophysical Colloquium.
Mitchell, K. J., Howell, S. B., and Usher, P. D. 1987, IAU Colloquium No. 95.
Mitchell, K. J., Warnock, A. III, and Usher, P.D. 1984, Ap.J. Lett., 287, L3.
Purgathofer, A. T. 1969, Lowell Obs. Bull. No. 147, Vol. VII, 98.
Sandage, A., 1965, Ap.J., 141, 1560.
Sandage, A. 1972, Ap.J., 178, 25.
Sandage, A. 1978, (private communication).
Schmidt, M. and Green, R. F. 1983, Ap.J., 269, 352.
Usher, P. D. 1978, Ap.J., 222, 40.
Usher, P. D. 1981, Ap. J. Suppl., 46, 117.
Usher, P. D., Green, R. F., Huang, K.-L., and Warnock, A. III 1983, in Quasars and Gravitational Lenses: Proceedings of the 24th Liege International Astrophysical Colloquium, p.245.
Usher, P. D., Mitchell, K. J., and Warnock, A. III 1988, Ap. J. Suppl., 66, No. 1.
Warnock, A. III and Usher, P. D. 1980, Pub.A.S.P., 92, 799.
Warnock, A. III, Usher, P. D., Mitchell, K. J., and Howell, S. B. 1986, M.N.R.A.S., 218, 445.

SURVEYS OF ULTRAVIOLET EXCESS QUASAR CANDIDATES IN THREE LARGE FIELDS

J.P. SWINGS, J. SURDEJ and E. GOSSET*
Institut d'Astrophysique, Université de Liège,
Avenue de Cointe, 5, B-4200 COINTE-OUGREE (Belgium).

Abstract : A general description is given of the surveys performed in the fields around NGC 450 and NGC 520, and in the ESO field n° 300 with the aim of detecting ultra-violet excess quasar candidates.

INTRODUCTION

In order to give further evidence for or against the location of quasars in superclusters, in the vicinity of irregular galaxies or nearby the companions thereof (cf. e.g. a review by Arp, 1983), we initiated a long term program for collecting new and extensive observing material on the background density, distribution and luminosity function of quasars in fields near bright galaxies as well as far from them. A first report was given for the field of NGC 5334 (Surdej *et al*, 1982) where we reported on spectroscopic observations of 13 optically selected QSOs in a large ($\simeq 25$ deg^2) field around that SB galaxy. In the present paper we describe some results concerning the fields of NGC 450 (see a preliminary report in Swings *et al*, 1983) and NGC 520, and in the ESO field n° 300 ($\alpha \sim 3$ h , $\delta \sim - 40°$) void of any particularly bright or "active" galaxy. Statistical methods applied to these results are briefly sketched in Gosset, Surdej and Swings (1988; Gosset *et al*., these proceedings).

---

* as of January 1988 : European Southern Observatory, Garching, FRG.

OBSERVATIONS : Search for QSO candidates

The first goal consisted in the detection of ultraviolet excess objects on dual exposure Schmidt plates obtained for our searches at both the Palomar 1.2 m and ESO La Silla 1 m Schmidt telescopes : the ultraviolet exposure was performed behind a UG 1 filter, and the blue one, offset by several arc seconds from the former, was obtained on the same emulsion behind a GG 13 filter. By visual inspection of the double image of a single object, quasar-candidates were chosen on the basis of their U-B color index. In this selection, we expect the ultraviolet threshold to be around U-B = - 0.4, and the limiting magnitude B ≃ 20.0 mag.

The large field ($\simeq 25$ deg^2) plates were scanned twice, slowly, systematically (and tediously !) by two persons on an X-Y machine built for that purpose in Liège : lists of primary and secondary candidates were then drawn for those objects that are common to the two independent double surveys. Our aim was to minimize any bias in the selection of the candidates, so that the latter could subsequently be used for the statistical programs developed by Gosset (1987) and that are briefly described by Gosset *et al.* (1988). It is only at the time of the spectroscopic observations that some subjective selection may take place : brightness of the object, association of targets, proximity to an interesting object, ...

The equatorial coordinates, accurate to about 1-2 arcsec, were measured on the Uccle-Liège digitized Zeiss "Blink" comparator. Direct plates of the field were also used either for coordinate measurements, or for finding chart purposes.

OBSERVATIONS : Spectroscopy of QSO candidates

The bulk of the spectroscopic data has been gathered at the Shectograph (Boller and Chivens spectrograph + intensified Reticon) attached to the Cassegrain focus of the Irénée Dupont 2.55 m telescope (Las Campanas, Chile) : with an entrance slot of 2 x 2 arcsec, and a dispersion of 114 A mm^{-1}, the resolution is of order 3 A. Additional data have been obtained at the :
(i) Palomar 5 m Hale telescope, with either a photon counting system or the "2 D-Frutti" double spectrograph;
(ii) ESO 3.6 m telescope, with an Image Dissector Scanner

(entrance slot 4 x 4", 224 or 171 A mm^{-1}, resolution ⩽ 12 A);

(iii) ESO Max PLanck 2.2 m telescope, with a Reticon Photon
      Counting System, giving about 10 A resolution at 220
      A mm^{-1}.

PHOTOMETRIC OBSERVATIONS

Photometry in areas of the different fields has been performed in the U, B, V bands at the Las Campanas 2.5 m and ESO
1 m telescopes, and CCD frames have been obtained at the 1.5 m
danish telescope at La Silla as well. Once reduced, the data
will lead to accurate limiting magnitudes, which in turn will
enable us to derive meaningful values of quasar densities (in
a specific redshift range).

SOME RESULTS

A brief summary of the presently available results is
given below, in tabular form :

TABLE 1 : Summary of the survey results

Field	Candidates		Objects observed spectros- copically	Number of QSOs	Notes
NGC 450	primary	95	91	59 + 3 assim.	1
	secondary	45	1	(Seyf. or HII)	
NGC 520	primary	86	86	58	
	secondary	59			
3 H, - 40°	primary ~ 450		35	21	2
	secondary ~ 400				

Notes : 1 : see Gosset (1987) for list of objects, spectra,
            line identifications and values of equivalent
            widths, redshifts, etc.
        2 : deeper plate from ESO Schmidt.

It is to be remembered that the redshift range of the QSOs detected here is located between 0.0 and 2.25.

Some associations of objects have been analyzed in more detail, and have led to individual publications, e.g. Q 0107-025 A, B and 3 "nearby" quasars (Surdej et al., 1986), Q 0118-031 A, B, C (Robertson et al., 1986) in the field of NGC 450.

A few results from the statistical tests applied to the objects of our surveys are :
(i) clustering on a   scale of about 10 arc minutes has been detected in the fields of NGC 450 and NGC 520 (see e.g. Gosset, 1987; Gosset et al., 1986, 1988);
(ii) a lack of any meaningful alignment of quasars near NGC 520 has been clearly demonstrated on the basis of our sample (Gosset et al., 1987).

CONCLUDING REMARKS

On the basis of UV excess objects detected on dual exposure plates it appears that two thirds of the primary candidates turn out to be quasars (in the redshift range $0 < z < 2.2$). Statistical tests applied to the candidates and/or to the QSOs, and their results concerning clustering(s) of objects are presented in Gosset et al. (1988, these proceedings).

REFERENCES

Arp, H.C., 1983, Proceedings of 24th Liège Astrophysical Colloquium "Quasars and Gravitational Lenses", p. 307.
Gosset, E., 1987, Ph.D. Dissertation, Univ. of Liège.
Gosset, E., Surdej, J., and Swings, J.P., 1986, Proceedings of IAU Symposium 119 "Quasars", eds. Swarup G. and Kapahi, V., p. 45.
Gosset, E., Surdej, J., and Swings, J.P., 1987, Proceedings of IAU Symposium 124 "Observational Cosmology", eds. Hewitt, A., Burbidge, G., and Fang, L.Z., p. 499.
Gosset, E., Surdej, J., and Swings, J.P., 1988, in "Optical Surveys for Quasars", these proceedings.
Robertson, J.G., Shaver, P.A., Surdej, J., and Swings, J.P., 1986, Monthly Notices Roy. Astron. Soc., **219**, 403.
Surdej, J., Swings, J.P., Arp, H., and Barbier, R., 1982, Astron. Astrophys., **114**, 182.

Surdej, J., Arp, H., Gosset, E., Kruszewski, A., Robertson, J.G., Shaver, P.A., and Swings, J.P., 1986, Astron. Astrophys., **161**, 209.

Swings, J.P., Arp, H., Surdej, J., Henry, A., and Gosset, E., 1983, Proceedings of 24th Liège Astrophysical Colloquium "Quasars and Gravitational Lenses", p. 37.

ACKNOWLEDGEMENTS

Part of this research has been supported by NATO grant n° 0161/87.

A SURVEY FOR FAINT VARIABLE OBJECTS IN SA 57

DARIO TREVESE
Istituto Astronomico, Universitá degli Studi "La Sapienza"
Via G.M. Lancisi, 29, 00161 Roma

RICHARD G. KRON
Yerkes Observatory, The University of Chicago
Box 0258, Williams Bay, WI 53191

ABSTRACT   Nine Mayall prime focus IIIa-J plates spanning an eleven-year baseline are analysed for the detection of variable objects to B = 22.6. Techniques are developed that succeed in independently finding a sample of known QSOs, but few additional objects have yet been identified with high certainty.

1. INTRODUCTION

We present preliminary results of a survey for variable objects at the North Galactic Pole (SA 57) that goes fainter than the Schmidt survey of Hawkins (1983). The high incidence of variability for QSOs found by Kron and Chiu (1981) and by Koo, Kron, and Cudworth (KKC, 1986) encourages the notion that a variability criterion would be moderately efficient as an independent search technique. The connection between degree and character of variability, and redshift, could also yield important clues and constraints to the QSO phenomenon.

2. DATA AND PROCEDURES

The data are derived from digital microphotometer scans of a homogeneous collection of plates of SA 57 (Table I). The scanned area was about 0.37 deg^2 (8000 x 8000 pixels), centered at α(1950) 13^h 06^m 05^s, δ(1950) +29° 39' 30". The 8000 x 8000 rasters for each of the nine plates of Table I were reduced to form catalogues of image parameters for detected sources using the software tools developed at the Observatory of Rome and described in Koo *et al.* (1986) and Pittella and Trevese (1987).

TABLE I   Plate Journal of SA 57
IIIa-J + GG385, UBK-7 corrector, Mayall prime focus

MPF	UT date	exposure (min)	HA(end)	weight
1053	1974 Mar 21	4 5	0W33	1.0
1561	1975 Apr 04	4 5	1E08	0.8
1562	1975 Apr 04	4 5	0E16	0.6
2176	1976 Dec 01	4 5	3E17	1.0
3133	1980 May 17	2 0	1E02	0.6
3622	1982 Jan 31	5 5	0W42	1.0
3919	1984 Apr 05	5 0	1E57	0.6
3921	1984 Apr 05	7 0	2W01	0.8
3977	1985 Apr 26	6 0	3W12	0.8

For present purposes we have restricted the scope of this study as follows. First, we select objects such that the completeness limit is relatively bright, namely J = 22.5. This can be justified primarily on the grounds that the candidate variables ultimately should be studied spectroscopically. Second, we choose only stellar images in order to avoid the complications of variable seeing. The price we pay for the exclusion of the extended objects is that it is possible that the most interesting new discoveries would be among them. Finally, we consider one plate only as defining the original finding list of objects (which means that objects like supernovae that appear only on one of the other plates would not be identified). We chose MPF 1053 as this fiducial plate because it is one of the best plates in terms of overall quality.

We compute for each object on each plate, the dispersion in the magnitude differences Δm compared to MPF 1053, with respect to the mean. Since some plates are known to be better than others on account of seeing, air mass, guiding, *etc.*, we gave relative weights according to Table I for purposes of calculating the standard deviation. We call this fundamental measure of variability $\sigma*$. Any single occurrence of $|\Delta m| > 0.67$ mag was excluded to reject glitches due to dust, *etc.* Only one such exclusion (the largest) per object was allowed. Also, we assume that there are no true variables between MPF 1561 and MPF 1562, and between MPF 3919 and MPF 3921; these pairs were assigned the weighted average Δm at the respective epochs.

3. ANALYSIS AND CONCLUSIONS

Figure 1 shows $\sigma*$ *vs.* the J magnitude on MPF 1053, measured within 1.1 arc sec radius, for all of the stellar objects. The variability criterion is arbitrarily $\sigma* > 0.15$ mag. Since the average measured $\sigma*$ of course increases with increasing magnitude, this criterion will be conservative for brighter objects (the survey is relatively reliable), and it will be liberal for the fainter objects (the survey is relatively complete).

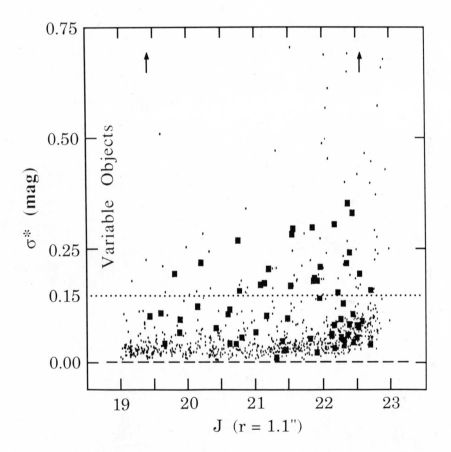

Fig. 1   The weighted *rms* deviation of the magnitude differences of stellar objects, over the 11-year baseline, as a function of magnitude. The variability criterion is $\sigma* > 0.15$. The black squares indicate objects in common with KKC.

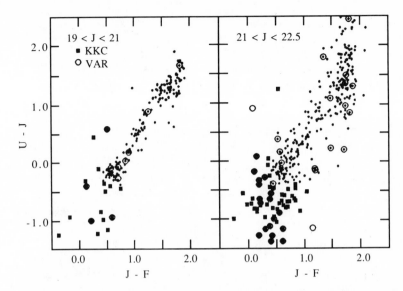

Fig. 2   Color-color diagram for stellar objects in the field. Objects in the KKC study are squares, and objects passing the variability criterion are circles. Circles without dots have J > 22.5. Many of the KKC objects not found to be variable are known spectroscopically to be stars or narrow emission-line galaxies, and are not expected to be variable (*e.g.* the region devoid of variables centered at U-J = -0.7, J-F = 0.7 is rich in narrow emission-line galaxies).

A sensitive diagnostic of systematics in the selection is a plot of the x, y distribution of the objects. It turns out that many of the candidate variables are near one edge of the survey area. This bias is apparently due to a slightly incorrect transformation of the coordinates from MPF 1053 to some of the other plates, so that the photometric subrasters were not properly centered. When this problem is corrected in our next iteration, the appearance of Figure 1 and Figure 2 may change somewhat.

We cross-correlated the catalogue of candidate variables with the photometric catalogue containing UJFN measurements for stellar images (Koo 1986) in order to investigate the color characteristics of the objects selected as variables (Figure 2). Since the QSO sample of KKC was taken from this same photometric catalogue, we can also check the extent to which these were independently discovered. We restrict the analysis area to correspond to the region where the photometry is evidently reliable at all epochs based on good correspondence with KKC. A circle of radius 3300 pixels contains a total of 22 candidate variables, 18 of which were already identified by KKC. Of the four objects not selected by KKC, the brightest object (J = 21.6) has the colors of a subdwarf, and is a good case of a possible QSO that would have been missed by the color-selection criterion of KKC. The other three objects are all fainter than J = 22.5, and so by definition do not appear in the KKC list. One of these three is visible in the right panel of Figure 2 at U-J = 0.92, J-F = 0.10. The probability is very small that it has such peculiar colors and at the same time is selected as a variable, unless it really is a QSO. (Moreover, this part of the color-color diagram is generally populated by high-redshift QSOs.) The other two objects have colors consistent with stars, one red and the other blue.

Hence it appears that in the central area our list is quite reliable, but the completeness is harder to judge. One way is to look at those objects called variables by KKC, *i.e.* those detected by them to have varied at more than one epoch. (KKC used essentially the same plate material as analysed here, but they used independent PDS scans and completely different reduction software.) Of the total of 75 objects studied for variability by KKC, 31 were claimed to be variable. Of these, 20 are independently discovered in this work. Of the remaining 11 objects, all except one is brighter than J = 21.1. It is thus likely that our criterion $\sigma^* >$ 0.15 is too conservative at the brighter magnitudes. Figure 1 shows that the lower bound of σ^* for the KKC QSO candidates is larger than the mean σ^*, at least for J < 21, which implies that the selection threshold could be lowered to achieve greater completeness with little corresponding loss in reliability.

ACKNOWLEDGMENTS

We thank G. Pittella for much assistance with this project, and D. Koo for discussions. RGK thanks the University of Rome for a travel subsidy.

REFERENCES

Hawkins, M.R.S. 1983, *M.N.R.A.S.*, **202**, 571.
Koo, D.C. 1986, *Ap. J.*, **311**, 651.
Koo, D.C., Kron, R.G., and Cudworth, K.M. 1986, *Pub. A.S.P.*, **98**, 285.
Koo, D.C., Kron, R.G., Nanni, D., Trevese, D., and Vignato, A. 1986, *A. J.*, **91**, 478.
Kron, R.G. and Chiu, L.-T. G. 1981, *Pub. A.S.P.*, **93**, 397.
Pittella, G. and Trevese, D. 1987, in *IAU Symp. 121, Observational Evidence of Activity of Galaxies*, eds. E. Khachikian, G. Melnick, and K. Fricke (Dordrecht:Reidel), in press.

A SAMPLE OF QUASARS AT FAINT MAGNITUDES[1]

GIANNI ZAMORANI
Osservatorio Astronomico
Via G.B. Tiepolo 11, 34131 - Trieste - ITALY

VALENTINA ZITELLI
Dipartimento di Astronomia, Università di Bologna
Via Zamboni 33, 40126 - Bologna - ITALY

BRUNO MARANO
Osservatorio Astrofisico, Città Universitaria
Viale Andrea Doria, 95125 - Catania - ITALY

ABSTRACT   We present here a brief summary of the results of
a quasar survey down to J $=$ 22.0 in a field of about 0.7 $sq.deg.$
The list of quasar candidates has been obtained by applying three
different methods of selection: color, presence of emission lines and
variability. Spectroscopic work is presently complete for candidates with J $\leq$ 20.9 and is in progress for fainter objects. A more
complete and detailed description of this work is given in Marano,
Zamorani and Zitelli (1988)

INTRODUCTION

Samples of quasars as complete as possible and without biases with respect to their intrinsic properties (as, for example, redshift, luminosity,
emission line intensity, color) are necessary if one wants to study quasar
evolution and luminosity functions. Well defined and controlled selection criteria are particularly important when conclusions about quasar
properties are derived by comparing with each other results from different surveys, employing different selection techniques (see, for example,
Schmidt 1988). On the other hand, the simultaneous use of various samples is necessary to conduct such a study, because it is well known that
no selection criterion, by itself, can assure not only 100% completeness
but also absence of biases in the selection of the quasar candidates (see
Peterson 1988 and references therein). For this reason, in order to minimize possible biases and losses of objects during the selection of the

---

[1] Work based on observations collected at the European Southern
Observatory, La Silla, Chile.

candidates we have conducted a survey in which three different selection criteria have been applied in the same field. The final aim of this work is to provide a new sample of spectroscopically confirmed quasars with $J \leq 22.0$.

DESCRIPTION OF THE SURVEY

We have selected quasar candidates with $J \leq 22.0$ in an area of about 0.7 *sq.deg.* centered at $\alpha(1950) = 3^h 13.8^m$ and $\delta(1950) = -55°25'$. The final list of candidates results from the application of three different selection criteria (multicolor selection, grism plates and variability analysis).

Multicolor Selection
The multicolor selection has been based on the analysis of two sets of U, J and F plates taken at the ESO 3.6m telescope in November 1981 and November 1982. The three photometric bands are essentially the same as those used by Koo and Kron (1982). Each of the six plates has been scanned with the ESO PDS micro-densitometer in raster mode with an aperture of 50x50 μm. These scans have been used to create a working list of about 6000 objects, in such a way that all the objects with $J \leq 22.5$ are included in this list. Small regions (11.4x11.4 arcsec) around these 6000 objects have later been scanned using a window of 25 μm with a 20 μm step. These higher resolution scans have been used to improve the photometric precision (typical random errors on the average magnitudes are of the order of 0.05 - 0.07) and to better separate extended from point-like objects. Such a separation, based on three different morphological parameters of the images, appears to be extremely reliable for objects with $J \leq 22.0$. Quasar candidates have then been selected from the color - color diagrams (U-J vs. J-F) of both the point-like and the slightly resolved objects. Following the extended color criterion suggested by Koo and Kron (1982), we have considered quasar candidates all the objects whose colors are different from those of normal stars. In order to minimize losses we have included in this list also objects whose colors are at or near the edges of the color distribution of normal stars. The total number of objects selected in this way is of the order of 130. Most of them are objects with ultraviolet excess (with an expected redshift smaller than about 2.2), and very few have the red colors which are typical of high redshift quasars.

Selection From Grism Plates
Searches for emission line objects and/or objects with blue continuum on grism or grens plates are usually highly efficient in finding high redshift quasars. Because of the complementarity of the color and grism selections, we obtained some grism plates at the ESO 3.6m telescope and analyzed the two best of them. The dispersion is about 2200Å/mm and the wavelength coverage is 3400 - 5400 Å. The IIIa-J long wavelength cutoff is such that the Lyα line would be detected up to a redshift of the order of 3.3. The two plates were visually inspected several times

searching for quasar candidates. The criteria adopted for this broad selection were either the presence of possible emission lines or the shape
of the continuum. All the objects in this preliminary list have then been
scanned with the PDS machine at ESO on the best of our plates and
a tighter selection has been performed. At this stage we considered as
grism candidates only the objects showing probable or possible emission
line features. Objects simply showing a blue continuum spectrum were
no longer selected as grism candidates, in order not to duplicate, with
poorer information and on a subjective basis, the color selection. The
total number of grism candidates is of the order of 50. Nine of them
were not selected on the basis of the colors.

Selection From Variability Analysis

In addition to color and grism techniques, it is known that variability
studies can also be useful in searching for quasars. For this reason,
despite the short time baseline covered by our direct plates (one year),
we have searched for variable objects all the stellar objects in our list. For
each band the objects have been divided in various magnitude ranges, in
such a way that each of them contains approximately the same number
of objects (of the order of 250). For each magnitude range we have first
computed the average and the dispersion of the distribution of m_1-m_2,
where m stays for J, F, or U and the subscripts 1 and 2 refer to the
first and second year plates, respectively. Finally, we have considered as
candidate variable objects all the objects whose magnitudes in the two
different years were found to be discrepant in at least two different bands.
This sample is constituted by 30 objects; because of the way these objects
were selected, about five of them are expected to be included in this list
on the basis of purely random errors of the magnitude measurements.

RESULTS

At the time of this writing we have spectroscopic data for all the candidates brighter than J = 20.9 and for a relatively small fraction of
the candidates with $20.9 < J < 22.0$. Our main results (see Marano,
Zamorani and Zitelli 1988) can be summarized as follows:

a) Our counts of quasars and quasar candidates are in excellent
agreement with the results of similar surveys in the magnitude range
$20.0 < J < 22.0$ (Koo, Kron and Cudworth 1986; Boyle et al. 1987;
Crampton, Cowley and Hartwick 1987) and confirm the flattening of the
log N - m curve beyond J = 20.0.

b) We have spectroscopic confirmation for 30 quasars. Twenty-
three of them constitute a complete sample of quasars with J < 20.9.

c) The redshift distribution of the confirmed quasars is approximately flat in the redshift range $0.6 < z < 2.8$. No high redshift quasar
($z \geq 2.8$) has been found so far in this field of 0.7 sq.deg. This absence
of high redshift quasars at faint magnitudes is in qualitative agreement
with the results of other similar surveys (see, for example, Koo, Kron
and Cudworth 1986; Crampton, Cowley and Hartwick 1987; Schmidt,

Schneider and Gunn 1986 and 1987) and with the suggestion of a significant flattening of the luminosity function of quasars at $z \geq 3.0$ (see Warren 1988).

d) Within the magnitude limit of our complete spectroscopic sample, the color and the grism selection were about equally successful. At fainter magnitudes ($21.0 < J < 22.0$), instead, the efficiency of selection from our grism plates is significantly lower than that based on the multicolor technique.

e) The variability analysis yielded 8 confirmed quasars out of 9 variable candidates with $J < 20.9$; all of them were already discovered through the color and/or the grism selection. Additional variable objects, without spectroscopic observations yet, have been found at fainter magnitudes.

REFERENCES

Boyle, B.J., Fong, R., Shanks, T., and Peterson, B.A. 1987, M.N.R.A.S., 227, 717.

Crampton, D., Cowley, A.P., and Hartwick, F.D.A. 1987, Ap.J., 314, 129.

Koo, D.C., and Kron, R.G. 1982, Astr. Ap., 105, 107.

Koo, D.C., Kron, R.G., and Cudworth, K.M. 1986, P.A.S.P., 98, 285.

Marano, B., Zamorani, G., and Zitelli, V. 1988, M.N.R.A.S., in press.

Peterson, B. 1988, This Volume.

Schmidt, M. 1988, This Volume.

Schmidt, M., Schneider, D.P., and Gunn, J.E. 1986, Ap.J., 306, 411.

Schmidt, M., Schneider, D.P., and Gunn, J.E. 1987, Ap.J.(Letters), 316, L1.

Warren, S. 1988, This Volume.

AQD SURVEYS

ROGER G. CLOWES

Royal Observatory, Blackford Hill, Edinburgh EH9 3HJ, Scotland.

ABSTRACT   This paper divides into two parts. The first part is a description of the various types of optical surveys for quasars, the purpose here being to emphasise complementarity and context. The second part is concerned with the progress and results of AQD (Automated Quasar Detection — Clowes *et al.* 1984, Clowes 1986) surveys. It includes descriptions of the AQD technique, AQD surveys, astronomical projects for which AQD is particularly well suited, and the results that have so far been obtained.

1   OPTICAL SURVEYS FOR QUASARS

1.1   Introduction

There are four main types of optical surveys for quasars:

- Ultraviolet excess
- Slitless spectroscopy
- Variability
- Multicolour

Some simple generalisations can be made: all of them work, have selection effects, are best when automated, favour particular subsets of quasars, and are appropriate to particular subsets of astronomical projects.

All of them do indeed work, although in the scramble for funds and telescope time there may be exalted claims for one technique and unedifying economies with the truth to diminish another. Selection effects are often portrayed as awful, but are rarely worse than simple conceptually and a little difficult practically: ignoring them is the real problem. Automation is essential for quantitative data and, except perhaps for the most able practitioners of slitless spectroscopy, is essential for objectivity too. Selection effects imply that particular subsets of the quasar population are favoured; the overlap between techniques is not generally excessively large and complementary information will be obtained. The favoured subsets are appropriate to particular subsets of astronomical projects: for example, ultraviolet excess may be suited to projects on the luminosity function and clustering for $z <\sim 2.2$ but will have nothing to contribute for larger redshifts; an objective-prism survey may be ideal for investigating large-scale structure at intermediate redshifts but have too small a magnitude range to be useful, by itself, for work on the luminosity function.

The following small sections summarise the features of each of these types of surveys. The choice of references is biased first to automated surveys, because automation is the modern and preferred way, and then to those from which information on selection effects and success rates can be extracted. Note that the ubiquity of UK work is due to the unique combination of the UK Schmidt Telescope, the COSMOS (Royal Observatory, Edinburgh) and APM (Institute of Astronomy, Cambridge) fast plate-measuring machines, and the computer software associated with these machines.

1.2   Ultraviolet excess

Originally, ultraviolet-excess surveys were visual searches of separate U and B plates or offset U and B exposures on a single plate. The technique has now been revitalised by automation, which introduces quantitative data and objectivity.

An ultraviolet excess survey would be contained by a multicolour survey. Thus ultraviolet excess would not provide information complementary to multicolour, but multicolour would provide information complementary to ultraviolet excess.

Telescopes
   UK Schmidt.

Detectors
   Photographic.

Plates per field
   A minimum of 1 U, 1 B plates.

Measuring machines
   COSMOS, APM.

Selection effects
   There is a commonly held but mistaken belief that the ultraviolet-excess technique is independent of emission lines. A $U - B$ limit of -0.5 would be sufficient to select all purely power-law continua $f_\nu \propto \nu^\alpha$ $[f_\lambda \propto \lambda^{-(\alpha+2)}]$, with $\alpha >\sim -1.8$, and at all redshifts. The standard power law $\alpha \sim -0.5$ gives $U - B \sim -0.8$. Adding emission lines and the Ly-α forest of absorption lines leads to large changes in $U - B$ as redshift varies. In particular, when Ly-α moves into the B band at $z \sim 2.2$ the ultraviolet excess disappears. Furthermore, as the plots of $U - B$ against z (< 2.2) show (Véron 1983, Boyle *et al.* 1987), there are redshift ranges ($z \sim 0.5$–0.9, 1.5–1.7, 1.9–2.2) in which the $U - B$ colour reaches particularly red values and incompleteness may be substantial — $\sim 30\%$ for $z \sim 0.5$–0.9 (Boyle *et al.* 1987). Quasars with $z <\sim 0.4$ may appear to be rare because their images may be classified as galaxy-like whereas the technique often specifies that images must be star-like.

Further selection effects are possible. In confirming spectroscopy the detectability of emission lines will decrease with increasing magnitude: there may therefore be a tendency for equivalent widths to be larger at fainter magnitudes. If plates are not taken contemporaneously then procedures to avoid spoiled and spurious images could cause variable quasars to be omitted.

Success rates

Boyle *et al.* (1987) confirm as quasars $\sim 40\%$ of their ultraviolet excess objects for $b < 21$. Spectroscopic confirmation of candidates is essential: the inefficiency of observation implied by the moderate success rate can be overcome at faint magnitudes ($\rightarrow$ high surface densities) by multi-object spectroscopy. Note that multi-colour surveys can improve the success rate for "ultraviolet excess" to $\sim 90\%$, at least for the brighter quasars (Miller & Mitchell, this conference), but at the expense of more plates and work.

Projects

Major projects include:

1. The luminosity function for $z <\sim 2.2$ (Boyle *et al.* 1987).

2. Clustering of quasars and large-scale structure of the universe for $z <\sim 2.2$ (Shanks *et al.* 1987).

References

Boyle *et al.* (1987).

1.3 Slitless spectroscopy

Originally, slitless-spectroscopy surveys were visual searches of objective-prism, grism and grens plates. Automation has now introduced quantitative data, guaranteed objectivity, and made feasible very large surveys.

The technique usually incorporates selection by one or both of emission lines and ultraviolet excess. Note that the ultraviolet excess may refer to the continuum only, and then avoids the effects of emission lines. In general, the technique is restricted to a narrower range of magnitudes than direct-plate techniques, and cannot extend so faint. Large areas of sky may be surveyed and large numbers of quasars discovered quite easily.

Telescopes

UK Schmidt (objective prism), CTIO 4m (grism), CFHT (grens).

Detectors

Usually photographic, may be CCD.

Plates per field

A minimum of 1 objective-prism/grism/grens plate. A sky-survey direct plate may also be essential.

Measuring machines

COSMOS, APM, PDS.

Selection effects

The selection effects for slitless spectroscopy in particular have had a bad reputation for no good reason. Several papers show and reiterate that the selection effects are conceptually simple and well understood (Clowes 1981, Schmidt *et al.* 1986a,b, and Gratton & Osmer 1987).

The fundamental selection effect is that limiting equivalent width is a function of magnitude or, equivalently, limiting magnitude is a function of equivalent width. This result should not be surprising and is, of course, also true of confirming spectroscopy for any technique. In principle the limits are also a function of wavelength but usually this dependence is minor. Each emission line is available only for a restricted range of redshifts: for example, for IIIa-J, MgII $\lambda2798$, CIV $\lambda1549$, Ly-α $\lambda1216$ $z \sim 0.14$–0.92, 1.07–2.47, 1.63–3.42 respectively (in practice the ranges are slightly more restricted than these). Ly-α quasars with $z \sim 1.8$–3.0 are very easy to detect.

For ultraviolet excess, the selection effects are a little simpler than for the conventional ultraviolet-excess technique: emission lines are avoided but the Ly-α forest is not.

A further selection effect, usual with automated surveys, is the omission of overlaps, which causes statistically predictable losses of quasars. If star-like images are specified then quasars with $z <\sim 0.4$ may appear to be rare.

Success rates

The success rate may vary greatly depending on how the sample is defined. For example, a sample that is defined by strong lines, or a sample that is defined by both moderately strong lines and ultraviolet excess, may have a success rate $\sim$ 80%, whereas one that is defined by ultraviolet excess and disregards emission lines will have the usual $\sim$ 40% success rate for ultraviolet excess.

Projects

Major projects include:

1. The ROE/ESO large-scale AQD survey (Clowes *et al.* 1987).

2. Clustering of quasars and large-scale structure of the universe for $z >\sim 1.8$ (Clowes *et al.* 1987).

3. Discovery of quasars in preparation for the ROSAT X-ray survey (Beuermann & Clowes 1987).

4. Frequency of occurrence of gravitational lenses (Webster *et al.* 1987).

References

Clowes *et al.* (1984), Clowes (1986); Hewett *et al.* (1985); Schmidt *et al.* (1986a,b); Borra *et al.* (1987); Crampton *et al.* (1987).

1.4   Variability

This is a fairly new technique, which is practicable only with automation. The photographic requirements are very demanding.

Plates are taken in groups: each group comprises plates separated by short intervals, $\sim$ weeks, and groups are separated by larger intervals, $\sim$ years (Hawkins 1986).

Selection requires that an object varies between groups of plates but does not vary significantly within groups. This last condition that there should not be short-term variations minimises contamination by stars (e.g. RR Lyrae stars) and galaxies (from photometric difficulties).

Telescopes
UK Schmidt.

Detectors
Photographic.

Plates per field
E.g. $\sim$ 20 B plates.

Measuring machines
COSMOS.

Selection effects
Very little detailed information is presently available on the selection effects. The technique is often portrayed as having no major redshift biases, but, as Keable (1987) emphasises, this claim is not yet justified. It is made only because quasars have been discovered at both low and intermediate redshifts — but the same is also true of slitless spectroscopy, which does have redshift biases. Certainly, redshift biases are plausible first because of time dilation and secondly because the intrinsic time-scale of variability might depend on cosmological epoch.

The condition that there should not be short-term variations minimises contamination by stars but, of course, also excludes quasars which vary on a time-scale $\sim$ weeks.

Success rates
The success rate may vary greatly depending on how the sample is defined. Hawkins (1986) achieved a success rate of 100% for an appropriately defined sample of 11 candidates with $17.0 < B < 18.5$. Once contamination begins, however, it increases rapidly (Hawkins, private communication). Note that two of these 11 quasars have redshifts of 1.56 and 1.58 (two of three in the range 1.5–1.7 — see the section on ultraviolet excess) but are too red for selection by ultraviolet excess. There are no firm statistics on the fraction of quasars that are detectable as variables, but Hawkins (1986) estimates 40–70%.

Projects
An example is:

1. The galaxy components of quasars (Hawkins & Woltjer 1985).

References
Hawkins (1986); Koo *et al.* (1986).

1.5 Multicolour

This too is a fairly new technique, which is also practicable only with automation. The photographic requirements are demanding.

Plates are taken for at least three passbands (e.g. Marano *et al.* 1987) and frequently for as many as five (e.g. Warren *et al.* 1987a,b). Selection is by identifying outliers on colour–colour plots or in multi-dimensional passband space. Two plates per passband allow selections of spoiled and spurious images to be avoided.

Telescopes
 UK Schmidt, ESO 3.6m, KPNO 4m.

Detectors
 Photographic.

Plates per field
 E.g. 2 U, 2 B, 2 V, 2 R, 2 I plates.

Measuring machines
 COSMOS, APM, PDS.

Selection effects
 Very little detailed information is presently available on the selection effects. The fundamental selection effect is that it is difficult to define the classes of quasars that will be lost in the stellar locus (Keable 1987 and Miller & Mitchell, this conference). The quasar candidates are selected by some algorithmic definition of outliers in multi-dimensional passband space, and so form a heterogeneous set with no simple defining limits. The difficulty then is that quasars otherwise similar to members of this heterogeneous class but having one or more of different redshifts, spectral indices, emission-line strengths and broad-absorption troughs may be lost in the stellar locus. The complexity of the quasar loci as these quantities vary may causes the losses to remain unknown. The difficulty is minimised by using a large number of passbands to reduce the proportion of passband space occupied by the stellar locus. Even with five passbands, however, quasars in the redshift range $\sim$ 2.2–3.2 often lie within the stellar locus.

Further selection effects are possible. The technique specifies star-like images and so quasars with $z <\sim 0.4$ may appear to be rare. If the two plates per passband are not taken contemporaneously then procedures to avoid spoiled and spurious images could cause variable quasars to be omitted.

Success rates
 Statistics on success rates are presently quite scarce. Miller & Mitchell (this conference) achieved a success rate $\sim$ 90% for bright "ultraviolet-excess" quasars ($z <\sim 2.2$) using five passbands. Warren *et al.* (1987a) achieved 2/12 or $\sim$ 20% for $z > 3.3$ again using five passbands.

Projects
 Major projects include:

1. High-redshift quasars (Warren *et al.* 1987a,b).

2. A large-scale multicolour survey for bright quasars (Miller & Mitchell, this conference).

References
 Koo & Kron (1982); Koo *et al.* (1986); Marano *et al.* (1987); Warren *et al.* (1987a,b).

1.6 Other types of optical surveys

Kron & Chiu (1981) (see also Koo *et al.* 1986) have used absence of proper motion as a selection criterion. The photographic requirements are very demanding, and all of the quasars in their small sample of 10 would have been discovered using conventional techniques. The technique may well prove to be a valuable addition in future.

1.7 Complementarity

The above sections describing the main types of optical surveys for quasars should show that there is more reason for emphasising complementarity than competition. No single technique subsumes the others as the following table of imprecise quantities summarises for current surveys based on the UK Schmidt Telescope:

Technique	Magnitude	Area	Redshift	Plates per field
Ultraviolet excess	Faint	Small	Low	Minimum of 2
Slitless spectroscopy	Intermediate	Large	Intermediate, low	Minimum of 2
Variability	Bright	Small	Low, intermediate	E.g. $\sim$ 20
Multicolour	Bright	Large	All	Minimum of 10
Multicolour	Faint	Small	All	Minimum of 10

The most productive approach to quasar surveys must be to use combinations of techniques (e.g. Koo *et al.* 1986, Marano *et al.* 1987). Ideally perhaps, these combined surveys should be applied to *selected areas*, but if the universe is isotropic then this restriction is not strictly necessary.

2 AQD SURVEYS

2.1 Introduction

Section 1 was intended to show the complementarity of the main types of optical surveys for quasars. The remainder of the paper concentrates on the AQD technique, which is in the category of slitless spectroscopy.

AQD, for Automated Quasar Detection, is the generic name for the system of software and procedures at the Royal Observatory in Edinburgh (ROE) that allows quasar candidates to be discovered from measurements of objective-prism or similar plates. It is based on the COSMOS fast plate-measuring machine (MacGillivray & Stobie 1984) at ROE, and typically uses plates from the UK Schmidt Telescope (UKST).

AQD was developed to remove from an inherently powerful technique the drawbacks of visual scanning, which include unknown and time-varying selection criteria, wastage of information, and tedium. A long period of software development was necessary, but the end result realises the full potential and allows large numbers of quasars to be discovered in large areas of sky, with pre-defined and constant criteria, recording of all important data, and relatively little tedium. The initial version of AQD is described by Clowes *et al.* (1984) and the current version by Clowes (1986).

The following options are available for finding quasars or, indeed, for finding other classes of objects with appropriately distinctive spectra:

- Emission lines
- Absorption lines
- Continuum discontinuities
- Ultraviolet excess
- Red excess

Of course emission lines and ultraviolet excess are the most productive options for quasars, and the others are intended for the rarer types such as BAL and high-redshift quasars.

The limiting values of the selection criteria are deliberately set quite low. Consequently many candidates are selected, most of which will not be quasars. The purpose here is to make all potential candidates available so that grading of candidates and re-selections can then be performed on this much smaller database according to the needs of particular applications.

Some important features of AQD are these:

1. The usual magnitude range is B $\sim$ 17.0–20.5.

2. AQD excels at detecting quasars in the range $z \sim 1.8$–3.0, but can, in fact, detect quasars at all $z <\sim 3.0$.

3. Many quasar candidates are so obvious that while spectroscopy is important for establishing the identifications of lines and for accurate redshifts it is not strictly necessary for their *confirmation.*

4. AQD is well suited to projects that require large numbers of quasars and/or coverage of large areas of sky.

5. The photographic requirements are easily satisfied. A minimum of one objective-prism plate and a sky-survey direct-plate are required. The prism plates must be of the highest quality but, even so, a collection of $\sim$ 100 suitable UKST plates exists.

6. The CPU requirements are quite small on a small VAX, (usually $\sim$ 40 hours on a VAX 11/780 for one UK Schmidt field) but processing in real time can be rather more lengthy at present because of the very heavy loading of the computer resources at ROE.

7. Losses from overlapped spectra occur at a rate $\sim$ 15%, but may be made negligible by also processing a second prism plate with the dispersion direction rotated by 90° relative to the first.

8. The maximum area that can be measured by COSMOS in a single measurement is 286.72 × 286.72 mm^2, which for a UKST plate is $\sim$ 28.6 deg^2.

A few simple statistics readily illustrate that AQD is a very efficient way of finding quasars in large numbers over large areas of sky. The Hewitt & Burbidge (1987) compilation lists 3681 quasars that have been published in a 23 year period since the first quasars were discovered. In a period of just eight months the ROE/ESO large-scale AQD survey (Clowes *et al.* 1987) yielded $\sim$ 1500 high-grade candidates from the five best plates ($\sim$ 140 deg^2); these are candidates for which the success rate is $\sim$ 80%. Including other surveyed fields the total number of high-grade candidates is $\sim$ 3000. Additionally $\sim$ 3000 quasars are expected to be present in the lower grades, for which confirming observations will be required.

2.2 AQD surveys

The following table lists the UKST fields involved in AQD surveys:

Field	Field centre	Survey
294	00 26 00 −40 00 00	ROE/ESO
295	00 52 00 −40 00 00	ROE/ESO
296	01 18 00 −40 00 00	ROE/ESO
297	01 44 00 −40 00 00	ROE/ESO
351	00 48 00 −35 00 00	ROE/ESO
352	01 12 00 −35 00 00	ROE/ESO
411	00 46 00 −30 00 00	ROE/ESO
119	05 04 00 −60 00 00	ROSAT
120	05 42 00 −60 00 00	ROSAT
416	02 41 00 −30 00 00	
899	01 20 00 +05 00 00	
927	10 40 00 +05 00 00	
N1	02 53 00 +00 20 00	
SGP	00 53 00 −28 03 00	

The ROE/ESO survey and the survey for ROSAT are described by Clowes *et al.* (1987) and Beuermann & Clowes (1987) respectively; the other fields are for various special-purpose, single-plate surveys. A new, major survey of $\sim$ 10 fields is planned.

2.3 Projects and results

AQD is well suited to projects that require large numbers of quasars and/or coverage of large areas of sky. A valuable feature is that one often has a reasonable estimate of the probability that any particular candidate will be a quasar.

The following is a list of some AQD projects:

1. Clustering of quasars and large-scale structure of the universe for $z >\sim$ 1.8 (Clowes *et al.* 1987).

   Clowes *et al.* (1987) analysed the clustering properties of high-grade candidates from the best five ($\sim$ 140 deg^2) of the seven fields of the ROE/ESO survey. Redshifts were assigned on the assumption that the typically single lines were

Ly-α , and were then restricted to the range 1.8–2.4. High-grade candidates were estimated to be $\sim$ 80% quasars, and the assumption of Ly-α was estimated to be correct in 70% of cases. Given these estimates the analysis was sensitive to very weak clustering (to a level of 7% of quasars occurring in pairs on scales $\sim 5h^{-1}$ Mpc) but none was found.

2. Discovery of quasars in preparation for the ROSAT X-ray survey (Beuermann & Clowes 1987).

The ROSAT satellite is expected to conduct an all-sky survey in soft X-rays. The expected density of AGNs is typically $\sim$ 2–3 deg^{-2}, but reaches $\sim$ 30 deg^{-2} at the ecliptic poles because of the repeated exposures there. AQD surveys are in progress to discover quasars in some of these deep-survey areas in advance of ROSAT. The fields chosen are 119 and 120, close to the SEP, with an expected source density of $\sim$ 10 deg^{-2}, but not so close to the LMC that overcrowding is too serious. The project is still in its early stages, but the first set of confirming spectra has now been obtained.

3. Common and associated absorption in close pairs of quasars.

AQD has overcome the previous shortage of close pairs of quasars for studies of common and associated absorption. It can find $\sim$ a few good pairs with separation $<\sim$ 2 arcmin in each field. Two pairs, of separation 1.34 and 1.80 arcmin, have been examined for MgII absorption but none was found (presently unpublished data).

4. Discovery of gravitational lens candidates.

The project is still in its early stages. High-grade candidates have been examined for evidence of gravitational lensing. One candidate lens has been observed: the brighter object was confirmed as a quasar, but the second object was too faint for satisfactory spectroscopy.

REFERENCES

Beuermann,K. & Clowes,R.G., 1987. In: *Observational and Analytical Methods Relating to Large-scale Structures in the Universe, Workshop of the Astronomisches Institut der Universität Münster,* in press.

Borra,E.F., Edwards,G., Petrucci,F., Beauchemin,M., Brousseau,D., Grondin,L. & Beaulieu,A., 1987. *Publs astr. Soc. Pacif.,* **99,** 535.

Boyle,B.J., Fong,R., Shanks,T. & Peterson,B.A., 1987. *Mon. Not. R. astr. Soc.,* **227,** 717.

Clowes,R.G., 1981. *Mon. Not. R. astr. Soc.,* **197,** 731.

Clowes,R.G., 1986. *Mitteilungen der Astron. Ges.,* **67,** 174.

Clowes,R.G., Beard,S.M. & Cooke,J.A., 1984. *Mon. Not. R. astr. Soc.,* **207,** 99.

Clowes,R.G., Iovino,A. & Shaver,P.A., 1987. *Mon. Not. R. astr. Soc.,* **227,** 921.

Crampton,D., Cowley,A.P. & Hartwick,F.D.A., 1987. *Astrophys. J.,* **314,** 129.

Gratton,R.G. & Osmer,P.S., 1987. *Publs astr. Soc. Pacif.,* **99,** 899.

Hawkins,M.R.S., 1986. *Mon. Not. R. astr. Soc.,* **219,** 417.

Hawkins,M.R.S. & Woltjer,L., 1985. *Mon. Not. R. astr. Soc.,* **214,** 241.

Hewett,P.C., Irwin,M.J., Bunclark,P. Bridgeland,M.T., Kibblewhite,E.J., He,X.T. & Smith,M.G., 1985. *Mon. Not. R. astr. Soc.,* **213,** 971.

Hewitt,A. & Burbidge,G., 1987., *Astrophys. J. Suppl.,* **63,** 1.

Keable,C.J., 1987. *PhD thesis, University of Edinburgh.*

Koo,D.C. & Kron,R.G., 1982. Astr. Astrophys., **105,** 107.

Koo,D.C., Kron,R.G. & Cudworth,K.M., 1986. *Publs astr. Soc. Pacif.,* **98,** 285.

Kron,R.G. & Chiu,L.-T.G., 1981. *Publs astr. Soc. Pacif.,* **93,** 397.

MacGillivray,H.T. & Stobie,R.S., 1984. *Vistas Astr.,* **27,** 433.

Marano,B., Zamorani,G. & Zitelli,V., 1987. *Mon. Not. R. astr. Soc.,* in press.

Schmidt,M., Schneider,D.P. & Gunn,J.E., 1986a. *Astrophys. J.,* **306,** 411.

Schmidt,M., Schneider,D.P. & Gunn,J.E., 1986b. *Astrophys. J.,* **310,** 518.

Shanks,T., Fong,R., Boyle,B.J. & Peterson,B.A., 1987. *Mon. Not. R. astr. Soc.,* **227,** 739.

Véron,P., 1983. In: *Quasars and Gravitational Lenses, 24th Liège Astrophysical Symp.,* p. 210.

Warren,S.J., Hewett,P.C., Irwin,M.J., McMahon,R.G., Bridgeland,M.T., Bunclark,P.S. & Kibblewhite,E.J., 1987a. *Nature,* **325,** 131.

Warren,S.J., Hewett,P.C., Osmer,P.S. & Irwin,M.J., 1987b. *Nature,* **330,** 453.

Webster,R.L., Hewett,P.C. & Irwin,M.J., 1987. Preprint.

DISCUSSION

Marshall : What kind of statistics do you have on the
selection efficiencies for overlapping redshift ranges, for
example, $1.8 < z < 2.2$, where it is possible to check against
the UVX method?

Shanks : We did a check in our UVX areas by doing an eyeball
prism survey. We went down to about 19-19.5 in the B band on
UK Schmidt prism plates and found Lyman alpha emission in 60%
of the UVX objects.

SPECTROSCOPIC CCD SURVEYS FOR QUASARS AT LARGE REDSHIFT

MAARTEN SCHMIDT
Palomar Observatory, California Institute of Technology,
Pasadena, CA 91125

DONALD P. SCHNEIDER
Institute for Advanced Study, Princeton, NJ 08540

and

JAMES E. GUNN
Department of Astrophysical Science, Princeton University,
Princeton, NJ 08544

ABSTRACT   We discuss the first results from a 4-Shooter
CCD grism transit survey for large redshift quasars. We
find that the co-moving space density of quasars with
Lyman-α line luminosity exceeding 10^{45} erg/sec at redshift
3.3 is about 7 times smaller than that at redshift 2.2.

INTRODUCTION

Our present knowledge about the space distribution of quasars
at large redshift is characterized by two features. First,
while for redshifts below 2.2 the co-moving space density for
quasars of given luminosity rises dramatically with redshift
(see Schmidt and Green 1983), at high redshifts there appears
to exist a decrease in quasar density or redshift "cutoff",
based on Osmer's failure to find quasars in a search covering
redshifts in the range 3.7-4.7 (Osmer 1982). Second, there is
considerable uncertainty about the completeness of photographic
prism or grism surveys in which the candidate quasars are
selected by visual inspection. This was illustrated in
particular by Osmer (1980) who showed that Curtis-Schmidt
objective prism surveys and 4-meter grism surveys yield numbers
of quasars at the same magnitude differing by a factor of up
to 10.
    There is not much doubt that the redshift cutoff does
exist. Schmidt, Schneider, and Gunn (1986a,b) failed to find
any quasars with redshift exceeding 3 in two surveys with
well-defined completeness limits, whereas dozens were expected

based on a smooth extrapolation of numbers observed at lower redshifts. Various hypotheses have been proposed to explain the redshift cutoff. The main hypotheses are (a) that the quasar phenomenon was activated only several billion years after the Big Bang, (b) that quasars suffer spectral evolution such that those at very large redshift are not easily recognized as quasars, and (c) that dust absorption causes the observed drop in numbers, as proposed by Ostriker and Heisler (1984).

In order to test these hypotheses, it is essential to have available substantial samples of high redshift quasars that are statistically complete to well-defined limits. In particular, the criteria used for selection of quasars should be explicit and objective in order to avoid the uncertainty accompanying selection by eye. We will describe the essential features of our current survey program and give a preliminary evaluation of the first complete sample of high redshift quasars obtained recently.

CCD GRISM TRANSIT SURVEYS

We are using 4-Shooter (Gunn et al. 1987) with the 200-inch telescope at Palomar in transit mode (Schmidt, Schneider, and Gunn 1987). In this mode, the CCDs are read out continuously at the sidereal rate. The area covered by the four CCDs is 8.5x8.5 arcmin, so a transit survey yields about 2 square degrees per hour of observation at the celestial equator. Each 4-Shooter survey strip is observed twice: once through color filters to provide a catalog of objects in the survey area, and once through grisms to produce low-dispersion spectra. Photometric calibration is provided by the observation of standard stars. Accurate flat fielding for CCD columns of constant declination is obtained from sky readings taken during the entire survey. For details of the reduction procedure, readers are referred to Schmidt, Schneider, and Gunn (1986b).

Emission-line candidates are selected on the basis of an algorithm requiring (1) that the emission-line flux exceed 7 times the noise of the underlying background, and (2) that the equivalent width of the line be at least 50 Å. Slit spectra of all emission-line candidates are obtained to confirm and identify the emission line and to determine the redshift.

As an example, we show in Figure 1 the grism spectrum and the slit spectrum of PC 0751+5623, which has a redshift of 4.30, the third-largest redshift known at the time of this Workshop.

PRELIMINARY EVALUATION OF A COMPLETE SAMPLE

Observational work on the first 4-Shooter survey has been completed. The survey area covers RA = 20^h48^m-4^h54^m at declination

1°29!8-1°38!2 (epoch 1985.8). The useable area is 14.3 degrees.
Among around 100,000 grism spectra recorded, about 180 objects
were selected as having a grism emission line fulfilling the
abovementioned criteria. Slit spectra of all quasar candidates
have been observed. A total of 143 line detections were confirmed, yielding a list of 44 quasars and 85 galaxies.

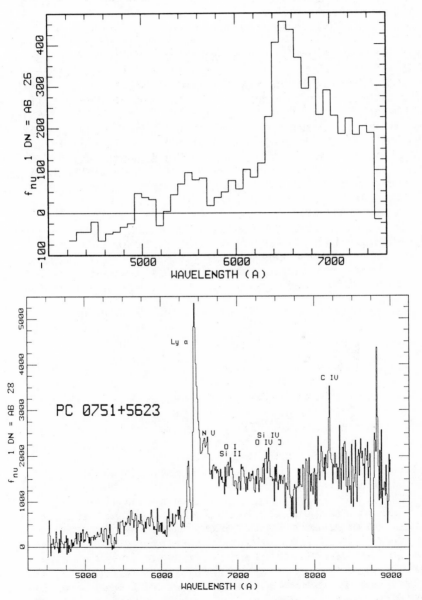

Fig. 1 Grism spectrum (top) and slit spectrum (bottom)
of quasar PC 0751+5623, redshift 4.30.

The limit of the survey is essentailly determined by the detection of an emission line on the sky background. The photometric calibration of the survey, together with the seeing profile and the sky brightness, set the limiting line flux as a function of wavelength. Since it is line flux rather than continuum magnitude that defines the survey limit, we will discuss the results in terms of line fluxes and line luminosities.

For all the emission-line objects found in the survey, we perform a modified V/V_m test as follows. We hypothetically move each of the objects along the line of sight until its emission line becomes undetectable given the survey critera. Suppose this happens at redshift z_{min} if we move the object toward us, and at z_{max} if we move it out. If $V=V(z)$ is the co-moving volume out to redshift z then, in obvious notation, $V_e/V_a=(V-V_{min})/(V_{max}-V_{min})$ is a measure of the object's location in the volume available to it within the survey limits. We have derived the average of this quantity separately for objects detected by the different emission lines. The results are given in Table I.

TABLE I

Line	# of lines	$\langle z \rangle$	$\langle V_e/V_a \rangle$
Hα	37	0.06	0.46±0.05
[O III]	58	0.15	0.52±0.04
Mg II	9	1.10	0.68±0.10
C III]	9	1.96	0.57±0.10
C V	15	2.47	0.47±0.07
Lyα	15	3.30	0.36±0.07

The first two entries refer to Hα or [O III] detections of compact galaxies or extragalactic H II regions. Since these objects are not active galaxies and are at small to moderate redshifts, their space distribution at given luminosity is expected to be uniform. This is borne out by their $\langle V_e/V_a \rangle$ values which do not differ significantly from 0.50.

The next two entries refer to quasars with redshifts below 2. The large observed $\langle V_e/V_a \rangle$ values are qualitatively consistent with the strong evolution exhibited by these quasars (Schmidt and Green 1983). For the Lyman-α detections at a mean redshift of 3.3, we find that $\langle V_e/V_a \rangle$ is substantially below 0.50, corresponding to a decline of co-moving density with increasing redshift. While this determination is only significant at the two sigma level, we will see below that at redshift 3.3 the density is declining sharply with increasing redshift.

Before we derive a space density of quasars detected by
their Lyman-α emission, it is of interest to check the effect
of the limiting value of 50 Å for the observed equivalent
width. This limit corresponds to a rest equivalent width of
12 Å at the mean redshift of the Lyman-α lines observed in this
survey. The median rest equivalent width of recorded Lyman-α
lines in quasars is 75 Å with a total range of about a factor
of 10. Hence, our equivalent width criterion is not expected
to exclude any quasars detected by their Lyman-α emission.

Since the volume surveyed depends on the absolute Lyman-α
luminosity, some care is required in interpreting the observational results in terms of the luminosity function. Based
on a preliminary evaluation of the slope of the luminosity
function and of the density decline, we find that the Lyman-α
sample produces the following value of the luminosity function:

$$\Phi(L(Ly\alpha) > 10^{45} \text{ erg/sec}, z = 3.3) = 56 \text{ Gpc}^{-3},$$

where we have used H_o = 50 km s^{-1} Mpc^{-1}, and Ω = 1.

It is of interest to see whether there is information on
the Lyman-α luminosity function at somewhat lower redshifts
than our mean redshift 3.3. Both Osmer (1980) and Crampton et
al. (1987) give information on line strengths for surveys based
on grism or grens plates. These surveys list 30 and 18 quasars,
respectively, with redshifts in the range 2.0-2.5. Somewhat
fortuitously, the Lyman-α luminosity limit of 10^{45} erg/sec
corresponds to a rather conservative flux limit for the two
surveys: 9 and 7 quasars, respectively, are above the limit.
Hence, we may hope that these restricted subsamples are complete, even though the full samples were selected by visual
inspection of the plates. The numbers observed lead to the
following values of the luminosity function:

$$\Phi(L(Ly\alpha) > 10^{45} \text{ erg/sec}, z = 2.2) = 397 \text{ Gpc}^{-3} \text{ (Osmer survey)},$$
$$= 384 \text{ Gpc}^{-3} \text{ (Crampton survey)}.$$

On the basis of this preliminary discussion, we conclude that
the co-moving density of quasars with the indicated luminosity
decreases by a factor of 7 between redshifts 2.2 and 3.3.

For redshifts below 2 all information on the luminosity
function is based on continuum luminosity, usually in the
B-band. The relation between B-band luminosity and Lyman-α
luminosity depends on the equivalent width of the Lyman-α line,
and on the ratio of continuum flux between 1216 and 4400 Å.
Since the Lyman-α equivalent width distribution shows a total
range of about a factor of 10, the proper statistical relation
to be used is subtle, since selection bias similar to Malmquist
bias will enter. We have used at this stage the simple
equivalence of M_B = -25.9 and $L(Ly\alpha)$ = 10^{45} erg/sec, based on

a median equivalent width of Lyman-α of 75 Å and a spectral
slope of -0.5, but this surely will have to be examined more
carefully.

We have used the luminosity functions given in Boyle et
al. (1987) and in the review by Boyle (this Workshop) to derive
the luminosity function $\Phi(M_B < -25.9)$ for redshifts between 0
and 2. This is plotted in Figure 2 together with the z = 2.2
value from the Osmer and Crampton surveys, and the z = 3.3
value based on our 4-Shooter survey. Even while the match
between the continuum luminosity function and that based on
Lyman-α luminosity may be subject to revision, it is clear
that the co-moving density of these quasars reaches a peak at
a redshift of around 2 to 2.5 and then declines sharply. The
data in Figure 2 were derived for $\Omega = 1$. For $\Omega < 1$, the
decline in space density for z > 2 would be steeper yet.

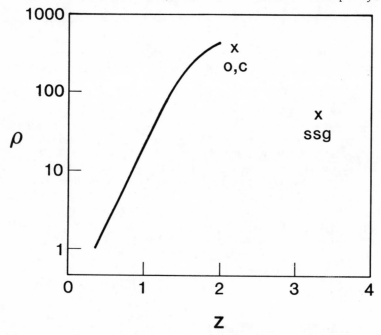

Fig. 2 Co-moving density of quasars with M_B <-25.9 (line)
or $L(Ly\alpha) > 10^{45}$ erg/sec (crosses) as a function of redshift. The crosses are based on the Osmer and the
Crampton et al. surveys (O,C) and the survey discussed in
this paper (SSG).

DISCUSSION

These first results based on a well-defined sample of high
redshift quasars are encouraging. We are planning to enlarge

the survey material by a factor of at least five in order to reduce the statistical uncertainty which at the larger red-shifts is still the dominating source of uncertainty. With an enlarged sample, we should also get information about the shape of the luminosity function, the shape of the continuum energy distribution, the distribution of line strengths, etc., which will be useful in evaluating the various hypotheses put forward to explain the redshift cutoff.

REFERENCES

Boyle, B. J., Fong, R., Shanks, T., and Peterson, B. A. 1987, M.N.R.A.S., 227, 717.
Crampton, D., Cowley, A. P., and Hartwick, F. D. A. 1987, Ap. J., 314, 129.
Gunn, J. E., et al. 1987, Opt. Engineering, in press
Osmer, P. S. 1980, Ap. J. Suppl., 42, 523.
Osmer, P. S. 1982, Ap. J., 253, 28.
Ostriker, J. P., and Heisler, J. 1984, Ap. J., 278, 1.
Schmidt, M., and Green, R. F. 1983, Ap. J., 269, 352.
Schmidt, M., Schneider, D. P., and Gunn, J. E. 1986a, Ap. J., 306, 411.
Schmidt, M., Schneider, D. P., and Gunn, J. E. 1986b, Ap. J., 310, 518.
Schmidt, M., Schneider, D. P., and Gunn, J. E. 1987, Ap. J. (Letters), 316, L1.

DISCUSSION

Peacock : Do you get less of a cutoff for $\Omega = 0$?

Schmidt : I have not yet tried anything but $\Omega = 1$. It would be worth repeating for $\Omega = 0$.

Turner : If you used a scale linear in time instead of in redshift, you would get a very narrow peak.

Schmidt : Yes, it would be incredibly narrow, leading one to say that there was only one epoch of quasar formation (according to my perhaps biased view).

Kron : am surprised that the points don't bunch up more at the faint limit. Is that because you have the equivalent width criterion?

Schmidt : I don't think so. There are too few points to be able to say much, but I have a suspicion that the shape of the luminosity function is involved.

Osmer : Clearly the completeness of slitless spectrum surveys is a critical question. Yet, with an observed limit of 50Å or 12Å in the emitted frame for the minimum detectable equivalent width, I doubt if any significant number of Lyman alpha quasars are being missed, because it is such a strong line. I think the burden is on those with doubts to show that there is a trend of line strength with redshift or that there is a significant population of very weak lined quasars.

Schmidt : If you wish, you can really think of this study as one of the evolution of the broad lined clouds that surround quasars, not of the quasars themselves. If the two evolve differently, then that becomes interesting in itself. In view of the uncertainties to date in this field, I think it better to take a well-defined sample, which in this case is based on the emission line luminosities. In the end, however, it will be necessary to tie the line and continuum results together.

Marshall : A comment and then a question. I applaud the use of methods with strict and definitive limits for the data, and you have already commented on the only objection I would have, namely, how can the line and continuum data be tied together. My question is, are not the line data also consistent with no evolution over the whole redshift range, at least from the V/Vm test?

Schmidt : Ahh, look here. I'm not going to prove from this

survey that there is quasar evolution which initially
increases steeply with redshift and then declines. I do
consider that the evolution at low to moderate redshift is
well established from other work, and I am pleased that these
data are consistent with that result.

The quasar luminosity function at redshifts $z > 3$

Stephen J. Warren, Paul C. Hewett
Institute of Astronomy, Madingley Road, Cambridge CB3 OHA, UK

Patrick S. Osmer
Kitt Peak National Observatory, National Optical Astronomy Observatories,
Tucson, Arizona 85726, USA

Abstract

We describe some of the important features of a relatively deep (m_R=20.0) wide-field multicolour survey designed to measure the luminosity function of quasars at high redshift. Low-resolution spectra of candidates in the first 30 square degree field confirm 19 of the candidates as new quasars of redshifts $z > 3.0$. These quasars possess a wide range of colours and a great variety of spectral type. By a preliminary quantification of the sample incompleteness and combining the new quasars with the 5 previously known high-redshift quasars in this field we are able to make a first estimate of the quasar luminosity function at $z = 3$ and $z = 4$, over a range of two magnitudes. The results confirm and quantify a decline in co-moving space density as indicated by earlier surveys covering small areas of sky.

1 Introduction

The implications for cosmology of the apparent decline in co-moving space density of high-redshift ($z >\sim 3.5$) quasars reported by Osmer (1982), if real, are of great significance. However this feature in the redshift distribution of quasars remains unlikely to figure prominently in the debate on theories of galaxy formation until the decline in number density is confirmed and mapped in a quantitative manner, and the question of incompleteness satisfactorily addressed. The more recent null detections of a number of deep searches for high-redshift quasars over small areas of sky (Koo, Kron and Cudworth 1986, Schmidt, Schneider and Gunn (hereafter SSG) 1986 a,b), and the contrasting success of Hazard, McMahon and Sargent (1986) in finding several bright high-redshift quasars in wide-field searches, have stimulated the development of objective techniques for conducting searches over large areas of sky (tens of square degrees), to relatively faint magnitude limits, and for which the incompleteness of the samples may be quantified. In particular the grism/CCD transit survey of SSG (1987 a,b, 1988), and the wide-field photographic plate multicolour technique described here, are both designed to detect high-redshift quasars in significant numbers. The notable recent successes in this field, with the detection of several quasars of redshift $z > 4$, are attributable largely to improvements in technology (CCDs, measuring machines, making available very large quantities of digitised source material), and the employment of sophisticated computer routines for the selection of candidates.

In the following section we discuss some of the important features of the multicolour survey underway at Cambridge, which employs scans by the Automated Plate Measuring machine (APM) of United Kingdom Schmidt Telescope (UKST) direct plates. In section 3 we describe the results of follow-up spectroscopy of candidate quasars observed in the 1986 and 1987 seasons, for the first 30 square degree field surveyed. A preliminary quantification of the incompleteness of this sample allows an estimate of the luminosity function to be derived, in section 4. We close with a brief discussion of the results in section 5.

2   Multicolour survey

The rationale behind the Cambridge multicolour survey, and some of the more spectacular results have been presented by Warren *et al.* (1987 a,b,c). These references include a rather general outline of some of the survey procedures adopted. In the following we provide a brief summary of the method and go on to discuss some of the notable features of the survey.

The multicolour technique, as applied to searches for quasars, entails the identification of star-like objects with colours that are unusual in comparison with the colours of common galactic stars (see e.g. Koo, Kron and Cudworth 1986). Follow-up spectroscopy is required to confirm the nature of these candidate quasars. The technique has the significant advantage that it is not necessary, in principle, for a quasar to possess strong emission lines for it to be detectable. We select candidates over as broad a range of colours as possible as we find that quasars of similar redshifts can have widely different colours.

Our photometric data come from APM scans of high-quality UKST plate pairs in each of five passbands U, B_J, V, R and I. By extending the number of passbands from 3 (produces 2-colour diagrams) to 5 (produces four-dimensional (4-D) multicolour space), our survey will be sensitive to a wider range of spectral type of quasar and to a broader redshift band.

There are three novel features of this survey, which we consider in more detail below. Firstly, we allow for the possibility that some objects will be absent from some of the plates. For an object to be included in the data set we require only that matching images appear on both the R plates. Secondly, we dispense with 2-colour diagrams in the selection of candidates. Instead we employ a computer search to identify outliers in the 4-D multicolour parameter space. Thirdly, we will quantify the sample incompleteness through the derivation of the 'selection function', which describes numerically the whole of the detection procedure i.e. the difference between the intrinsic and the observed luminosity functions.

In the long term the survey will cover 60 square degrees and the redshift range $2.2 < z < 4.5$, to a limiting magnitude of $m_R = 20.0$. We report here on a preliminary analysis of the results from the first 30 square degree field, the South Galactic Pole (0 h 53 min., $-28° 03'$ (equinox 1950.0)), and the redshift interval $3.0 < z < 4.5$, for which follow-up spectroscopy is largely complete.

2.1   "missing" images

A sample of stellar images that appear on all ten of the U, B_J, V, R and I plates will exclude many of the interesting objects which it is the very purpose of the survey to identify, because they are absent from one or more of the plates. In particular, high-redshift quasars missing from the shorter-wavelength passbands will not be found. On the other hand the condition that objects appear on more than just one plate is an effective means of removing spurious images, which would have apparently interesting

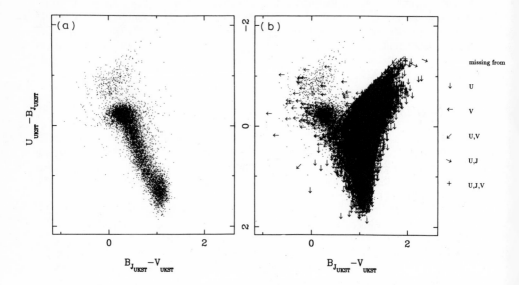

Figure 1: UB$_J$V 2-colour diagrams a.) all stellar images in the SGP field brighter than $m_R = 20.0$ that appear on at least one plate in each of the U, B$_J$ and V passbands, b.) the same diagram including objects missing from one or more passbands.

colours and which, therefore, would swamp the candidate list. The R plates are the deepest of the long-wavelength passbands, and we have compiled a set of broad-band magnitudes for all the objects that match on the two R plates, with no requirement that images appear on any of the U, B$_J$, V and I plates. This maximises the area of the redshift – magnitude diagram that can be investigated effectively. For objects that are absent from both plates of a particular passband, upper limits to their brightnesses corresponding to the limit of the deeper plate, provide the optimal magnitude information. A sample apparent-magnitude limit m_R=20.0 is set, beyond which the photometric errors become excessive. Fig. 1 illustrates the large number of missing objects brighter than the survey limit that are included in the sample.

2.2   candidate selection

Candidate quasars are selected by identifying star-like objects that have unusual colours. The five broad-band magnitudes (or limits to magnitudes) define four colours (conventionally U-B$_J$, B$_J$-V, V-R, R-I although we use 'orthonormal' axes as defined by Warren *et al.* 1986 b) which are the coordinates of each object in a 4-D multicolour parameter space. Candidate quasars are selected by identifying objects that lie in low-density regions of this 4-D space, away from the locus of normal stars. By increasing the number of colour dimensions from two to four the range of redshifts over which quasars are detectable and the variety of spectral type of quasar that can be found are greatly increased.

Rather than search through a series of 2-colour diagrams to identify candidate quasars we have developed a density-search algorithm that selects outliers by measuring the 4-D distance an object lies from the stellar sequence. There are two principal advantages of this approach, neither of which is widely appreciated. Firstly, the method is much more effective at identifying objects of unusual colour than an inspection of 2-colour diagrams, primarily because the 4-D vector defining the shortest distance from an outlier to the stellar locus will, at best, appear foreshortened in any of the 2-colour diagrams used, and, more often than not, will be completely hidden within the normal stars. Secondly, the density-search technique provides a means of quantifying the incompleteness of the survey, because, as described below, it allows the calculation of the probability of detection for any type of quasar, making it possible to derive the 'selection function' that defines the difference betwen the observed and the true luminosity functions. In addition the density search has the advantages of reducing the labour involved in the selection of candidates, of allowing the incorporation of the photometric errors, and, further, it can provide a ranking of candidate quasars ordered by how far from the stellar locus they lie.

With regard to the choice of algorithm for the selection of outliers, potentially a considerable body of work on topics in cluster analysis and pattern recognition bear on this area. Nevertheless we have found that the use of a simple 'isolation criterion' is effective in picking out known high-redshift quasars and finding new ones. For each object in the catalogue we calculate a 4-D distance (with appropriate subtraction of the photometric errors) to all the neighbouring points that are of similar magnitude or brighter (note that faint normal stars could envelop a bright outlier as the stellar locus swells with the increasing photometric errors). We define an isolation criterion of the distance to the nth nearest neighbour, where $n = 6$ (Miller and Mitchell 1988 employ a similar approach, with $n = 10$): this gives an estimate of how different an object is from the nearest population of common galactic stars. If the value of n is too small outlying groups of two or three objects of similar colours might be missed, while if n is too large normal stars in a slightly underdense region of the stellar locus will become selected in preference to single outliers.

While it is possible that a more sophisticated approach would prove more effective, we have found that the large majority of the previously known and the newly found quasars lie at the very top of the ranking list output from the density search. The success in terms of the wide range of spectral type of quasar identified, and the broad range of colours over which detected quasars of similar redshift are spread has vindicated the adoption of a general approach in which no prespecification of the regions of the multicolour space to be searched has been made.

The incorporation of limits to colours for objects missing from different passbands can be achieved relatively simply and in an optimal manner. For images absent from only one passband, the colour limit defines a semi-infinite line in 4-D space along which the object might lie. These objects are selected as candidate quasars if the line does not intersect the stellar locus (the distance to the 6th nearest neighbour is calculated at different positions along the line), and further, if the object has colours disimilar to all the other objects in the same category that likewise do not intersect the stellar locus. For images missing from two or three passbands the colour limits define respectively an area bounded by two semi-infinite lines and a volume bounded by three such lines, and in the same manner the object is eliminated if the area/volume intersects the stellar locus. Over two-thirds of our sample of quasars of redshift $z > 3$ fall into the category of 'missing' objects. Indeed one of the quasars in our sample appears on only three of the ten plates. Note that the upper limits provide successively less information than the real magnitudes as one goes fainter, and this contributes to the search procedure becoming progressively less efficient at faint magnitudes. The drop-off in efficiency is established in deriving the selection function.

2.3   incompleteness

The use of a density-search algorithm for selecting candidate quasars allows the calculation of the selection function for the survey. From a knowledge of the photometric errors in the determination of the object magnitudes in the different passbands it is possible to calculate precisely the probability of detecting a quasar of any particular combination of the three variables, *vis.* restframe spectral type, redshift and apparent magnitude. The colours of the object are computed and run several times through the density search, varying them in a statistical manner as appropriate for the known errors. The selection function is fully described by the array of computed probabilities for each combination of the above three variables.

The selection function is a necessary requirement for the analysis of the quasar luminosity function. In principle the intrinsic luminosity function (and its evolution over the redshift range of interest) may be derived by deconvolving the selection function from the observed luminosity function, limiting the conclusions to the region of the spectral type – redshift – magnitude parameter space for which the probability of detection is significantly non-zero. Note that this procedure requires that candidate quasars have been observed in a uniform manner i.e. all possible regions of the multicolour parameter space have been investigated, down to some uniform threshold value of the isolation criterion. Where, as in our case, the number of quasars detected is relatively small the calculated intrinsic luminosity function will be subject to large errors so that it is preferable to invert the procedure (Peacock 1985) and to identify the range of intrinsic luminosity functions which, convolved with the selection function, are compatible with the observations. We are currently undertaking such an investigation. Meanwhile we present here a preliminary calculation of the intrinsic luminosity function, based on a simplified estimate of the sample incompleteness.

3   Observational results

Follow-up spectroscopy of candidate quasars was undertaken during August and September 1987 at the 4-metre telescope at Cerro Tololo Inter-American Observatory and at the Anglo-Australian Telescope (AAT) (Warren *et al.* 1987c), following a preliminary assessment of the candidate list at the AAT in August 1986 (Warren *et al.* 1987 a). We observed candidates of widely different colours, while concentrating on the volume of multicolour parameter space in which we believed it conceivable that quasars in the redshift interval $3.0 < z < 4.5$ could lie, and follow-up spectroscopy in this region is largely complete. We excluded objects with colours $U-B_J < -0.35$, for which the numbers of high-redshift quasars are known from the very large ultra-violet excess survey of Boyle *et al.* (1988), as well as the region R-I>1.5, bearing in mind that at $z = 4.5$ Ly$-\alpha$ is still within the R band. In all we observed 157 candidates of which 65 proved to be quasars (21 of these have redshifts $z < 2.2$). The success rate was generally higher for the bluer objects observed, and noticeably higher for candidates at the head of the density-search ranking.

In the SGP field a total of 19 new quasars with $z > 3.0$ have been identified, including three with $z > 4.0$. The five quasars with $z > 3.0$ already known in this field were readily detectable, making a total sample of 24. The number-redshift distribution for these quasars is shown in Fig. 2, and the locations of the quasars in a B_JRI 2-colour diagram of all the stellar images in the SGP field down to the survey limit are shown

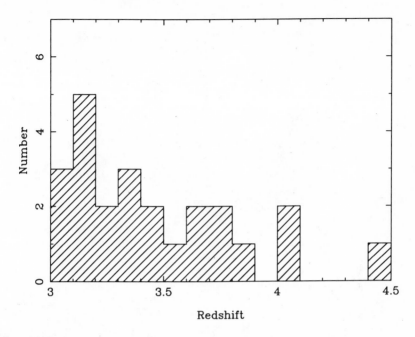

Figure 2: The number – redshift distribution for the 5 previously known and 19 new quasars of redshift $z > 3$ in the SGP field

in Fig. 3. Note that many of the quasars lie in the stellar locus in this diagram, but have been successfully picked out in the 4-D density search. The quasars in our sample exhibit a very wide range of colours, and include broad-absorption-line quasars and quasars with weak emission lines. The surface density of this sample is similar to that achieved by SSG (1988) for the same redshift range, but it is likely that the multicolour survey detects a somewhat higher fraction of the total quasar population — including objects without strong emission lines — but to a brighter limiting magnitude. Using parameters, $H_0 = 50$ km s^{-1} Mpc^{-1}, $q_0 = 0.5$, $\alpha = -0.5$, the absolute magnitudes of these quasars lie predominantly in the range $-28.0 < M_B < -26.0$.

4   Luminosity function

We have undertaken an analysis of the luminosity function over the redshift interval $3.0 < z < 4.5$, on the basis of a number of simplifying assumptions, pending an in-depth assessment of the incompleteness of the sample through the measurement of the selection function. This preliminary analysis provides an indication of the shape and normalisation of the high-redshift luminosity function, and how it evolves. We have based our calculations on m_R, which we assume provides a reasonable estimate of the continuum magnitude at this wavelength i.e. we have made no correction for emission or absorption within the passband. In order to reduce the large errors associated with the coarse binning of a small number of data points, we have adopted the assumption of simple density evolution over the absolute magnitude – redshift range defined by our sample i.e. we suppose that the luminosity function retains its shape but that the normalisation varies as a function of redshift. Choloniewski (1987) provides an optimal

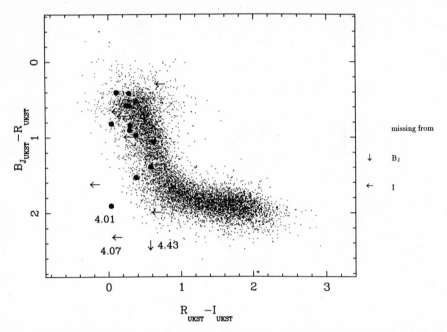

Figure 3: The $B_J RI$ 2-colour diagram for stellar images in the SGP field brighter than $m_R = 20.0$, showing the 24 quasars of redshift $z > 3$. The large number of normal stars missing from the B_J and I passbands have been omitted for clarity.

technique for calculating the orthogonal luminosity $\phi(M_B)$ and density $\rho(z)$ functions that describe this form of evolution, with what amounts to an extended version of the 'C' method of Lynden-Bell (1971). The functions $\phi(M_B)$ and $\rho(z)$ are derived in the form of a series of weighted delta functions, corresponding to the individual data points, which may be smoothed to produce continuous functions. The smoothed functions define the co-moving space density of quasars at any point in the redshift – absolute magnitude plane $N(M_B, z) = \phi(M_B).\rho(z)$. Dividing by the selection probability in the appropriate manner gives the estimated true space density.

In estimating the sample incompleteness we have firstly calculated the detection limiting apparent magnitudes, m_R, by making each of the quasars in the data set artificially fainter until the measured density-search distance becomes less than the adopted threshold value for the survey. We note from our data firstly that most of the quasars in our sample lie well away from the stellar sequence, which suggests that the range of spectral type observed is a reasonable representation of the true range. Secondly we find that the distribution of detection limits, for this small number of data points, shows no obvious strong dependence on redshift, at least out to $z=4.0$. If we assume that the observed quasar spectra are representative of the intrinsic range, and that the selection probabilities are redshift-independent then we can simply use the distribution of detection limits to give us the incompleteness as a function of apparent magnitude and adopt this as a first estimate of the selection function. We have incorporated the incompleteness into the calculation of the luminosity function by weighting each data point by the reciprocal of the completeness at that apparent magnitude.

The results of the calculation are illustrated in Fig. 4, which compares the estimate

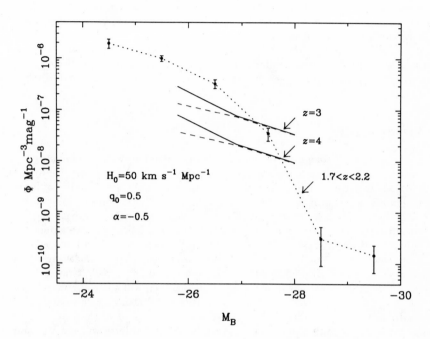

Figure 4: The derived luminosity function at $z = 3$ and $z = 4$ compared to the result of Boyle *et al.* (1988) for $1.7 < z < 2.2$. The dashed line is the result of our calculation with no allowance for incompleteness, the solid line is the result allowing for incompleteness, in the manner explained in the text

of the luminosity function at redshifts $z = 3.0$ and $z = 4.0$, with that measured by Boyle *et al.* (1988) for the redshift interval $1.7 < z < 2.2$. The noticeable features are that, over the absolute magnitude range sampled, the average slope of the luminosity function at high redshifts is rather flat, similar to the slope found for the fainter quasars at lower redshifts, and that there is a marked decline in the co-moving space density of the fainter quasars with increasing redshift. For an absolute magnitude $M_B = -26.0$, at redshift $z = 3.0$ the co-moving space density is down by a factor of nearly three, compared to redshift $z = 2.0$, and by an order of magnitude by redshift $z = 4.0$. There is some indication of an excess of bright $M_B \leq -28.0$ high-redshift quasars, but this conclusion is based on only a handful of objects, and remains to be confirmed by a systematic survey covering a larger area of sky (such as that of Miller and Mitchell 1988, currently underway).

5   Discussion

The data from the multicolour survey confirm and in a preliminary manner quantify the apparent decline in co-moving space density of quasars at high redshifts implied by the null detections of earlier surveys. Indeed despite the seemingly large number of high-redshift quasars in our sample the derived luminosity function is broadly compatible with the majority of the earlier searches. We stress that our conclusions regarding the high-redshift luminosity function are very uncertain in detail and rest on a number of assumptions particularly regarding the selection function, but also, to a lesser extent, on the choice of quasar spectral index, K corrections, cosmological model etc. The completion of the survey over 60 square degrees and over the full redshift range $2.2 < z < 4.5$, and a detailed assessment of the selection function, will allow a more reliable measurement. While we expect that our conclusions will change in a quantitative way, we argue that the qualitative picture of a change in sign of the evolution of quasars and a steady decline in numbers towards higher redshifts will remain, and deserves the closer attention of cosmologists. The extension to both brighter and fainter magnitudes will provide important clues in distinguishing between two different plausible explanations for this decline, of obscuration (Heisler and Ostriker 1988) or quasar formation (Efstathiou and Rees 1988).

Acknowledgements

We are grateful to the staff of the UKST for the excellent plate material, and to the staffs as AAO and CTIO for assistance with the spectroscopic observations. We thank the members of the APM group for their help. S.J.W. is supported by an Isaac Newton studentship, and P.C.H. by the Royal Society.

References

Boyle, B.J., Fong, R., Shanks, T. & Peterson, B.A., 1988, Mon. Not. R. astr. Soc., *in the press.*

Choloniewski, J., 1987, Mon. Not. R. astr. Soc., **226**, 273.

Efstathiou, G. & Rees, M.J., 1988, Mon. Not. R. astr. Soc., **230**, 5P.

Hazard, C., McMahon, R.G., & Sargent, W.L.W., 1986, Nature, **322**, 38.

Heisler, J. & Ostriker, J.P., 1988, Ap. J., *in the press.*

Koo, D.C., Kron, R.G. & Cudworth, K.M., 1986, P.A.S.P., **98**, 285.

Lynden Bell, D., 1971, Mon. Not. R. astr. Soc., **155**, 95.

Miller, L.A. & Mitchell, P.S., 1988, *this workshop*.

Osmer, P.S., 1982, Ap. J., **253**, 28.

Peacock, J.A., 1985, Mon. Not. R. astr. Soc., **217**, 601.

Schmidt, M., Schneider, D.P. & Gunn, J.E., 1986 a, Ap. J., **306**, 411.

Schmidt, M., Schneider, D.P. & Gunn, J.E., 1986 b, Ap. J., **310**, 518.

Schmidt, M., Schneider, D.P. & Gunn, J.E., 1987 a, Ap. J., **316**, L1.

Schmidt, M., Schneider, D.P. & Gunn, J.E., 1987 b, Ap. J., **321**, L7.

Schmidt, M., Schneider, D.P. & Gunn, J.E., 1988, *this workshop*.

Warren, S.J., Hewett, P.C., Irwin, M.J., McMahon, R.G., Bridgeland, M.T., Bunclark, P.S. & Kibblewhite, E.J., 1987 a, Nature, **325**, 131.

Warren, S.J., Hewett, P.C. & Irwin, M.J., 1987 b, in **Proc. IAU Symp. 124 Observational Cosmology**, eds Hewitt, A. & Burbidge, G. Reidel.

Warren, S.J., Hewett, P.C., Osmer, P.S. & Irwin, M.J., 1987 c, Nature **330**, 453.

DISCUSSION

Weedman : How many candidates in your survey turn out to be high-redshift quasars?

Warren : About 1 in 5.

Weedman : Does that mean that the other 80% are not quasars at all, are they quasars of different redshift, or are they objects you can't identify?

Warren : At faint magnitudes, the high redshift objects are contaminated by subdwarfs and by compact galaxies which appear stellar on the plates.

Marshall : If you consider the volume occupied by quasars in your four-dimensional space, what fraction is occupied by stars?

Warren : I only have rough estimates, but it may be as low as 5%. For three dimensions, I have estimated it to be 10%.

Smith : It is interesting that nature appears to have conspired once again to help us in the area of Schmidt telescopes and automatic measuring machines. It is only the very luminous objects which seem not to participate in the rapid decrease in space density at high redshifts. This leads to precisely the situation where only a Schmidt telescope can really enable you to bring up very large numbers of objects, which has been very fortunate.

Warren : Yes. I think one of the main concerns in recent years has been the possibility of selection effects blocking the discovery of high redshift quasars. I have tried to show you that the multicolor technique is really very sensitive to a wide variety of quasar spectra.

REDSHIFT AND LUMINOSITY CHARACTERISTICS
FOR APM OBJECTIVE-PRISM SURVEY QUASARS

GORDON M. MACALPINE and STACY S. MCGAUGH
Department of Astronomy, University of Michigan,
Ann Arbor, Michigan 48109

SCOTT F. ANDERSON and RAY J. WEYMANN
Mount Wilson and Las Campanas Observatories,
813 Santa Barbara Street, Pasadena, California 91101

DAVID A. TURNSHEK
Space Telescope Science Institute, Homewood Campus,
Baltimore, Maryland 21218

PAUL C. HEWETT
Institute of Astronomy, Madingley Road,
Cambridge CB3 0HA, England

FREDERIC H. CHAFFEE, JR. and CRAIG B. FOLTZ
Multiple Mirror Telescope Observatory,
University of Arizona, Tucson, Arizona 85721

INTRODUCTION

The APM objective-prism survey will ultimately produce roughly
1000 quasars with $m_J \leq 18.5$, detected from UK Schmidt direct and
objective-prism IIIa-J plates which are scanned with the Cambridge Automated Plate Measuring facility. In the first of a
planned series of papers, spectroscopic and photometric data
have been reported for 192 quasars in a 102 square degree area
centered on the Virgo cluster (Foltz et al. 1987). Although a
thorough analysis of selection effects has not yet been completed, we present a preliminary discussion of the source-
count, redshift, and luminosity characteristics for this
initial quasar sample. Our purpose here is to demonstrate the
potential of the new survey, and no attempt to derive conclu-
sive results is intended at this time.

QUASAR CHARACTERISTICS

Figure 1 shows integral quasar surface densities, comparing

the APM Virgo field sample with results from various UVX
surveys. The error bars represent statistical uncertainties;
and we have adopted B=m$_J$+0.067. The latter comes from
m$_J$=B-0.28(B-V) (see Harris and Smith 1981) with B-V=0.24
assumed for typical APM quasars (see Barbiere et al. 1982).
Objective-prism surveys have generally been significantly less
complete down to a given limiting magnitude than UVX samples;
for instance, Impey and He (1986) found a factor of two deficiency of quasars to V=20 toward Virgo, using eye searches of
UK Schmidt plates. However, the success of the APM survey
should not be surprising because the low-dispersion spectra on
the discovery plates have good signal-to-noise characteristics
considerably fainter than the chosen magnitude limit of 18.5
and because the automated search employs color as a primary
selection criterion, along with emission or absorption lines
and spectral discontinuities.

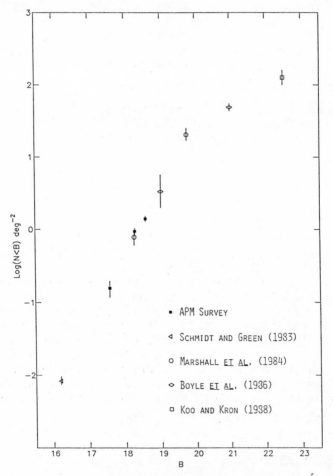

Fig. 1   Quasar surface densities.

In Figure 2, we illustrate how the APM quasars fill in a
previously sparsely populated region of the Hubble diagram,
complete coverage of which will facilitate the derivation of
luminosity function evolution. The diagonal line represents
M_B=-23 for H_0=50 km s^{-1} Mpc^{-1} and q_0=0.5. The Curtis Schmidt
and 4-M Grism data come from several surveys (e.g., MacAlpine,
Smith, and Lewis 1977 or Osmer and Smith 1980 for the former
and Hoag and Smith 1977 or Sramek and Weedman 1978 for the
latter). The new Durham UVX survey quasars (not shown) would
be distributed primarily in the upper right part of the diagram in the magnitude range 19<B<21 (see Boyle et al. 1987).

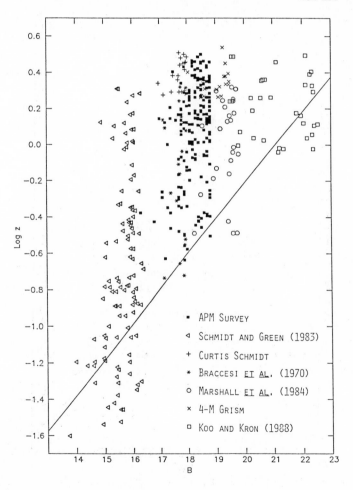

Fig. 2   Quasar Hubble diagram.

Figure 3 compares a broadly-binned part of the redshift
distributions for APM survey quasars (solid-line histogram)

and UM Curtis Schmidt survey quasars (dotted histogram; see
Lewis, MacAlpine, and Weedman 1979). The solid-line and dotted curves represent the same dN/dz evolution model, with the
latter having been corrected for several potentially important
selection effects which are inherent in eye searches of objective-prism plates and which (in this case) result primarily
from the wavelength sensitivity of the IIIa-J emulsion. It is
evident that redshift-dependent selection effects are considerably alleviated in the APM automated survey. Biases that
remain will be recognized and quantified by testing the quasar
selection software using appropriate synthetic spectra.

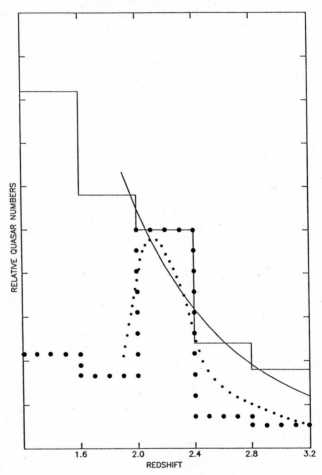

Fig. 3   Partial redshift distributions for APM survey
quasars (solid-line histogram) and UM Curtis Schmidt survey quasars (dotted histogram), normalized at z=2.1.   The
solid-line and dotted curves represent the same evolution
model, with the latter including selection effects (see
text).

     Figure 4 demonstrates the value of the APM quasar sample
for investigating luminosity functions. We show a luminosity
function diagram, adapted from Koo and Kron (1988) for q_0=0.5
and the redshift bin 2.51<z<3.16. The APM survey is useful
for delineating the bright end of the luminosity function at
all redshifts and particularly important for z>2.1 where UVX
surveys are known to suffer from incompleteness due to shifting of strong emission lines into measured wavelength bands.

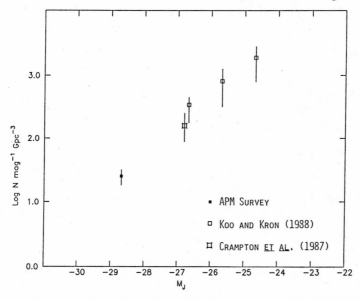

Fig. 4   Composite quasar luminosity function for q_0=0.5
and 2.51<z<3.16.

ACKNOWLEDGMENTS

Part of this work is supported by NSF grants AST-8615485 and
AST-8700741.

REFERENCES

Barbiere, C., Capaccioli, M., Cristiani, S., Nardon, G., and
    Omizzolo, A. 1982, Mem. S. A. It., 53, 511.
Boyle, B. J., Fong, R., Shanks, T., and Peterson, B. A. 1986,
    in Structure and Evolution of Active Galactic Nuclei, ed.
    G. Giuricin (Dordrecht: Reidel), p. 491.
------. 1987, M.N.R.A.S., 227, 717.
Braccesi, A., Formiggini, L., and Gandolfi, E. 1970, Astr.
    Ap., 5, 264.
Crampton, D., Cowley, A. P., and Hartwick, F. D. A. 1987,
    Ap. J., 314, 129.

Foltz, C. B., Chaffee, F. H., Jr., Hewett, P. C., MacAlpine, G. M., Turnshek, D. A., Weymann, R. J., and Anderson, S. F. 1987, A. J., 94, 1423.

Harris, W. E., and Smith, M. G. 1981, A. J., 86, 90.

Hoag, A. A., and Smith, M. G. 1977, Ap. J., 217, 362.

Impey, C. D., and He, X.-T. 1986, M.N.R.A.S., 221, 897.

Koo, D. C., and Kron, R. G. 1988, Ap. J., in press.

Lewis, D. W., MacAlpine, G. M., and Weedman, D. W. 1979, Ap. J., 233, 787.

MacAlpine, G. M., Smith, S. B., and Lewis, D. W. 1977, Ap. J. Suppl., 34, 95.

Marshall, H. L., Avni, Y., Braccesi, A., Huchra, J. P., Tananbaum, H., Zamorani, G., and Zitelli, V. 1984, Ap. J., 283, 50.

Osmer, P. S., and Smith, M. G. 1980, Ap. J. Suppl., 42, 333.

Schmidt, M., and Green, R. F. 1983, Ap. J., 269, 352.

Sramek, R. A., and Weedman, D. W. 1978, Ap. J., 221, 468.

DISCUSSION

Marshall : Is there any overlap of your fields with
traditional UVX surveys, such as Mitchell's?

MacAlpine : We plan to cover 20 to 25 fields which are
accessible to the followup telescopes available to us. We do
intend to overlap with other surveys so we can make
comparisons.

Chaffee : In our work to date we have searched every catalog
of known quasars to see if we have found everything. So far,
this is the case except for some objects in which our spectra
were contaminated by overlaps.

THE EFFICIENCY OF QUASAR SELECTION BY BROAD-BAND COLOURS

L. MILLER
Royal Observatory, Blackford Hill, Edinburgh EH9 3HJ, U.K.

P. S. MITCHELL
Dept. of Astronomy, University of Edinburgh, Edinburgh EH9 3HJ, U.K.

ABSTRACT. We discuss the efficiency with which quasars may be selected by their broad-band UBVRI colours. The multicolour method is assessed from tests using prism-selected quasars and from colours generated from models of quasars. Three regimes of redshift are considered: the traditional UVX regime for $z < 2.2$, the range $2.2 < z < 3.5$, and $3.5 < z < 4.5$. It is concluded that the addition of extra colours to the UVX technique can reduce contamination in bright samples and increase completeness. In the intermediate redshift range the technique appears to be about 80 percent complete, quasars with the strongest emission lines being most susceptible to incompleteness at redshifts around 3. A combination of the technique with the slitless-spectrum technique may prove very effective. At high redshifts quasars tend to have very extreme colours, and the effectiveness of the technique is limited mainly by the depth of the multicolour plate material. Weak-lined objects may be most at risk in this redshift range.

1. INTRODUCTION

The last year has seen some spectacular developments in quasar selection, with the discovery of six quasars at redshifts $z \gtrsim 4$. Four of these quasars were found by virtue of their peculiar broad-band colours (Warren *et al.* 1987a & b). But how *complete* is the multicolour technique? Are there any types of quasar which cannot be selected, and what are the redshift dependences of any selection effects? We shall present some preliminary answers to these questions below.

We shall assume that the selection method is familiar to you: see the work of Warren *et al.* (op. cit. & this workshop) if it is not. Basically, the technique involves measuring the broad band colours of every stellar image on UBVR and I band Schmidt plates with a fast measuring machine and looking for those objects whose colours are peculiar. The selection is achieved by calculating for each object in turn the distance in the four-dimensional multicolour space to the nth nearest other object, that distance being a measure of how isolated the object is in multicolour space. The value of n is not critical: in this work $n = 10$. A value of $n = 100$ gives similar results, and the only important criterion is that n should not be so small that small isolated groups of objects are missed. Quasars can have very extreme colours, and they may be undetected in one or more bands. It is thus important to include objects which only have limits on their colours in the quasar search. Unfortunately there are usually a large number of red stars which are not detected in U, and these can only be dealt with by assigning the U magnitudes to values as close as possible to the main, dense, stellar locus within the constraint of the U plate limit. For red objects the U constraint is usually not very restrictive and hence real quasars in this part of colour space can only be selected if they are peculiar in the other colours not involving U. This restriction is included in the tests below.

The results described below are based on the multicolour survey currently under-way at Edinburgh, which comprises COSMOS measurements of UBVR and I plates in 13 U.K. Schmidt fields. Colours have been derived for about 2×10^6 objects and datasets chosen at B < 18.5 and R < 18.5. We first discuss the results of a spectroscopic study of a UVX sample of quasars selected from the multicolour survey, and then discuss the multicolour technique for redshifts z > 2.2.

2. IMPROVING UVX SAMPLES

The UVX technique has been the most successful at providing us with some physical information on the quasar population: the recent studies of Boyle *et al.* (1987) and Marshall (1985) combined with previous UVX surveys have been able to demonstrate that the quasar luminosity function appears to evolve in a manner which can be de-scribed by pure luminosity evolution. To improve further on those conclusions we need to increase the numbers of quasars in complete samples at intermediate magnitudes to improve the statistics, and to remove the residual selection effects in those samples, as errors in the luminosity function due to these will dominate over errors due to the random statistical uncertainties.

Both these aims can be aided by adding further colours to the UVX technique. At bright magnitudes (B $\lesssim$ 17) UVX samples are dominated by hot stars, a source of contamination which increases the amount of telescope time needed. Adding one or more extra colours can discriminate very effectively between quasars and hot subdwarfs and the most extreme hot white dwarfs. The reason for this is obvious: most low redshift quasars and all hot stars are extremely blue in the U and B bands: but quasars have underlying power-law continua whereas stars have underlying black-body continua, so that including a red passband should be able to discriminate between them. A similar UBV technique has been advocated by Usher *et al.*.

Fig. 1 shows the results of a spectroscopic survey of a UVX sample at B < 18 in two U.K. Schmidt fields. The main stellar locus for stars later than type F is clearly defined, and the A type horizontal branch stars in the Galactic halo can also be seen, with a few hotter stars continuing to follow the main sequence track. Slit spectra of stars on that track show them all to be hot stars. Conversely the quasars are all found at much redder B-I colours. The division is very distinct, and the use of the third colour in this way can greatly reduce contamination of bright UVX samples.

But note also the two quasars which have been discovered redder than the usual UVX limit, with colours U-B $\sim$ -0.25. It would not normally have been possible to select these without excessive contamination from normal stars, but they were selected, using the nearest-neighbour method, on the basis of their UBVRI colours which clearly show them to be non-stellar. They both have redshifts z $\sim$ 0.6, and illustrate the redshift hole in the UVX method (e.g. Wampler & Ponz 1985). In fact these two quasars were selected from a somewhat fainter sample, so the incompleteness of the pure UVX technique is not as great as implied by Fig. 1, and is probably only of order 3 percent, but nonetheless such objects need to be found if the present determination of the form of the luminosity function at low redshifts is to be significantly improved upon.

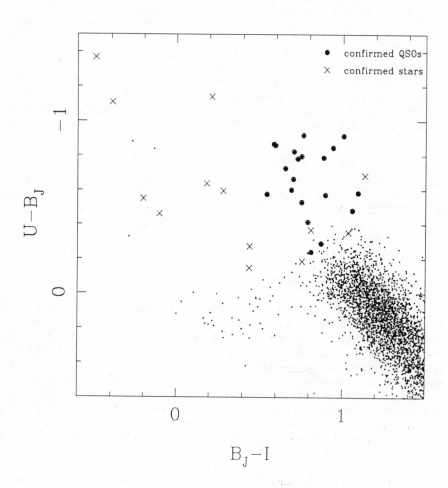

Fig. 1. The U-B/B-I colour diagram from the Edinburgh survey.

3. TESTS AT HIGHER REDSHIFTS

We employ two methods to investigate the technique at higher redshifts. One is to see how efficiently previously-known quasars can be selected. To achieve this we have compiled the U.K. Schmidt colours of as many confirmed quasars as we could find at $z > 2.2$. These are all prism-selected quasars so although this is a vital comparison to make we have to be aware that weak-lined quasars are liable to be missing from this test. To investigate whether this is important, and to extend the test to yet higher redshifts, we have also derived the predicted colours of some toy model quasars with a wide range of parameters. These models cannot tell us the completeness fraction of our surveys, since we don't know the relative numbers of any particular spectral type, but they can tell us if there is a particular class of quasars (such as weak-lined objects) which may be missed by the technique.

Before describing the results let us describe the models used. The basic model comprises a power law continuum of variable slope, on which is superimposed a blue bump and an emission-line spectrum of variable equivalent width (for simplicity the line ratios are fixed at those given by Wilkes 1986 and Baldwin 1975). Absorption by the Ly-α and Ly-β forests is incorporated as a variable blanket suppression of the continuum shortward of Lyman-alpha, an adequate approximation as we are only interested in the broad-band colours of these models. The absorption is assumed to be due to clouds whose number as a function of redshift increases as $(1+z)^{2.2}$ so that higher redshift quasars are more strongly absorbed. An independently-variable amount of absorption shortwards of 912 Angstrom is included: the variation of this absorption with wavelength is assumed to be due the combined effect of the variation in the hydrogen ionization cross-section with wavelength and the same $(1+z)$ dependence for the number of absorbing clouds as for the Ly-α forest. Finally, a small amount of dust extinction is superimposed on the continuum, equivalent to $E_{B-V} \sim 0.02$. This only has a noticeable effect at the very shortest wavelengths.

Five of these parameters were varied in the tests described below: the continuum slope, the emission-line equivalent width, the Ly-α/β and Ly continuum absorption, and the quasar redshift. The other parameters were fixed at values typical for quasars. Fig. 2 illustrates some of the ranges of model quasar spectra produced. The top three models show the range in emission-line strength for a quasar at $z = 3$, the next three the range in continuum slopes and the last three the range in absorption at short wavelengths for a quasar at $z = 3.75$. The broad-band colours were derived by integrating the model fluxes in the U.K. Schmidt passbands, the magnitude zero point being derived from the same procedure applied to the spectrum of an A0 star.

4. THE DATASET

Fig. 3a shows the distribution of nearest-neighbour distances for a dataset selected from the Edinburgh UBVRI survey at B $\lesssim$ 18.5. Objects would be selected for follow-up observations if they lie at large nearest-neighbour distances: the vast majority of objects at small distances ($\lesssim$ 0.1 mag.) are not plotted. The figure shows that use of the distance to the 100th nearest neighbour provides much the same information as the distance to the 10th nearest neighbour: there is no clustering of data outside the main stellar locus. Before looking at where quasars lie on this diagram, it is interesting to ask what the most extreme objects apparent in this particular dataset are. The filled dots are those stars which lie in the A-star region of the colour space - the horizontal branch stars. Slit spectra and objective-prism spectra confirm that these are indeed HB stars. Unfortunately we cannot simply remove all these objects from the dataset because

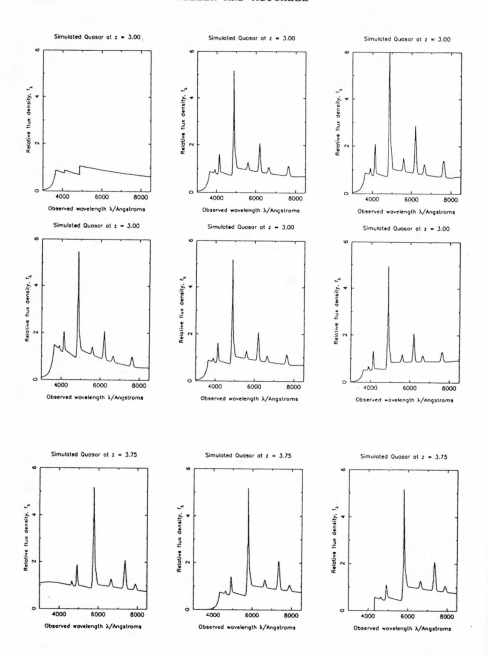

Fig. 2. Models of quasar spectra showing (*top*) the range of assumed emission-line strength, (*centre*) the range of continuum slopes and (*bottom*) the range of short-wavelength absorption.

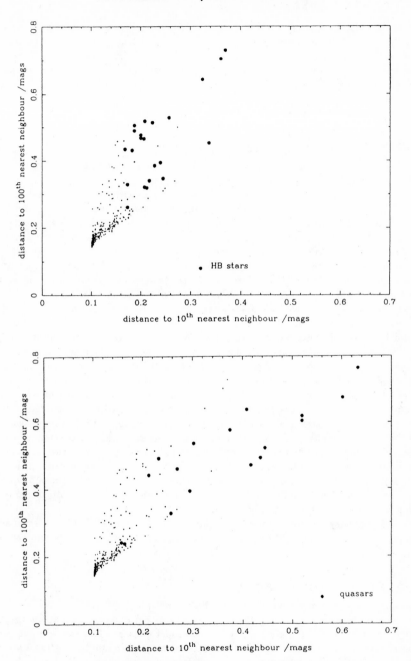

Fig. 3. (*top*) Nearest-neighbour distances in the Edinburgh multicolour survey show-ing the positions of the horizontal-branch stars and (*bottom*) showing the positions of objects with the colours of prism-selected quasars when added into the dataset at its faint magnitude limit.

quasars at redshifts around 3 may be found in that region of colour space. But including them in the dataset also greatly increases the contamination. This dilemma may be resolved by obtaining objective-prism spectra, since at bright magnitudes (B $\lesssim$ 18.5) such spectra can distinguish between these stars with their distinctive A-star spectra and genuine quasars. It is important to emphasise that the HB stars are less of a problem at fainter magnitudes, as we probe further out into the Galactic halo at fainter magnitudes and their apparent magnitude distribution tails off.

5. RESULTS IN THE RANGE 2.2 < z < 3.5

Fig. 3b shows the colours of the previously-known quasars superimposed on Fig. 3a, retaining the HB stars for the time being. If those stars can be excluded, then 15 out of the 17 known quasars would be selected by this technique without undue contamination, implying a completeness for this type of quasar of around 80 percent.

The model quasars reveal which type of quasar may be in the missing 15 percent or so. The redshift range most at risk is 2.7 $\lesssim$ z $\lesssim$ 3. The various parameters are correlated in the effect they have on the nearest-neighbour distance, and Figs 4 a - d show the effects of varying these parameters as a function of redshift. The most important are continuum slope and strength of absorption. Fig. 4a shows the results for quasars with red continua (α = -1, defined as $f_\lambda \propto \lambda^\alpha$) but typical absorption, for three values of the rest-frame emission-line equivalent width of Ly-α: 0, 60 & 120 Angstrom. At redshifts greater than 3.2 the effect of the absorption is to shift the quasar well away from the stellar locus, but at lower redshifts they lie only just above the selection limit shown as the dashed horizontal line, and at redshifts around 3 the **stronger**-lined quasars may be missed. Quasars with bluer continua (α = -2 in Fig. 4b) are more easily selected. At the lower redshifts the variation in continuum slope can successfully bracket the observed range of positions in colour space, but at higher redshifts we see one quasar which, although easily selected, lies closer to the locus than predicted by the models. This is because of a lack of strong absorption shortwards of 912 Angstrom in this quasar, and Fig. 4d shows the effect of a low amount of absorption on the red continuum model, compared with a large amount of absorption in Fig. 4c (in this model there is essentially no U-band flux remaining). Fig. 4c is almost identical to Fig. 4a, but it can be seen that Fig. 4d correctly reproduces the position in colour space of the z = 3.67 quasar. But now at redshifts around 3 only the very strongest-lined quasars are in danger of being missed, contrasting with Fig. 4b.

In summary, around 15 percent of quasars in the range 2.2 $\leq$ z < 3.5 may be missed, those at risk being the stronger-lined quasars with red continua and redshifts around 2.7 to 3.

6. RESULTS FOR z > 3.5

We can conduct the same experiment for higher redshifts, except that now we do not have enough measured UBVRI colours of prism-selected quasars, so that for the time being we will have to be guided solely by the model quasars. Thus we will not yet be able to give a completeness estimate. It is usual for high-z candidates to be selected from somewhat fainter datasets, and here we have chosen a limit of R = 18.5. It turns out that the limit adopted is the most critical single factor, because the technique loses power as objects disappear below the plate limits in each band: quasars at high redshifts always have extreme colours, but we are limited by our ability to measure those colours on the plates. In the Edinburgh survey the plate limits are such that the technique rapidly becomes ineffective at magnitudes fainter than R = 18.5. By using deeper

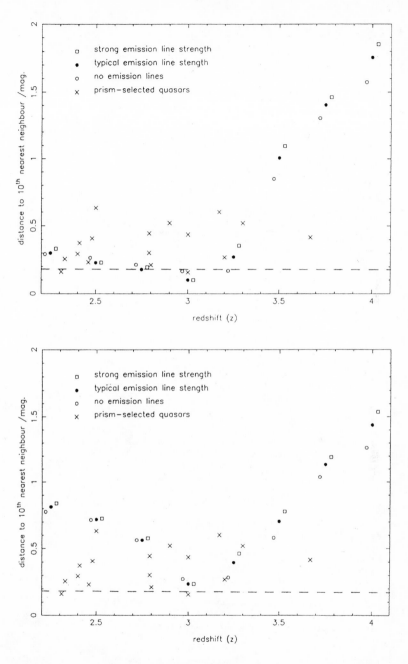

Fig. 4. Nearest-neighbour distances as a function of redshift, for (a) red quasar models with typical short-wavelength absorption, (b) blue quasar models, (c) red quasar models with high short-wavelength absorption, (d) red quasar models with little short-wavelength aborption.

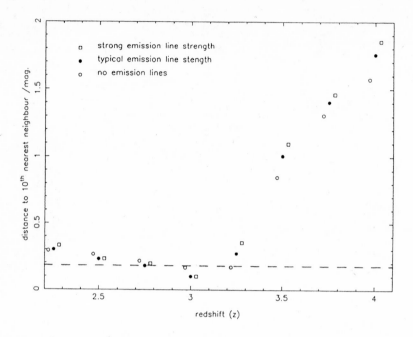

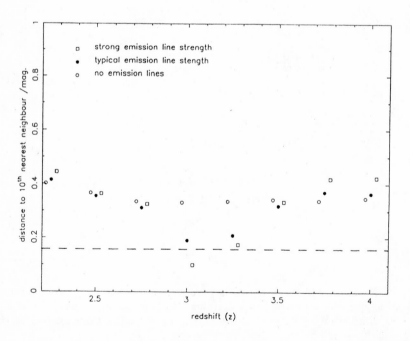

material from the U.K. Schmidt, fainter samples such as that of Warren *et. al.* (op. cit.) may be investigated.

It was found that all the models with rest-frame Ly-α stronger than about 25 Angstroms would be selected over the redshift range $3.5 \leq z \leq 4.5$. Again in this redshift range increasing the amount of absorption leads to larger distances from the main stellar locus. Some of the weaker-lined models may be missed, however, although even here only certain combinations of parameters lead to problems, and we cannot say whether in fact any of these models will be observed in practice. The general feature which the problem models have is that they mimic the colours of either cool stars or faint compact galaxies. At redshifts higher than 4.5 the colours continue to depart from the stellar locus but these extreme colours can no longer be measured on our plate material and several of the models have small nearest-neighbour distances.

7. CONCLUSIONS

By using a combination of the colours of known quasars and the synthesised colours of model quasar spectra it is possible to quantify the effectiveness of the multicolour selection method for any particular dataset. It is important to conduct such tests on each individual dataset as plate limits and measurement errors vary, and the effectiveness of the technique depends largely on these factors. In the Edinburgh UBVRI survey it is found that non-UVX quasars at low redshifts can be found, and that quasars with redshifts around 3 can be selected with around 80 percent completeness. Strong-lined quasars with red continua may be most at risk of being missed, and the multicolour technique should be regarded as complimentary to the technique of searching for emission-line objects.

At higher redshifts it is not yet possible to estimate completeness, but we can say that almost all types of quasar should be selected in the redshift range 3.5 to 4.5 provided that we work sufficiently brighter than the plate limits. Such quasars tend to have rather blue (R-I) colours but red colours at shorter wavelengths, so it important to be able to set good constraints on colours such as (B-R).

ACKNOWLEDGEMENTS

We are grateful to the U.K. Schmidt unit and the COSMOS group for providing the data on which the Edinburgh UBVRI survey is based, to STARLINK for computing facilities, and to Paul Hewett and Steve Warren for providing some of the information used in this study.

REFERENCES

Baldwin, J.A., 1975. Astrophys. J., **201**, 26.
Boyle, B.J., Fong, R., Shanks, T. & Peterson, B.A., 1987. Mon. Not. R. astr. Soc., **227**, 717.
Marshall, H.L., 1985. Astrophys. J., **299**, 109.
Wampler, E.J. & Ponz, D., 1985. Astrophys. J., **298**, 448.
Warren, S.J. et al. 1987a. Nature **325**, 131.
Warren, S.J., Hewett, P.C., Osmer, P.S. & Irwin, M.S., 1987b. Nature **330**, 453.
Wilkes, B.J., 1986. Mon. Not. R. astr. Soc., **218**, 331.

DISCUSSION

Warren : Is there a strong effect of the Lyman continuum at z = 4, or of the Lyman limit?

Miller : Yes. Objects that would be a problem to detect were ones with perverse combinations of various parameters, such as weak lines and strong 912 absorption. However, other quasars with redder continua and strong absorption were easily detectable. No one parameter could be singled out as a problem in itself; the problems came from the different combinations.

Kron : Did you say which photometric errors are required to get the 85% limits?

Miller : I didn't show the proper diagram, but in fact the errors have a negligible effect on the selection process.

Smith : Presumably the use of infrared photometry may enable you to improve the selection at higher redshifts.

Miller : Indeed. The strength of the multicolor technique increases with the wavelength range covered. Infrared measurements will help greatly; the main problem is in achieving coverage over a large enough sample of objects.

A MULTICOLOR CCD SURVEY FOR FAINT QSOS

SCOTT F. ANDERSON AND PAUL L. SCHECHTER
Mount Wilson and Las Campanas Observatories
813 Santa Barbara St., Pasadena, CA 91101

ABSTRACT   In order to further study the faint end of the QSO luminosity function, we are engaged in a multicolor survey for faint QSOs. CCD images are taken through B, g, r, and i filters with the Four-Shooter on the Palomar 5m. Two fields have been surveyed, allowing for multicolor selection of QSOs to $m < 22.5$. The first field overlaps with the SA 57 field previously surveyed for faint QSOs by Koo, Kron, et al., and will provide tests of sample completeness and contamination by non-QSOs. A second, *new* field covering 0.3 sq. deg. has also been surveyed. Initial results are encouraging, and an extension of the survey to $m \approx 24.5$ is planned.

INTRODUCTION

An understanding of the QSO luminosity function (hereafter, LF) and its redshift dependence is central to a number of issues: It constrains global models (e.g., luminosity-dependent density evolution, pure luminosity evolution, etc.) for the the QSO population as whole. Further, the evolution of the LF may provide insights into how "typical" QSOs evolve (e.g. Cavaliere et al. 1983, Koo and Kron 1988). In addition, a comparison of the LF for QSOs (defined here to be AGN with $M_B < -23.0$, for $q_0 = 0$, $H_0 = 50$ km/sec) with that of Seyferts provides information on the link between QSOs and lower luminosity AGN. Finally, even without a detailed knowledge of the redshift distribution itself, constraints on the surface densities of AGN are relevant to such issues as the contribution of AGN to the diffuse X-ray background radiation.

Although a number of recent and ongoing QSO surveys are efficiently probing the nature of the QSO LF at high luminosities and/or low redshifts using "bright" ("bright" is here defined to be $m < 21$) objects, surveys at fainter magnitudes are (for obvious reasons!) comparatively much more rare (e.g., Koo and Kron 1988, Marano et al. 1987). In order to further study the faint end of the QSO luminosity function, we recently obtained multicolor CCD images in pseudo-Johnson

B and Gunn g, r, and i covering ≈ 0.5 sq. deg. at high galactic latitude. These multicolor images are currently being analyzed to obtain QSO samples to $m < 22.5$. Described here are some very early results from our program.

THE FOUR-SHOOTER MULTICOLOR SURVEY

The multicolor selection approach is similar to the *photographic* technique used by Warren et al. (1987) to select several $z > 4$ QSOs. The multicolor technique, however, can be used to select "complete" QSO samples over a wide redshift range. Normal stars occupy a small, well-defined volume in multicolor space, while QSOs, having power-law continuous spectra and emission lines, occupy regions away from the stellar sequence. A primary limitation of the use of photographic plates is the difficulty in obtaining high photometric accuracy at very faint magnitudes: photometric errors artificially broaden the intrinsic "width" of the stellar sequence in multicolor space, making it increasingly difficult to separate QSOs from stars. In addition, the low signal to noise of the photographic plates at the faint end makes the morphological separation of stellar-like objects (QSOs and stars) from galaxies increasingly difficult. On the other hand, CCDs are capable of achieving high photometric accuracy to quite faint magnitudes. The Four-Shooter (Gunn et al. 1987) on the 5m is especially well-suited to the task: exposures of a less than a few minutes allow for photometric accuracies of a few percent at 21st magnitude; each exposure covers a relatively large field of $\approx 9'X9'$ (the Four-Shooter detector consists of a 2-by-2 mosaic of four 800X800 TI chips); and, the small pixel size (1/3" per pixel) is useful in morphologically separating stars and QSOs from galaxies. The multicolor filters we use are a pseudo-Johnson B, and Gunn g, r, and i with central wavelengths of 4340, 5000, 6500, and 7800 Å, respectively.

In our first run for this program in the summer of 1987, we obtained Four-Shooter multicolor data for two fields. The first, overlaps with the SA 57 field surveyed for faint QSOs by Koo, Kron, and collaborators (hereafter, KK et al.). Time limitations did not allow us to cover the entire region in SA 57 surveyed by KK et al. However, the area imaged with the Four-Shooter includes all 11 of the $B < 21.1$ QSOs spectroscopically confirmed by Koo, Kron, and Cudworth (1986). The region of SA 57 surveyed with the Four-Shooter includes an additional 50 objects to $B < 22.6$ originally classified by KK et al. as QSO *candidates*. Many of the objects in this latter category have also been followed-up spectroscopically (Koo and Kron 1988). Our Four-Shooter data in SA 57 will be used to investigate and empirically calibrate the utility and "completeness" of accurate (few percent at $m \approx 21$), multicolor photometry in distinguishing QSOs from stars at faint magnitudes. By using SA 57 as such a test field, we take advantage of the already extant follow-up spectroscopy.

Early results are encouraging. Shown in Figure 1 are the reduced data from one full Four-Shooter frame covering a ≈9'X9' area of sky in SA 57. Warren, Hewett, and Irwin (1987) have discussed the use of orthonormal color axes; however, for simplicity, we merely display r-i versus B-g in Fig. 1. Instrumental colors are shown. The objects shown are those detected in all four colors, and having accurate photometry (estimated errors < 0.05 mag.) in either of the two displayed colors; effectively this means that the objects are brighter than ≈ 22 in B and g, or in r and i. The spread in colors at the red end of the stellar sequence is largely due to the uncertainty in the B magnitude: typically, the reddest objects in the diagram have $B > 23$. The object marked 'Q' is a $B = 21.5$ magnitude QSO candidate from the KK et al. survey; it is the only object within this Four-Shooter frame selected as a QSO candidate by KK et al. that also satisfies our photometric accuracy criteria. It is easily distinguished from the stars in this two dimensional color-color diagram. These first results are encouraging, but it is clear that further reductions are required before we can quantitatively address such issues as the completeness of our multicolor approach and contamination of the sample by stars and other non-QSOs.

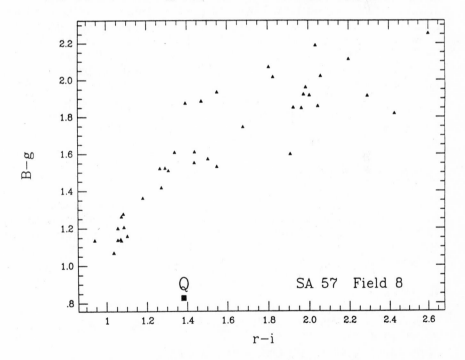

Fig. 1   Multicolor data from one Four-Shooter frame in SA 57. The object flagged 'Q' is a QSO candidate with $B = 21.5$.

The data of Fig. 1. were reduced using a batchable software package, *onesie/twosie*, recently developed by one of us (P. Schechter) which takes flattened data, and, with very little further human interaction, automatically locates objects, determines the stellar point spread function, morphologically separates galaxies from stellar like objects, and produces magnitudes for both stars and galaxies.

The Four-Shooter was also used to survey a second field to obtain a *new* CCD multicolor-selected sample of QSOs to $m < 22.5$. This second field covers ≈ 0.3 sq. deg. of sky previously imaged in X-rays with the Einstein X-ray Observatory. In collaboration with S. White (Univ. of Arizona), this new field will also be used to study the contribution of faint QSOs to the diffuse X-ray background radiation.

FUTURE EXTENSION OF THE SURVEY TO $m \approx 24.5$

Additional time has been granted on the Four-Shooter at the Palomar 5m early in 1988 for an extension of this survey to even fainter magnitudes. In collaboration with D. Koo (UC Santa Cruz) and R. Windhorst (Az. State Univ.), we will use the same multicolor CCD approach to conduct a survey to $m \approx 24.5$. This limiting magnitude will allow us to nearly fully sample the faintest end of the QSO LF, even for objects at quite high redshifts. We intend to survey ≈ 0.25 sq. deg. of sky, a portion of which will be in SA 57. In a long-term program, we intend to spectroscopically confirm QSO candidates from this upcoming fainter survey; R. Kron (invited paper, this conference) has discussed the feasibility of such faint spectroscopic follow-up using even existing instrumentation.

ACKNOWLEDGEMENTS

The observations discussed here were made at Palomar Observatory as part of a collaborative agreement between the California Institute of Technology and the Carnegie Institution of Washington. David Koo, Rogier Windhorst, and Simon White are collaborating on various aspects of this program.

REFERENCES

Cavaliere, A., Giallongo, E., Messina, A., and Vagnetti, F. 1983, *Ap. J.*, 269, 57.
Gunn, J. E., et al. 1987, *Opt. Eng., in press.*
Koo, D. C., Kron, R. G., and Cudworth, K. M. 1986, *PASP*, 98, 285.
Koo, D. C. and Kron, R. G. 1988, *Ap. J., in press.*
Marano, B., Zamorani, G., and Zitelli, V. 1987, *MNRAS, in press.*

Warren, S. J., et al. 1987, *Nature*, <u>325</u>, 131.
Warren, S. J., Hewett, P. C., and Irwin, M. J. 1987, in *Observational Cosmology, Proceedings of IAU Symposium 124*, eds. A. Hewitt et al., Dordrecht: Reidel, p. 661 ff.

AUTOMATED SURVEY OF CFHT GRENS PLATES

E.F. BORRA, M. BEAUCHEMIN AND G. EDWARDS,
Département de Physique, Université Laval,
Québec, QC CANADA G1K 7P4

ABSTRACT We have developed automated software to process slitless spectra. The software, described in 2 published papers, automatically separates stars from extended objects and quasars from stars. Comparing the performance of the software with a plate taken in a region of SA 57 that has been extensively surveyed by others using a variety of techniques, we find that our automated software performs very well. The complete automated analysis of all the spectra on a plate and computer simulations are used to calibrate and understand the characteristics of our data.

INTRODUCTION

We have developed automated software to process slitless spectra. The software automatically separates stars from extended objects and quasars from stars. Among the many parameters output by the program, we have: narrow-band and broad-band photometric indices, spectral discontinuities, emission and absorption lines. The lines are found with respect to an automatic continuum fit. All of these quantities are given with associated errors and signal to noise ratios. Considerable effort was made to calibrate the data and to understand systematic effects. We know the uncertainties associated with our data and strive to give all relevant quantities in terms of signal to noise ratios. We have also made extensive use of computer simulations to understand our algorithms and check systematic errors.The program does not depend on the measurement of direct plates. The software can be used to study stars, galaxies and quasars. Stars are very useful to understand the characteristics of a given plate. The software is described fully in Borra et al. (1987). In a second publication (Edwards, Beauchemin and Borra, 1987) we discuss our quasar search techniques in greater details. Table I summarizes the characteristics of the survey.

TABLE I Characteristics of the survey

Search technique : Automated
Area/Plate: 1 square degree (0.7 completeness)
Total Plates: 44 (> 31 Square degrees)
Limiting magnitude: B ~20.5 to 21.5 (B ~ 20 to 21 completeness)
Dispersion: 1000 Å/mm (~70 Å resolution)

Wavelength range: 3500 Å < λ < 5300 Å

GOALS AND PERFORMANCE OF THE SOFTWARE

The main goals of the project are to study the large scale (clustering) and very large scale distribution of quasars. In particular, we want to check, with quasars, the cosmological assumptions of homogeneity and isotropy . The fields observed (Borra et al. 1987) have been carefully chosen with these goals in mind. Because we will compare quasar properties among various fields, we must understand very well our data and pay attention to selection effects. The data will also be used for the usual work (evolution, luminosity function, etc...)

We can check the performance of our software with a plate taken in a region of SA 57 that has been extensively surveyed by Koo, Kron and Cudworth (1986; KKC), Hoag (1986), as well as others,using a variety of techniques: our automated software performs very well. Figure 1 shows quasar counts as a function of our probability classes for one of our plates taken in the SA57 reference region. The hatched region shows the quasars in common with KKC and others. We find all of their candidates to our completeness limiting magnitude (B=21), with the exception of 2 quasars contaminated by nearby spectra.

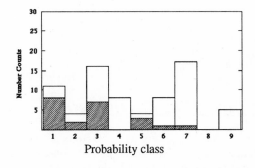

Probability class

Fig. 1   Quasar candidate counts as a function of our probability
class in a region in SA 57 extensively studied by others
The confirmed candidates, are shown in the hatched region.
The unshaded region represents counts of our candidates not seen
by other surveys. We may have picked up marginal quasars.

We also carried out an eyesearch of the same plate. The eyeselected
quasars were found by eye inspection by 2 people. We did not want to miss
any objects and flagged anything that looked abnormal. We flagged over 100
objects. Notwithstanding this overkill, the eye looses most of the faintest
confirmed quasars .

We observed some of our candidates with the AAT. The candidates
were originally found by eye inspection. Later, the plates were analyzed
automatically and the candidates ranked according to probability class.The
number of quasars candidates in an eyeselected subset of objects for which
follow-up spectroscopy was obtained at the AAT, is plotted as a function of
probability class in Fig. 2. We only include candidates confirmed by the
automated software. The hatched region shows the confirmed quasars. The
remaining objects are mostly blue subdwarfs or white dwarfs.

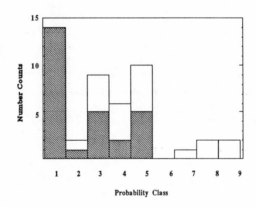

Fig. 2   The number of quasars candidates in the list of objects for
which follow-up spectroscopy was obtained at the AAT, as a function
of probability class. The hatched region shows the confirmed quasars.

USING STARS TO UNDERSTAND OUR DATA

Most of the spectra on a plate belong to stars. Stars are very useful to
understand our data and selection effects. For example, stellar counts as a
function of surface area as measured from the plate center are used to
determine the completeness area of a given plate.

In figure 3 we compare differential star counts obtained by Tyson and

Jarvis (1979) with the star counts derived from three of our plates in SA 57 . The Tyson and Jarvis counts were taken in a nearby region. The numbers shown represent true surface densities per square degree. Our counts agree very well with the TJ counts, showing that we do not miss too many spectra.

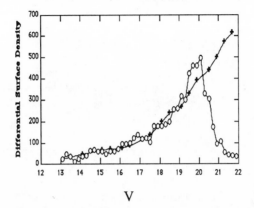

Figure 3: Comparison of the Tyson and Jarvis (1979) differential star counts (crosses) with the star counts derived from three plates in SA 57 (open circles). The numbers shown represent true surface densities per square degree.

The counts of all objects detected are plotted as a function of B magnitude for a plate are used to determine the limiting magnitude of each plate (Fig. 4) . Every plate contains faint photoelectric sequences set up with the Kitt Peak #1 0.9-m (Borra and Beauchemin, 1987; Borra, Edwards and Petrucci, 1985). We correct for vignetting and nonuniformities in the quantum efficiency of the plate with the maps of the sky intensity across a plate.

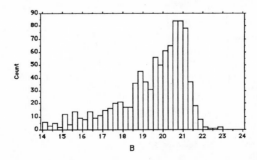

Figure 4   The counts of all spectra detected are plotted as a function of B magnitude for a plate in SA 57.

Plots of the Full Width at Half Maximum of all the objects on a plate (stars and galaxies) (Fig. 5) are used to measure the behavior of image quality within a plate. They also show seeing differences from plate to plate.

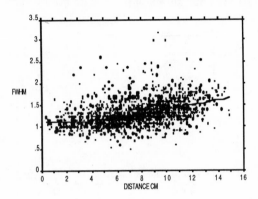

Fig. 5   The Full Width at Half Maximum of all the objects on a plate (stars and galaxies) are plotted as a function of distance from the plate center. The FWHM is ~ 1 arcsec at the center but gets worse at the edges.

REFERENCES

Borra, E.F., and Beauchemin, M. 1987, A.J. 1987 , 92, 421.
Borra, E. F., Edwards, G., Petrucci, F., Beauchemin, M., Brousseau, D., Grondin, L. and Beaulieu, A. 1987, Pub.A.S.P., 1987, 99, 535..
Borra, E.F., Edwards, G., and Petrucci, F. 1985, A. J. 90, 1529.
Edwards, G., Borra, E.F., and Beauchemin, M. 1987, Pub.A.S.P. in Press
Hoag, A.1986, I.A.U. symp. 119, D. Reidel Publishing Co, p. 47.
Koo, D.C., Kron, R.G. and Cudworth, K.M. 1986, Pub.A.S.P., 98, 285.
Tyson, J.A., and Jarvis J.F. 1979, Ap.J. Lett., 230, L153.

THE APM QSO SURVEY: A PROGRESS REPORT

FREDERIC H. CHAFFEE, JR. and CRAIG B. FOLTZ
Multiple Mirror Telescope Observatory, University of
Arizona, Tucson, Arizona 85721

PAUL C. HEWETT
Institute of Astronomy, Cambridge University, The
Observatories, Madingley Road, Cambridge CB3 0HA, England

GORDON M. MacALPINE
Department of Astronomy, University of Michigan,
Physics-Astronomy Building, Ann Arbor, Michigan 48109

DAVID A. TURNSHEK
Space Telescope Science Institute, 3700 San Martin Drive,
Homewood Campus, Baltimore, Maryland 21218

RAY J. WEYMANN and SCOTT F. ANDERSON
Mount Wilson and Las Campanas Observatories, 813 Santa
Barbara Street, Pasadena, California 91101

ABSTRACT   The APM QSO survey is aimed at obtaining a
sample of ~ 1000 QSOs brighter than V = 18.5 over a wide
range of redshifts ($0.2 \leq z \leq 3.3$) using well-defined and
consistently-applied selection criteria. Direct and
objective prism plate material is obtained with the UK
Schmidt; the plates are scanned at the Institute of
Astronomy's Automated Plate Measuring (APM) facility, and
candidate QSOs are selected using software developed by
Hewett and collaborators. Follow-up spectroscopy is
obtained primarily with the MMT and the Las Campanas 2.5
meter telescope where approximately 70% of the candidates
are confirmed as QSOs.

INTRODUCTION

Until recently, surveys to identify QSOs have been based on
continuum properties, as evidenced by optical color or radio
flux, or on subjective judgments of emission line characteristics, most frequently from optical objective prism plates.
With the recent advent of high speed, high accuracy plate

scanning machines, it has become increasingly possible to
objectify the QSO candidate selection process from plate material. A survey of objectively selected QSOs should allow us
to investigate many hitherto unavailable global QSO properties,
and its great strength is that it allows us to quantify the
probability of the selection of any QSO type.

We are now undertaking such a survey using plate material
from the UK Schmidt Telescope (UKST), Automated Plate Measuring (APM) facility at the Institute of Astronomy for plate
scanning, and the MMT and the 2.5 meter Dupont telescope to
obtain follow-up spectroscopic observations. This paper summarizes our progress as of the end of 1987 when 444 new QSOs
had been detected over approximately 400 square degrees of
plate material. Foltz et al. (1987) have discussed the APM
QSO survey in considerable detail, and we will provide only a
brief description here.

SCIENTIFIC OBJECTIVES

The final sample of ~ 1000 QSOs will be used for the following
investigations:
A.   To determine the emission line strengths and widths of a
      large sample of optically-selected QSOs as a function of,
      for example, luminosity and redshift.
B.   To investigate the frequency of occurrence of "rare" QSO
      types, e.g. broad absorption line (BAL) QSOs or QSOs with
      extremely narrow emission lines.
C.   To search for QSO pairs with separations less than 1 arc-
      minute to serve as probes of the size of intervening ab-
      sorbing clouds.
D.   To provide a list of bright QSOs displaying damped Lyman
      α absorption for further investigations at high spectral
      resolution.
E.   To provide an additional constraint on the space density
      of bright QSOs in the range $2.2 \leq z \leq 3.3$.

SURVEY CHARACTERISTICS

Our goal is to identify ~ 1000 QSOs brighter than $m_J = 18.5$ in
the redshift range $0.2 \leq z \leq 3.3$ using machine-scanned direct
and objective prism plates from the UK Schmidt telescope.
Plates are scanned at the Institute of Astronomy's APM facility. Algorithms select approximately 3 candidates per square
degree as possible QSOs. Follow-up spectroscopy at 6 Å resolution at the MMT and at Las Campanas is used to classify each
candidate.

Candidate Selection

Among the ~ 20000 objects per field with $16.0 \leq m_J \leq 18.5$, approximately 80 are selected as possible QSOs if they meet one or more of the following criteria: a) their colors are extremely blue relative to common Galactic stars, b) their spectra show evidence for strong emission or absorption features, c) their spectra show evidence for strong continuum breaks.

Magnitude Calibrations

CCD frames in two colors obtained at the McGraw-Hill 1.3 meter telescope or Las Campanas 1 meter will be used to provide magnitude calibrations for each field in the survey. Early tests suggest that APM magnitudes can be calibrated to a precision of ~ 0.1 magnitudes.

Spectroscopic Observations

A spectrum of each candidate is obtained with the MMT spectrograph at 6 Å resolution (FWHM) and S/N of ~ 15 in the continuum window closest to 4500 Å. On average, six candidates per hour are observed, of which approximately four are found to be QSOs. Figure 1 shows representative MMT spectra of confirmed QSOs over the redshift range of the survey.

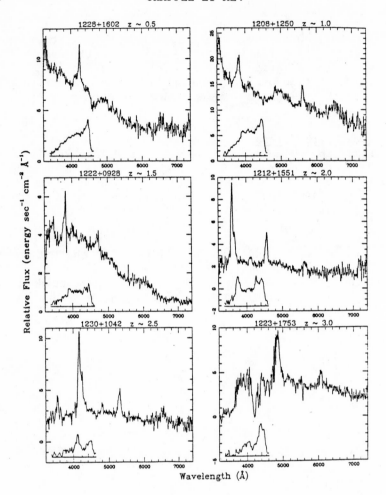

Fig. 1   Six-Angstrom resolution MMT spectra of representative QSOs found in Virgo. The insert in each frame
is the corresponding objective prism spectrum. MMT spectra are plotted on a linear wavelength scale; UKS spectra
are not. The spectra are aligned at $\lambda 4000$, and the tick
marks appear on the UKS spectra at 500 Å intervals. The
flux scale of the MMT spectra is in units of energy cm^{-2}
Å $^{-1}$ sec^{-1}.

For comparison we include tracings of the objective prism
spectrum for each object. Observations at comparable S/N and
resolution have now been obtained for three southern fields
using the 2.5 meter Dupont telescope.

SUMMARY OF OBSERVATIONS TO DATE

Virgo Cluster
Four overlapping direct objective prism pairs of UKS plates
were obtained centered on the Virgo cluster. The observations
in these fields are now complete and have been published by
Foltz et al. (1987). In the 102 square degrees covered in
Virgo, 192 QSOs were found, and when correction is made for
overlapping objective prism spectra, the surface density is
2.1 per square degree.

Other Fields
Table 1 shows the current status of all the survey observations; as of January 1, 1988, 444 QSOs had been identified in
11 fields.

TABLE I   APM QSO Survey Status - January 1, 1988

Plate Desig.	Plate Center R.A.	Dec.	No. of QSOs Confirmed	Status
UJ 6543 P	00^h 00^m	$+00^o$ $00'$	34	ID Spectroscopy complete.
UJ 11474 P	00 20	+00 00	35	Initial object selection complete. ID Spectroscopy underway.
SGP	00 53	-28 00	61	ID spectroscopy of high-probability objects complete over 36 $deg.^2$.
UJ 11757 P	11 40	+00 00	0	Object selection underway.
UJ 10732 P	12 16	+11 12	33	Virgo field. Spectra and magnitudes published (A.J. **94**, 1423, 1987).
UJ 10742 P	12 16	+15 27	42	As UJ 10732 P.
UJ 10738 P	12 34	+11 12	53	As UJ 10732 P.
UJ 10749 P	12 34	+15 27	64	As UJ 10732 P.
UJ 5853 P	12 40	+00 00	29	ID spectroscopy complete.
F 287	21 28	-45 00	43	ID spectroscopy of high-probability objects complete over 28 $deg.^2$.
MTF	22 03	-18 55	20	ID spectroscopy of high-probability objects complete over 18 $deg.^2$.
891	22 40	+00 00	30	ID spectroscopy complete.

Figure 2 presents the redshift distribution of these QSOs.
Contrary to the results from many ocularly scanned objective
prism surveys, the bias toward detecting QSOs near z = 2.2 is
not evident, and the redshift distribution is similar to that
seen in Parkes flat-spectrum ratio QSOs where emission line
biases are absent.

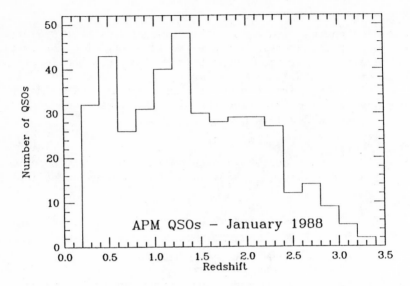

Fig. 2   The redshift histogram of the 444 QSOs found as
of January 1, 1988 is shown.

QUANTIFICATION OF THE APM SURVEY

One of the great advantages of the techniques being applied
here is that they allow us to estimate quantitatively the
probability of finding a QSO of any type at any redshift and
thus to estimate the completeness of our sample.   Figure 3
illustrates one approach with 1235+0857, one of the BAL QSOs
in the Virgo sample.

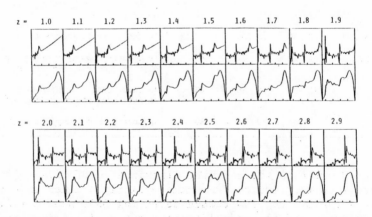

z = 3.0   3.1   3.2   3.3

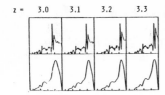

Fig. 3   An example of the transformation from a flux-
calibrated spectrum into a pseudo-APM spectrum. The MMT
spectrum of the newly-discovered BAL QSO 1234+0857 has
been truncated to the sensitivity window of the J emulsion, rebinned into the non-linear wavelength scale of
the APM spectrum, and multiplied by the empirically-
determined wavelength-dependent sensitivity function of
the UKST prism plates. The resulting spectra, transformed to a variety of redshifts, are shown in the upper panels. After degradation to the resolution of the APM
spectra, these are presented in the lower panels. Following the addition of noise, these spectra were passed
through the selection algorithms. **All were identified as
probable QSOs.**

In the upper strips of panels, we have extrapolated its MMT
spectrum beyond our observed window, compressed it, restricted
our attention to the $3400 \leq \lambda \leq 5300$ Å window of the J emulsion, and redshifted it in steps of 0.1 from z = 1.0 to z =
3.3. In the lower strips, we have degraded the resolution to
that of the objective prism spectra and convolved each spectrum with the J emulsion plate response. The result is a
spectrum of 1235+0857 of very high S/N as it would appear at
$1.0 \leq z \leq 3.3$ on a UK Schmidt objective prism plate. We can
then introduce noise into the spectra and pass each through
our selection algorithms to see at what magnitude and redshift
this object would have been selected as a probable QSO. In
the case of this object, our algorithms selected the object as
a probable QSO at all redshifts.

     This technique will be applied to a large grid of hypo-
thetical QSO spectra as well as a large number of real spectra. Multiple applications of noise into each spectrum will
allow us to calculate the **detectability** of an arbitrary QSO
type. Armed with the probability of detecting a QSO of given
type, we can calculate the completeness of the survey as a
function of redshift, magnitude, and QSO type.

FUTURE PLANS

Over the next 2 years we expect to complete our survey of ~

1000 QSOs. Numerical experiments such as described above will allow us to estimate the completeness of our samples and to examine for the first time the global properties of such a large QSO sample as a function of luminosity and epoch.

ACKNOWLEDGMENTS

The APM QSO survey is supported in part by grant AST 87-00741 from the National Science Foundation, for which we are very grateful. We gratefully acknowledge the cooperation of the United Kingdom Schmidt Telescope Unit, which provided the plate material, and of the APM group at Cambridge, where the plates were scanned. The APM is funded by the Science and Engineering Research Council as a National Facility. P.C.H. was supported by Corpus Christi College. We heartily thank John Salzer for his assistance with the photometry at MHO.

REFERENCES

Foltz, C. B., Chaffee, F. H. Jr., Hewett, P. C., MacAlpine, G. M., Turnshek, D. A., Weymann, R. J., Anderson, S. F. 1987, A. J., 94, 1423.

THE HAMBURG QUASAR SURVEY

D. ENGELS, D. GROOTE, H.-J. HAGEN and D. REIMERS
Sternwarte der Universität Hamburg, Gojenbergsweg 112,
D-2o5o Hamburg 8o, F. R. G.

ABSTRACT   Quasar candidates down to $B \sim 18.5$ are searched
on objective prism plates with a PDS 1o1oG microdensitometer. The Hamburg Quasar Survey (HQS) aims to provide a
sample of several hundred bright quasars suitable for
studies in several fields of quasar astronomy. This paper
describes the present status of the survey.

INTRODUCTION

New optical quasar surveys were often started with the aim to
study the luminosity function of QSO to fainter luminosities
and/or to shift the observed z-range to higher values. The
faint QSO found in these surveys, however, can be observed
spectroscopically only with a vast amount of observing time on
4 m -class telescopes.

Many topics of quasar research, like the physics of the
emission and absorption line regions, require high-resolution
spectroscopic data from a sample of objects which can be acquired only from brighter QSO ($B \leq 19$). New challenges as the
probing of the early universe by the analysis of absorption
lines in quasar light and the possibility that quasar light
may be severely affected by gravitational lensing, also demand
greater samples of bright quasars. Close pairs of bright QSO
of similar redshifts and angular separations less than $\sim 1^{\circ}$ are
of particular interest.

Thus a renewed interest in larger samples of bright
quasars has emerged. As the surface density of bright quasars
is low, only a wide angle survey with a Schmidt telescope can
provide such samples. To detect QSO candidates on a larger
area of the sky requires a quick search method on the plates.
If one aims for reasonable completeness down to a certain
magnitude only a machine-based search technique is useful.

The HQS is such a wide angle survey and uses a semi-
automated search technique to select QSO candidates. Our aim is to
provide several hundred bright quasars ($0.2 \leq z \leq 3.3$
and $B \leq 18.5$)

-    to search for particulárly interesting objects like,
     BALs, QSO pairs, gravitational lenses
-    to provide complete samples of quasars for selected
     fields (B (lim) $\stackrel{<}{\sim}$ 18.o).

These new bright quasars will give the base for follow-on
studies.

The plate material and processing capacity acquired in
the frame of the HQS is equally powerful to serve other astronomical applications. Peculiar objects showing emission or unusual absorption lines are easily picked-up by the search software. Also classification work for new sources coming from
satellite based surveys (e.g. ROSAT) is feasible.

OBJECTIVE PRISM PLATES

The survey plates are taken with the former Hamburg Schmidt
telescope, which is located since 198o at the Spanish-German
Astronomical Center on Calar Alto/Spain. The telescope is a
f/3 instrument with a mirror diameter of 1.2 m and a free
aperture of the correction plate of o.8 m. A 1.7 deg prism is
available, allowing to take prism plates with a dispersion of
$\sim$ 14oo $\AA$/mm at Hγ. The field size on the 24 x 24 cm plates is
5.5 x 5.5 deg.

The survey is carried out on KODAK IIIa-J plates which
are hypersensitized by baking in nitrogen (prior to 1987) or
forming gas. In about one hour integration time a limiting
magnitude of $\sim$18.5 mag is reached. For each field two prism
plates and one direct plate are taken.

The plates are not calibrated so that subsequent processing of the plates is made in densities. Calibration of the
direct plates is planned using stellar sequences obtained from
CCD-direct imaging on Calar Alto.

The survey region is concentrated on the northern sky
with $\delta > 25^\circ$ and galactic latitude b $> 2o^\circ$. As a first step
we try to cover 1oo fields ($\sim$ 25oo deg^2) of which $\sim$5o % are
finished. Additionally plates are available of several fields
on the southern sky with $\delta > o^\circ$.

SEMI-AUTOMATED CANDIDATE SELECTION

The plates are scanned in Hamburg by a PDS 1o1oG microdensitometer controlled by a PDP 11/24 computer. For data storage
currently a 1o MByte disc drive and a 16oo bpi magnetic tape
drive are available. Machine driver software and all data
processing programs have been written in Hamburg and are
running under RSX 11M plus on the PDP 11/24 (Hagen, 1987).

The machine-based search enables us to extract quasar
candidates with well defined selection criteria. We are not

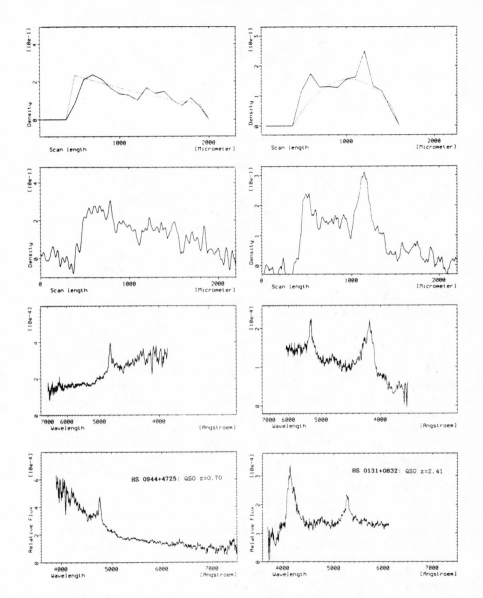

Fig.1: QSO spectra. Low- and high-resolution prism spectra, slit spectrum convoluted with prism dispersion and original spectrum

affected by the severe selection effects connected with eye-ball searches. The software is optimized for a quick extraction of quasar candidate spectra, for minimal data storage requirements and is thus especially suitable for rapid processing of a large quantity of plates.

The automated search proceeds in two major steps:
1. Extracting all spectra on a plate (full sample)

2. Classifying spectra to form a subsample of quasar candidates

Acquisition of the full sample

To extract the spectra, the microdensitometer works in a 'low-
resolution scan mode'. The entire plate is scanned perpendicu-
larly to the direction of dispersion (x), using a rectangular
slit of 2o x 1oo micron and a step-size of 1o micron in x and
1oo micron in y. The use of a 1oo micron wide slit reduces
noise, decreases the number of scan lines, and therefore scan
time and the amount of data. The slit width has been found
to be an optimal compromise between a sufficient spectral
resolution on one hand and saving of processing time on the
other hand. Fig. 1 shows that the spectral resolution is still
sufficient for the detection of typical quasar emission lines.

During the low-resolution scan the data of each scan-
line are processed immediately while the PDS-machine is
scanning the next line. The local background is determined by
searching for strings of length 1oo-2oo micron with alternating
differences between neighbouring pairs of pixel values. Between
these strings the background is interpolated linearly. Density
peaks in the scan-line have gauss-like profiles. They are
identified, if the density of several adjacent pixels exceed
a $n \times \sigma$-threshold above the local background. With $n = 2$ about
15o to 4oo peaks are found in one scan-line depending on the
surface density of the spectra on the plate. Position, height,
width of peaks, and the local background are stored on tape
(or hard-disc). The total scan time (on-line processing included) is at present ∿13 hours. A decrease to about 4 hours
will be achieved after changes in the PDS hardware, planned
in the near future. The amount of data for each plate is approximately 5 MByte, about 5 % of the space occupied by a full
map of the plate.

Images of spectra are built up from pixels of adjacent
lines by a subsequent code. Usually 15-2o pixels contribute
to each spectrum. This small number of pixels per spectrum
will result in short execution times for the forthcoming
classification procedure. About 2 hours computing time is
needed to extract the total number of 3o ooo-5o ooo spectra on
each plate.

Quasar candidate selection

The quasar candidate selection proceeds in three steps
i)   Selection of spectra from point-like sources with blue
      continua and/or emission features
ii)   Scan of the candidates on the direct plate for further
      star/galaxy discrimination, overlap detection and de-
      termination of the coordinates
iii) High-resolution scan of the reduced candidate sample and
      evaluation.

To evalute the spectra of the full sample, the selection code fits a polynomial to the continua of all spectra. Its slope at a given wavelength is taken as measure for the colour of the spectra. In case of deviations of more than 3σ above the adopted continuum, objects are flagged as emission-line candidates.

The candidates (including 2o-3o AGK-stars for coordinate reference) are then scanned on the direct plate. The scan-field for each object is 3ooo x 4oo micron and a 1o x 1o micron slit is used. A subsequent code determines accurate positions ($\lesssim 1"$), overlapping spectra and removes still remaining extended objects from the sample.

In a final step high-resolution scans of the remaining candidates are made on the prism plate. For each candidate a 3ooo x 2oo micron field is scanned with a 2o x 2o micron slit. To obtain a one-dimensional spectrum, the background is removed, followed by an integration perpendicular to the direction of dispersion. The spectra are evaluated interactively on a display. Objects with stellar features (Hydrogen lines, G-band, Ca-Break) are removed, while the rest is kept for examination by slit spectroscopy.

RESULTS FROM SLIT SPECTROSCOPY

Slit spectroscopy of prism selected objects has been performed with the 3.5 m-telescope on Calar Alto/Spain equipped with a B. & C. spectrograph and a CCD-camera. The spectral range was 36oo to 65oo $\overset{\circ}{A}$ and the dispersion 24o $\overset{\circ}{A}$/mm.

The objects were selected from several fields with the purpose to cover quasar candidates over the whole z-range up to 3.3 and all kinds of objects which might be confused with quasars, e.g. white dwarfs, subdwarfs and blue (active) galaxies. In total about 6o objects were observed and the success rate in confirming candidates for the different groups of objects was well above 75 %. In total 16 Quasars could be confirmed with o.23 $\leq$ z $\leq$ 2.41.

These first results show:
-     Using as selection criteria blue continua and emission lines we are able to find quasars in the range o.2 $\lesssim$ z $\lesssim$ 3.3. No selection effects in certain z-ranges are apparent yet. Completeness might be achieved down to $\sim$18.o mag.
-     The average number of quasars detected per field is 25-3o for plates of medium quality. This might increase to up to 5o quasars for excellent plates, leading to a surface density of quasars on our plates of 1-2 quasar/deg^2.
-     In addition, many interesting peculiar objects are found on our plates. For example, a rare magnetic white dwarf (Hagen et al., 1987) and another supergiant HII-region in a galaxy were discovered (Hagen et al., 1988).

REFERENCES

Hagen,H.-J.: 1987, Ph.D. thesis, University of Hamburg.
Hagen,H.-J., Groote,D., Engels,D., Haug,U., Toussaint,F.,
    Reimers,D.: 1987, Astron.Astrophys. <u>183</u>, L7 Letters.
Hagen,H.J., Groote,D., Engels,D., Toussaint,F., Reimers,D.:
    1988, Astron.Astrophys. (in press).

SELECTED AREA GRISM QSO SURVEY

ARTHUR A. HOAG and NORMAN G. THOMAS
Lowell Observatory, Flagstaff, AZ 86002

ALLAN R. SANDAGE
Mt. Wilson Observatory, Pasadena, CA 91101

ABSTRACT We photographed one square degree regions
centered on 10 Selected Areas through a grating prism at
the Mayall 4-m telescope prime focus. We have identified 246 QSO candidates at three different degrees of
certainty and have noted a number of other objects of
interest.

THE OBSERVATIONS

Details of the procedures we used are described by Vaucher
et. al. (1982). Candidate QSOs were selected visually then
evaluated by microdensitometry and computer analysis. Candidates are categorized in three groups as enumerated in Table
II. An object with two or more reasonably identified red
shifted emission lines is labeled Q. Q: denotes a candidate
having one emission line and either a vague indication of
another line or a continuum distribution that allows one to
guess at the single line identification. Finally, Q:: denotes
will-o'-the-whisp, borderline cases. Inconsistencies, apart
from intrinsic scatter, are to be expected, not only because
of variations in plate quality, but also because of variations
in judgement. Hoag and Schroeder (1970) labeled the technique of using a grating in the converging beam as "Nonobjective" grating spectroscopy and, unfortunately, our
application is nonobjective in more ways than one!
  Our sample is naturally biased in favor of QSOs having
strong emission lines. The redshift distribution illustrated
in Figure 1 shows the peaks attributable to 1216 (Lyα), 1549
(CIV), and 2798 (MgII) at redshift positions near the maximum
detectivity of our observing system ($\lambda\lambda$3800-4000 Å). This
grism selection bias has been thoroughly documented by Hewitt
and Burbidge (1987).
  This emission line biased sample is also magnitude limited. Our slitless spectograms are saturated at magnitudes

HOAG, THOMAS AND SANDAGE

TABLE I   Selected Areas

S.A.	α 1950	δ	l	b
28	$08^h43^m15^s$	44 45 46	176.1	38.7
29	09 41 42	44 38 57	175.4	49.1
55	11 32 28	29 42 12	200.7	73.1
57	13 06 15	29 39 03	64.7	85.6
71	03 14 24	15 13 26	167.1	-34.7
82	14 17 03	15 09 54	6.6	66.2
94	02 53 22	-00 18 53	176.1	-49.7
106	14 39 05	-00 13 15	351.3	51.8
107	15 36 28	-00 08 55	5.7	41.3
118	02 17 44	-14 38 49	185.8	65.7

TABLE II   QSO Candidates

S.A.	Q	Q:	Q::	SUM	BAL	Q only Found on N Plates N	1	2	3	4	5
28	15	6	6	27	1	5	3	3	–	–	
29	19	9	9	37	–	13	2	3	–	1	
55	12	14	7	33	–	4	–	7	1	–	
57	10	9		19							
71	3	5	5	13	2	–	–	1	2	–	
82	17	15	8	40	–	8	4	2	3	–	
94	17	4	4	25	–	13	1	3	–	–	
106	5	10	1	16	2	3	1	1	–	–	
107	1	9	4	14	–	1	–	–	–	–	
118	11	10	1	22	2	3	4	3	1	–	
Sums 10□°	110	91	45	246	7	50	15	27	7	1	

Q per □°   11.0 ± 5.9 RMS
SUM per □°   24.6 ± 9.9 RMS
if Q   90% correct, total Q   = 99
if Q: 80% correct, total Q: = 73
if Q::   60% correct, total Q:: = 27
then total emission line QSOs = 199 or ~20% per □°

b~17 making detection of emission difficult. Further, the
magnitude distribution, and the luminosity distribution at a
fixed value of Z, illustrated in Figure 2, probably reflect
incomplete detection at magnitudes greater than ~19.5 in addition to showing intrinsic properties of the sample.

    We found approximately 25% more QSO candidates on the
IIIa-J plates than on the IIIa-F plates for the eight central
fields photographed with both emulsions. We attribute this to
the narrower passband (hence longer exposure to the "sky

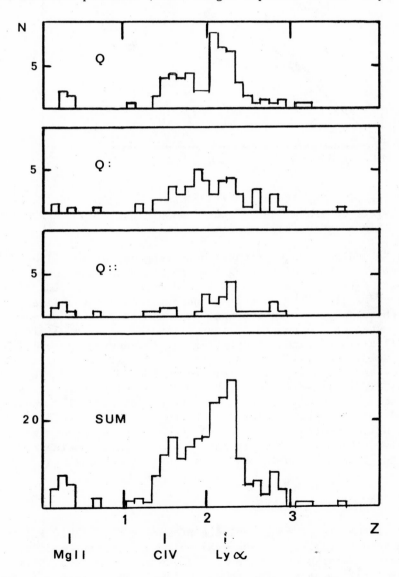

Fig. 1 Redshift distributions of QSO candidates

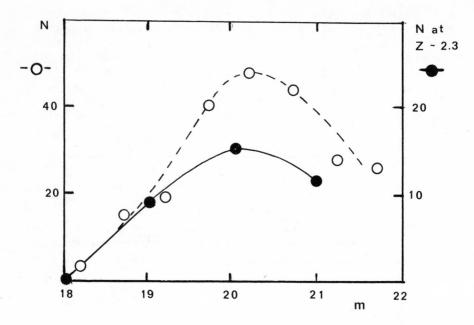

Fig. 2   Magnitude distributions of all QSO candidates
(left ordinate) and of the Q sample at 2.1 < Z < 2.4
(right ordinate)

limit") and more regular wavelength response of the IIIa-J
emulsion. A more disturbing systematic effect was found on
comparing the average number of candidates found on the IIIa-F
central plates with the average number found on the IIIa-F
peripheral plates. On average, we found half again as many
candidates on the central plates as on the peripheral plates.
The IIIa-J advantage, redundant coverage, unwitting memory of
the field, and small shifts in field orientation, thus changing overlap conditions, may all be factors. In any case,
these difficulties indicate that we have underestimated the
surface density distribution in ways as yet not fully understood. These difficulties illustrate the importance of im-
personal automated quasar search techniques which have been
widely discussed in the literature and which are being reviewed at this Workshop on Optical Surveys for Quasars.

OTHER OBJECTS OF INTEREST

We note numbers of other objects of interest in Table III including ultraviolet excess (UVX) objects, some of which may be
quasars. We also enumerate blue objects having large Balmer
jumps (lBj) emission line galaxies (eGal) and unresolved objects having spectra like those of H II regions.

TABLE III   Other Objects of Interest

S.A.	UVX	1Bj	eGal	H II
28	22	3	–	–
29	24	5	4	–
55	33	3	11	–
57	7	–	2	–
71	27	7	2	3
82	19	14	5	1
94	27	6	5	1
106	28	15	16	1
106	3	5	4	–
118	18	2	21	–

ACKNOWLEDGEMENTS

We are indebted to Bill Schoening, Barbara Schaefer, Doug
Tody, Saul Levy, Jim DeVeny, and Rob Hubbard for telescope
operations in obtaining the observations. Barbara Vaucher and
Tobias Kreidl, Lowell Observatory, devised and maintained the
computer program used to analyse the selected spectrograms.
Sister Mary Matthew, Mercyhurst College, worked with us, during one semester of sabbatical leave, on plate reviews and
preparations for astrometric work. We are also indebted to
Larry Wasserman and Barry Feierman, at Lowell, for help in
setting up procedures for identification astrometry of the
selected objects.

REFERENCES

Hewitt, A., and Burbidge, G., 1987 Ap. J. Suppl., 63, 1.
Hoag, A. A., and Schroeder, D. J., 1970, Publ. A. S. P. 82,
    1141.
Vaucher, B. G., Kreidl, T. J., Thomas, N. G., and Hoag, A. A.,
    1982 Ap. J. 261, 18.

CALAN TOLOLO SURVEY: BRIGHT QUASARS AT HIGH REDSHIFTS

JOSE MAZA, MARIA TERESA RUIZ, LUIS E. GONZALEZ and
MARINA WISCHNJEWSKY
Departamento de Astronomía, Universidad de Chile
Casilla 36-D, Santiago, Chile.

THE SURVEY

An objective prism survey has been started at cerro Tololo
Inter-American Observatory using the thin UV prism (1,360 Å/mm
at H_γ and 1,740 A/mm at $H\beta$) on the Curtis Schmidt telescope.
Unfiltered baked IIIaJ plates exposed 90 minutes have been obtained for 163 fields shown in Figure 1. Unwiden spectra taken
in good seeing reach B $\simeq$ 19. This survey is an extension of the
original Tololo survey (Smith 1975; Smith, Aguirre and Zemelman
1976).

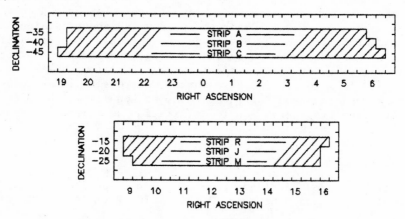

Fig. 1   Sky coverge of the Calán Tololo Survey. Three
parallel strips at declination $-35°$, $-40°$, and $-45°$, la-
belled A, B, and C, run from 19 to 6 hours. Strips R, J,
and M at declinations $-15°$, $-20°$, and $-25°$go from 9 to 16
hours. The survey covers $3,400$ deg^{-2} and is limited to
galactic latitude $|b| \geq 20°$.

Plates are searched at Cerro Calán using a binocular microscope at 15x. Two of us (LEG and MW) explore each plate

spending from 10 to 20 hours per plate; unusual spectra are
marked. The objective prism plate is then checked against the
ESO Quick Blue Sky Survey to discard overlapping spectra and to
evaluate the morphology of the objects; then a list of candidates is drawn for each field. The number of candidates varies
from 20 to 60 per plate, 40 being a typical number. Strips R,
J, and M have been looked at thoroughly; strips A, B, and C are
only 50% searched (January 1988). We plan to finish the candidate selection by the end of 1988.

We have learned to identify in this kind of photographic
plates a variety of interesting objects among which we can
point out Seyfert galaxies of type 1, emission line galaxies
(starburst type, liners and Seyfert 2's), carbon stars and a
group of emission line stars including T Tauri and cataclismic
variable stars (Ruiz et al. 1987).

HIGH REDSHIFT QUASARS

Objective prism plates are very suited for discovering high
redshift quasars, as found many years ago by M. G. Smith and
P. S. Osmer at CTIO (see, e.g., Osmer and Smith 1980). If a
strong $L\alpha$ line is present somewhere in the range from 3,500 Å
to 5,300 Å and if the apparent B magnitude is such that
$14.5 \leq B \leq 18.5$, then the probability of discovering such a
quasar is high. We have some 500 candidates, 93 of them confirmed during the spectroscopic follow up. List N°3 of the
Calán Tololo Survey shall present data for 80 new quasars (Maza
et al., in preparation). The rate of success for this quasar
survey is very high; near 90% of the candidates turn out to be
quasars. The number of high redshift quasar candidates varies
from plate to plate, from as low as 1 or 2 up to 25 per plate.
In some fields we have already confirmed more than 10 quasars
(each plate covers 25 square degree on the sky).

A preliminary estimate of the surface density of quasars
with $1.8 \leq z \leq 3.3$ and $14.5 \leq B \leq 18.5$ would be 0.4 deg^{-2}. This
is an average figure over a large area of the sky. The proba-
bility of discovering a quasar in this type of survey is a
strong function of the equivalent width of the $L\alpha$ line and the
apparent magnitude of the continuum (see, e.g., Gratton and
Osmer 1987). Quasar candidates are selected using several complementary criteria. On the first place they call the attention of the person searching the plate due to the presence of
an emission line (here the equivalent width of the line is crucial). As can be seen in Figure 2 the most common line found
in this survey is $L\alpha$ and the continuum usually changes crossing
$L\alpha$ being weaker to the blue; that detail is always taken into
account in the selection of a candidate. In addition a faint
broad emission somewhere from 3500 Å to 4100 Å if it is comple-
mented by a hint of another emission from 4500 Å to 5200 Å

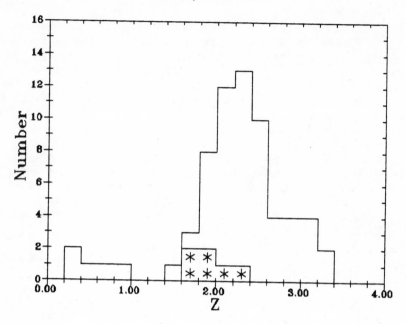

Fig. 2   Histogram presenting the redshift distribution of
66 Calán Tololo quasars observed with slit spectroscopy.
Six broad absorption line quasars (BAL's) are indicated by
an asterisk.   This is not a random sample of Calán Tololo
quasars and it might be biased toward high redshift quasars.

makes a good case for a quasar with Lα and CIV present in the
spectrum.   This effect tends to favour quasars with a redshift
z such that $1.9 \leq z \leq 2.4$ and a high CIV to Lα ratio. Selection
effects are difficult to evaluate because the selection of candidates is subjective and based in more criteria than the simple detection of an emission line.
     Figure 2 presents the redshift distribution of a sample of
66 Calán Tololo quasars.   Few (14%) have a redshift z < 1.8;
the majority, 65%, have a redshift z such that $1.8 \leq z \leq 2.6$
and 21 % have z such that $2.6 \leq z \leq 3.4$.   This could be biased
in favour of high redshift quasars and to quasars with strong
and broad absorption lines (BAL quasars) because we have observed mainly those candidates.   Only after observing every
single candidate in a large area of our survey could we produce
an unbiased redshift distribution.
     Finally we would like to stress here the importance of a
large area survey because it can provide a large number of
bright quasars very important for studies of the luminosity
function (and soace distribution) of quasars, to address the
problem of clustering of quasars, and to study the properties
of the intergalactic medium.

ACKNOWLEDGEMENTS

We would like to thank Cerro Tololo Inter-American Observatory
and Las Campanas Observatory for granting us the telescope time
that makes this project possible. This research is partially
supported through grants to JM and MTR from the D.I.B. from the
University of Chile.

REFERENCES

Gratton, R. and Osmer, P.S. 1987, Pub.A.S.P., 99, 899.
Osmer, P.S. 1980, Ap.J.Suppl., 52, 523.
Osmer, P.S. and Smith, M.G. 1980, Ap.J.Suppl., 52, 333.
Ruiz, M.T., Maza, J., Gonzalez, L.E., and Wischnjewsky, M.
     1987, A.J., 92, 1299.
Smith, M.G. 1975, Ap.J., 202, 591.
Smith, M.G., Aguirre, C., and Zemelman, M. 1976, Ap.J.Suppl.,
     32, 217.

AN UNBIASED LONG-SLIT SURVEY AND THE SERENDIPITOUS DISCOVERY OF A REDSHIFT 4.4 QSO[1]

PATRICK J. McCARTHY AND MARK DICKINSON

Department of Astronomy, University of California, Berkeley

[1]Based on observations obtained at the Lick Observatory, operated by the University of California.

ABSTRACT

We report the serendipitous discovery of an extremely high-redshift quasar (QSO), designated Q2203+29, found during long-slit observations of an unrelated object. The QSO spectrum contains strong emission lines of Lyα, N V λ1240, and C IV λ1549, as well as weaker Lyβ, O VI λ1034, Si II λ1262, and Si IV/O IV] λ1400. The derived redshift is 4.406 ± 0.005, one of the highest known. The r magnitude of the QSO is 20.8. This is the first serendipitously discovered QSO of any redshift in nearly five years of long-slit CCD spectroscopy of faint galaxies at Kitt Peak and Lick Observatories by H. Spinrad and collaborators. During this period ~ 0.01 square degrees of the sky has been observed down to very faint levels.

Key Words : Quasars–Redshifts–Surveys

I. INTRODUCTION

Long-slit spectroscopic observations of faint objects can be considered as nearly unbiased spectroscopic surveys of the sky. We are involved in such a survey using observations made with the Lick 3m and the KPNO 4m telescopes. The only significant bias contained in such surveys is introduced by the restricted wavelength coverage available to any CCD-disperser combination. An obvious and important shortcoming, of course, is that only a minute fraction of the sky can be sampled with this technique.

We report the discovery of an extremely high-redshift QSO from a survey of this type, carried out in parallel with spectroscopy of faint radio galaxies. It is quite surprising that such an unusual object should be discovered in a random survey of a very small area of the sky. Moreover, it comes at a time when intensive searches for $z > 4$ QSOs are for the first time yielding positive results.

II. The Long-Slit Survey

During the past five years a number of observers have been regularly obtaining long-slit spectra of very faint objects. The data set considered here was obtained in collaboration with H. Spinrad and S. Djorgovski using the Lick 3m and the Mayall 4m telescopes. These observations were usually targeted on faint $(V \sim 20 - 23)$ galaxies. These objects are sufficiently small such that they obscure a negligible fraction of the slit length. Integration times are typically 1 hour and only spectra taken in dark conditions with good $(\leq 2'')$ seeing and with exposure times > 1800 seconds are counted as part of the survey. A typical spectrogram will record 5 to 10 objects (other than the program source) with $R \sim 20 - 22$. At the faintest levels it is often difficult to distinguish faint stars from galaxies. Nearly 10% of the objects detected show strong or moderate emission lines, but only a modest fraction of these can be assigned unambiguous redshifts. While individual field galaxies with redshifts up to $z \simeq 0.8$ have been detected, no broad line objects had been detected prior to 1987. In table 1 we list the total number of hours of observation, the total area of sky surveyed, and the wavelength range covered for the various instruments involved in the survey.

TABLE 1

UNBIASED LONG-SLIT SURVEY

Telescope	Slit Size	Wavelength Coverage	Number of Spectra
Lick 3m	$2'' \times 1.'5$	3000 - 6000	31
Lick 3m	$2'' \times 2'$	4000 - 7800	66
Lick 3m	$2'' \times 2'$	6000 - 9500	20
KPNO 4m	$2.''5 \times 6'$	4500 - 7000	108
KPNO 4m	$2.''5 \times 6'$	6000 - 9500	43
Total	$0.0125\ deg^2$	3000 - 9500	268

III. The Discovery of Q 2203+29

Q 2203+29 was discovered (Dickinson and McCarthy 1987) on 1987 September 25 UT during the course of long-slit spectroscopic observations of the intermediate redshift radio galaxy 3C 441 ($z = 0.707$, Perryman *et al.* 1980). The observation was set up to obtain a spectrogram of 3C 441 and an object associated with it's northern radio hotspot. The discovery spectrogram was taken with the spectrograph slit centered on 3C 441 and oriented at position angle (P.A.) 145°, the radio P.A. of 3C 441. The combination of dispersing element and slit size used gave $\simeq 15$Å resolution in the wavelength range 4250 Å to 8600 Å with a slit length of 2.'2.

The QSO appeared along the slit displaced 51″ northwest of 3C 441. Direct images previously obtained with the same instrument (1987 August 23) revealed a faint stellar object at the approximate location of the QSO. The telescope was then offset to place this object on the slit, and an additional exposure was made with the slit at P.A. = 90°. This spectrogram confirmed our identification of the faint stellar object as the QSO.

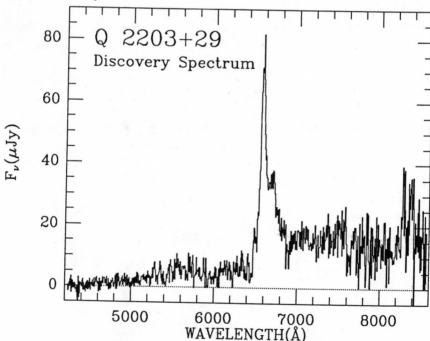

Fig. 1.—Discovery spectrum of Q2203+29, taken on 1987 September 25. The pixel size is 5 Å and the mean resolution is 12 Å. The total integration time was 1.5 hours. Note the strong break in the continuum shortward of Lyα emission.

Subsequent spectroscopic observations were made at Lick Observatory, both to observe the spectrum of the QSO over the entire visible spectrum ($\lambda\lambda 3000 - 10,000$Å) and to obtain higher resolution (5Å) spectra centered on Lyα. The details of these observations can be found in McCarthy *et al.* (1988).

An image of Q 2203+29 is presented in McCarthy *et al.* (1988) along with an astrometric position. The QSO is rather faint; we derive an magnitude of $r = 20.8$. The Lick r filter contains the Lyα emission line; thus, our r magnitude does not measure the continuum alone. Applying synthetic V and I filters to the spectrum yields continuum magnitudes of $V = 22.0 \pm 0.3$ and $I = 21.1 \pm 0.3$. This is considerably fainter than the high redshift QSOs identified by radio and objective prism techniques (Peterson *et al.*(1982); Hazard, McMahon and Sargent (1986), but is comparable to the QSOs recently discovered by CCD drift scanning (Schmidt, Schneider and Gunn 1987) and multi-color techniques (Warren *et al.* 1987a, b).

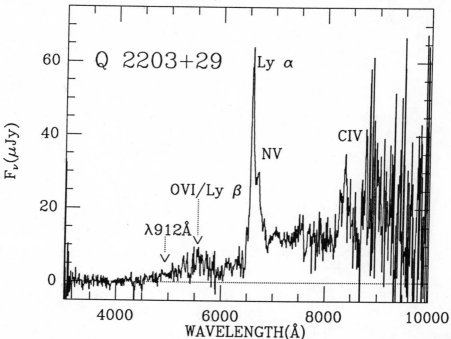

Fig. 2.—Average of all our spectra of Q2203+29. The total integration time was roughly 2 hours in most portions of the spectrum, except near Lyα where it was $\sim$ 4 hours. The redshift determined from the strong lines is 4.406 ± 0.005.

IV. DISCUSSION

As stated above, the QSO was discovered serendipitously during long-slit observations of an unrelated object.   This is the first serendipitously discovered QSO of *any* redshift during five years of long-slit CCD spectroscopy of faint radio sources at Kitt Peak and Lick Observatories by Spinrad and collaborators. As shown in table 1, we have sampled 0.01 square degrees of the sky. Based on the recent unbiased luminosity function of QSOs by Koo and Kron (1988), we would expect to find one QSO during these observations, as indeed we did. The APM multi–color survey of Warren *et al.*1987*b* resulted in the discovery of 3 QSOs with $z > 4$ in 30 square degrees surveyed to $R = 20.0$. From this we conclude that the probability of finding a $z > 4$ QSO in 0.012 sq. deg. of randomly selected sky (to $R \sim 22$) is a few times 10^{-3}, assuming that the surface density of $z > 4$ QSOs does not increase strongly from $R = 20$ to $R = 22$.

We wish to thank the staff of Lick Observatory, and Jim Burrous in particular, for assistance with the observations. The set of nearly 300 long-slit spectrograms used in the unbiased survey was obtained in collaboration with H. Spinrad, S. Djorgovski, M. Strauss, and D. Baker. We acknowledge support from NSF grants AST 85–13416 and AST 84–16177, and from the California Space Institute. Lick Observatory is operated by the University of California, and is funded in part by NSF grant AST 86–14510.

REFERENCES

Dickinson, M., and McCarthy, P. J. 1987, *Bull. A.A.S.*, *in press.*

Hazard, C., McMahon, R. G., and Sargent, W. L. W. 1986, *Nature*, **322**, 38.

Koo, D. C., and Kron, R. G. 1988, *Ap. J.*, **325**, *in press.*

McCarthy, P., J., Dickinson, M., Filippenko, A. V., Spinrad, H., and van Breugel, W. J. M. 1988, *Ap. J. Letters (in press).*

Perryman, M. A. C., Lilly, S. J., Longair, M. S., and Downes, A. J. B. 1984, *M. N. R. A. S.*, **209**, 159.

Peterson, B. A., Savage, A., Jauncey, D. L., and Wright, A. E. 1982, *Ap. J. (Letters)*, **260**, L27.

Schmidt, M., Schneider, D. P., and Gunn, J. E. 1987, *Ap. J. (Letters)*, **321**, L7.

Warren, S. J., Hewitt, P. C., Irwin, M. J., McMahon, R. G., Bridgeland, M. T., Bunclark, P. S., and Kibblewhite, E. J. 1987, *Nature*, **325**, 131.

Warren, S. J., Hewitt, P. C., Osmer, P. S., and Irwin, M. J. 1987,*Nature* **330** 453.

A STRIP SEARCH FOR QUASARS:
THE CCD/TRANSIT INSTRUMENT (CTI) QUASAR SURVEY

John T. McGraw, Michael G. M. Cawson,
J. Davy Kirkpatrick and Vance Haemmerle

Steward Observatory
The University of Arizona, Tucson, Arizona   85721

1.0 INTRODUCTION

Quasars are of great interest to astronomers principally because they are the most intrinsically luminous objects known. They can thus be seen to the greatest distance and can be used in a number of ways as probes of the early universe. The physical characteristics of quasars are, however, sufficiently heterogeneous that an investigation of the quasar phenomenon itself requires an unbiased picture of the various types of quasars and the distribution of quasars with redshift.

Unfortunately, the history of discovery of quasars is riddled with selection effects which have biased this picture. Initially, quasars were radio-loud, starlike objects. Then, quasars became starlike objects, only a fraction of which were radio-loud, but which showed an ultraviolet excess. Next, quasars became emission line objects. More recently, it was known that quasars suffered a "cutoff" in space distribution beyond $z = 3$. Each of these effects is now known to be either only partially true or incomplete, thus there still exists a need for surveys to find representative samples of quasars unbiased with respect to extrinsic selection effects, intrinsic physical differences, or redshift. The CCD/Transit Instrument (CTI) at Steward Observatory has underway such a survey involving multiple survey techniques to be used on the same strip of sky.

2.0 THE STRIP

The survey area is defined as the field-of-view swept out by the CTI in the course of a year. The CTI is a 1.8 m, f/2.2 telescope on Kitt Peak. It does not move, but is fixed in the meridian at a declination of +28 degrees. It utilizes two CCDs aligned east - west in the focal plane to observe the sky in two colors, V and one of (U), B, R or I, depending upon the sky brightness. The CCDs are operated in the "time-delay and integrate" (TDI) mode at the apparent sidereal rate. The resulting image is a strip 8.25 arcmin wide (N-S) with length determined by the length of the night. The CTI surveys about 15 square degrees per night or about 45 square degrees per year

to a limiting magnitude of V = 21. Each object in the strip
thus has (U)BVRI time-averaged colors and a V light curve with
a time resolution of one day.

In an effort to both select an unbiased sample of quasars
to a faint magnitude limit and to <u>measure</u> selection effects
among various techniques, we are conducting independent "blind"
surveys using as detection criteria variability and color. An
additional survey for radio-loud quasars utilizing the VLA has
been initiated.

3.0 THE SEARCH

The CTI quasar survey is multi-purpose by design. In
particular, it is designed to find quasars in a manner as
unbiased as possible and to monitor the light curves of the
discovered quasars. The emphasis of this survey is on precise,
continued <u>measurement</u> of quasars. We discuss selection on the
basis of color and variability. Each search is conducted
independently of all others.

3.1 Selection of Quasar Candidates on the Basis of Color

The principal advantages of selection by color are the
depth which can be reached and the relative ease of processing
data. The principal disadvantage is that the information
content per observation is lower than for objective
spectroscopy. We wish to exploit the advantages and minimize
the disadvantages of color selection.

Utilizing cluster analysis, selection of quasar candidates
on the basis of multiple broadband colors results in a sample
chosen relatively independently of redshift. We have
investigated the selection of quasar candidates using multiple
colors, but on the basis of arbitrary color-color diagrams in
an effort to more fully understand the information content of
multiple colors, intuitively understand the colors of quasars
of different types, discover quasars independent of redshift,
and select candidates in a less intensive computational
regime. Colors have been fabricated in linear combinations
which represent the projection of quasars onto arbitrary planes
in color space.

Our approach has been to first try to <u>predict</u> the colors of
quasars as a function of redshift and then to search for them,
utilizing the arbitrary color-color diagrams for the selection
of candidates. The colors we utilize are fabricated by
convolving the <u>measured</u> CTI filter bandpasses with the energy
distributions of real stars and quasars. Quasar energy
distributions were estimated from calibrated IUE and optical
spectroscopy. To avoid biases involved in producing an
"average" spectrum, we elected to utilized spectra from
multiple, independent quasars and to redshift the energy
distibution only over the range for which valid data existed.
Instrumental colors are calculated by convolving the quasar

enery distribution with the filter bandpass, the CCD response, the telescope efficiency and the atmospheric transmission. The result of this analysis should accurately represent the colors of quasars as directly measured by the CTI.

Our analysis includes two hypotheses graphically illustrated here:

* the best discriminant of a quasar is its power law energy distribution, not (necessarily) its emission line spectrum
* arbitrary color-color diagrams can always be constructed to either:
    1) minimize the area occupied by quasars
    2) minimize the area occupied by stars
    3) for a given redshift range, minimize contamination by stars.

The techniques we utilize to select quasar candidates on the basis of color are demonstrated in Figure 1.

In the four diagrams of Figure 1, each expressed in CTI instrumental magnitudes, can be seen the change in color induced by lines entering and leaving bandpasses. The main sequence is marked with letters denoting the spectral type. W and S are used for white dwarfs and subdwarfs, respectively. Colors for power laws are marked with open boxes with indicies noted. The quasar tracks are calculated empirically from selected "typical" emission line quasars with measured optical and uv spectra. Individual quasars are marked by line type over redshift ranges noted at the bottom of each plot. The redshift at integer values is noted where confusion of the tracks allows. In each of these diagrams the details of excursions below $z = 2$ are not considered significant - only the general locus is of real interest.

Figure 1A shows that quasar energy distributions are more like power laws than stars. A wide range in wavelength (U through I) is useful in maximizing the distance of the locus of most (low redshift) quasars from the main sequence.

Figure 1B shows a color projection which minimizes the area of the locus of quasars by requiring all power law colors to be identical. In this diagram the locus of quasars remains centered about the colors for power laws with principal contamination from white dwarfs and subdwarfs.

Figure 1C minimizes the locus of stars, with less stellar contamination, in general.

Figure 1D utilizes easily measured colors and indicates that, in this diagram, $z = 4$ quasars are (typically) separated from the main sequence by more than 0.2 magnitude. It is a pity all quasars are not "typical"!

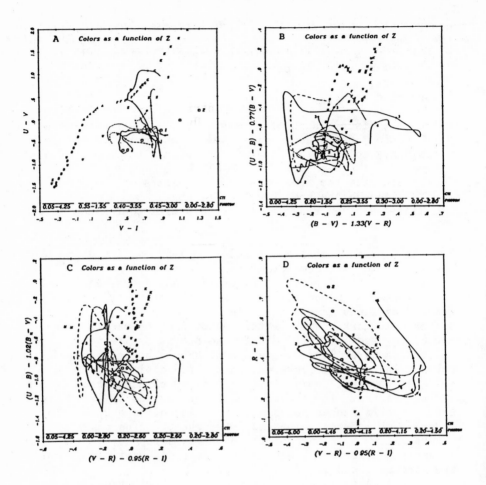

FIGURE 1. Typical excursions of quasars colors with redshift in color-color diagrams.

Analysis of the geometrical distance in color space (as a function of redshift) a quasar falls from the main sequence gives a measure of the probability of detection. This is shown in Figure 2. Assuming a uniform distribution of quasars with redshift to z > 5, it can be seen that the redshift region 3 < z < 4 has the lowest detection probability on the basis of color. This effect is a likely contributor to the deficiency of quasars in this redshift range. Overcoming this selection effect is dependent upon:

* multiple broadband colors spanning the optical wavelength range,
* precise measurement of colors, and
* survey depth.

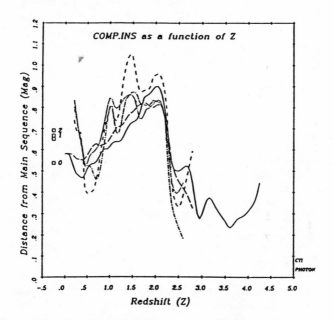

FIGURE 2. The absolute distance of quasar colors from the main sequence colors as a function of redshift. The quasar colors are those used in the above color-color diagrams. We hypothesize that the (former) paucity of quasars with z > 2.5 was contributed to by this effect. Photometric precision better than 0.1 magnitude will help investigate the redshift domain 2.5 < z < 4.0.

THE CASE LOW-DISPERSION NORTHERN SKY SURVEY: THE QSO
FINDINGS

N. SANDULEAK AND PETER PESCH
Warner and Swasey Observatory
Case Western Reserve University
Cleveland, Ohio 44106

ABSTRACT   Follow-up, higher resolution spectroscopic
studies of candidate objects found on objective-prism
plates from the Case Low-Dispersion Northern Sky Survey
have confirmed 50 new QSO's. These initial data suggest
that this survey will ultimately be found to contain
roughly 0.4 QSO's deg.$^{-2}$ to a limiting apparent magnitude
of B $\sim$ 18 over the range 0.06 $<$ z $<$ 3.3.

INTRODUCTION

The Case Low-Dispersion Northern Sky Survey utilizes
objective-prism plates taken with the 24-inch Burrell Schmidt
telescope at the Warner and Swasey Observatory Station on
Kitt Peak. The so-called thin-prism (1°.8 apex angle)
provides a dispersion of about 1300 Å mm^{-1} at Hγ and careful
guiding results in spectra having a width $<$ 0.03 mm. Kodak
IIIa-J plates (baked in forming gas) are given 75 minute
exposures which reach a limiting magnitude of B $\sim$18 for the
threshold detection of a stellar-like continuum. Because of
the ultraviolet transparency of both prism and corrector
plate, the spectra extend from λ5350 (i.e. the red cut-off
of the IIIa-J emulsion) down to about λ3300 in very blue
objects.

    The square plates measure 5°.1 x 5°.1 and have a scale of
about 100" mm^{-1}. The survey was originally intended to cover
that portion of sky lying north of declination +30° and above
+30° galactic latitude. This requires 227 fields and as of
this writing (Dec. 1987) usable plates have been obtained for
221 (97%) of these fields. We now plan to extend the coverage
southward to near the celestial equator and 31 (18%) of the
additional 174 fields have been photographed. The plate
acquisition phase will require several more observing seasons
to complete and ultimately should encompass about 8000 deg.^{2}
of sky.

The plates are scanned by both authors using a 15X
binocular microscope and a wide variety of objects of interest
are noted including blue and/or emission-line galaxies, QSO
candidates, blue subluminous stars, and distant horizontal
branch, carbon and late M-type stars in the galactic halo.
As the results become available they are published in an ongoing series of finding lists in the Astrophysical Journal
Supplements. In this report we consider only the QSO
findings but should stress the point that detection of QSO
candidates comprises only one (albeit highly important) goal
of this comprehensive high galactic latitude survey.

SPECTRAL CATEGORIES CONTAINING QSO CANDIDATES

The majority of QSO's can not be recognized conclusively as
such on our low-resolution objective-prism plates and hence
require follow-up observations for confirmation. Potential
QSO candidates are primarily to be found in the following two
of our spectroscopically defined natural groups.

Category II - Blue and/or Emission-line Unresolved Objects

Here we include two basic types of spectra both of which
appear stellar on our plates. The first involves objects
(only 2-3 per field on average) which actual show or are
suspected of having emission features characteristic of QSO's.
In a majority of cases, a solitary emission feature is seen
which is often (but not necessarily) redshifted Lyα in the
range $1.7 < z < 3.3$. These then are highly probable QSO's
but nevertheless require higher resolution spectroscopy for
confirmation and to establish the identification of two or
more emission lines so as to provide an accurate redshift. For
the purpose of this report, these will hereafter be designated
as CSO,em. objects.
    The other and more numerous representatives in this
category are only moderately blue (i.e. U-B $\sim$-0.5) and have
rather flat and featureless continua with no obvious emission
or absorption lines present. A majority of these should
prove to be blue galactic stars but many primarily low redshift ($z < 1.7$) QSO's having emission features too weak
(e.w. < 50 Å) to be readily detected on our survey plates are
to be found among these objects when follow-up spectroscopy
is performed. These will be called simply CSO objects.

Category III - Known and Probable Blue Stars

In this category we attempt to segregate out those blue
unresolved objects which we have reason to suspect, or know
to be galactic stars. These include objects which clearly

are stars because they show evidence of Balmer series hydrogen
absorption lines. One example would be DA-type white dwarfs
showing very broad hydrogen lines. Needless to say, any blue
stellar object independently selected by us, which we subsequently find to be a known star in the literature, is put
into this category. Likewise any blue object for which we
find a published sizable proper motion is considered a star
and included here. These are designated CBS objects.

   However, we initially decided to also put into this
category unresolved, extremely blue objects (i.e. U-B $<$ -1.0)
showing featureless continua with the expectation that these
will prove to be mainly sdO stars. Follow-up studies have
now shown that a small percentage of very blue, low z QSO's
is being included in Category III primarily among these
suspected sdO star candidates. Thus, in order to ferret out
all of the QSO's lurking in our finding lists, one should
subject these very blue, featureless objects from Category III
to higher resolution spectroscopy along with all of the
objects in Category II not already known to be QSO's.

RESULTS OF FOLLOW-UP SPECTROSCOPY

Thus far there have been four series publications giving
finding lists for objects in spectral categories II and III.
These are papers I. Pesch and Sanduleak (1983), II. Sanduleak
and Pesch (1984), III. Pesch and Sanduleak (1986), and V.
Pesch and Sanduleak (1987). Follow-up spectroscopic studies
of CSO and CSO,em. objects have been made by Zotov (1985) and
Wagner (1987). Follow-up spectroscopy of the CBS objects
has been done by Wagner et al. (1987). The results of their
findings as they pertain to QSO's are given in Table I.

TABLE I   Follow-up Spectroscopic Results

	CSO,em.	CSO	CBS
Total Number	76	325	160
Known QSO's, BL Lac's	5	13	0
Follow-up spectra	13	86	133
New QSO's	11	21	18

The first row gives the total number of objects listed in the
first four survey publications referenced above which cover a
contiguous area of 462 deg.^{2} The previously known QSO's and
BL Lac objects in row 2 were those we found listed in Hewitt
and Burbidge (1980). The figures given in row 3 show that
the follow-up studies are still incomplete particularly for
the CSO and CSO,em. objects. Based on these limited results,

it appears that a very large fraction of the CSO,em. objects
and about one-fourth of the CSO objects found by our methods
will prove to be QSO's. The numbers of new QSO's in the last
row include three objects originally found by us which are
now listed in Hewitt and Burbidge (1987) after having been
independently confirmed as QSO's by other observers. From
these findings one can project that our survey will ultimately
be shown to contain about 0.4 QSO's deg.$^{-2}$ to B$\sim$18 mag.

The redshift distribution of both the new and previously
known QSO's is given in Table II.

TABLE II   Redshift Distribution of QSO's

z	CSO,em.	CSO	CBS
0.06 - 0.50	4	11	12
0.51 - 1.00	2	9	2
1.01 - 1.50	2	9	3
1.51 - 2.00	2	1	1
2.01 - 2.50	4	0	0
2.51 - 3.00	2	1	0
3.01 - 3.30	0	0	0

Three of the CSO objects are known BL Lac objects for which
redshifts are unspecified. As expected the QSO's found among
the CSO objects and particularly among the CBS objects have,
on average, much lower redshifts than those detected on the
basis of very strong emission features. The detection of
these presumably lower luminosity QSO's constitutes one of
the major contributions of this survey. We strongly urge
interested observers to pursue the follow-up spectroscopy
necessary to identify the QSO's in our finding lists over this
wide range of z values. Preprints of our future lists will
be provided upon request.

ACKNOWLEDGMENTS

We wish to acknowledge the efforts of Rik Hill, our resident
Kitt Peak observer, and also express appreciation to the
National Science Foundation for its continuing support of this
project.

REFERENCES

Hewitt, A., and Burbidge, G. 1980, Ap. J. Suppl., 43, 57.
_____ 1987, Ap. J. Suppl., 63, 1.
Pesch, P., and Sanduleak, N. 1983, Ap. J. Suppl., 51, 171,
    (Paper I).

Pesch, P., and Sanduleak, N. 1986, Ap. J. Suppl., 60, 543
     (Paper III).
_____   1987, Ap. J. Suppl., (in press,
     Paper V).
Sanduleak, N., and Pesch, P. 1984, Ap. J. Suppl., 55, 517
     (Paper II).
Wagner, M.R., Sion, E.M., Liebert, J., and Starrfield, S.G.
     1987, Ap. J., (submitted).
Wagner, M.R. 1987, (private communication).
Zotov, N. 1985, Ap. J., 295, 94.

UNDETECTED GRAVITATIONAL LENSING EVENTS AND QUASAR POPULATION STATISTICS

EDWIN L. TURNER

Princeton University Observatory
Princeton, NJ 08544 USA

ABSTRACT

The possibility that straightforward interpretations of quasar population statistics are being distorted by undetected gravitational lensing events is reviewed. Two recent detailed studies by Ostriker and Vietri concluding that the effects are probably minimal and by Schneider concluding that they may well be large are discussed. The evidence, particularly the absence of the expected statistical signatures of lensing in the available data, favors the former conclusion. However, given the possibility of inadequacies in the present observations and (probably more importantly) the existence of incompletely explored theoretical complications, the latter conclusion can not be certainly excluded.

INTRODUCTION

Do undetected gravitational lensing events significantly perturb the intrinsic quasar luminosity function producing a false "apparent" luminosity function, false evolution, and so forth? Or equivalently, is the Robertson-Walker metric a sufficiently accurate approximation for routine use in large redshift observational cosmology? Or in yet other words, are luminosity distances only a function of H_o, q_o, and z, or do they also depend on RA and DEC?

This somewhat disturbing question was raised by the author (Turner 1980) in the context of modern observations of gravitational lenses and in a very extreme form much earlier by Barnothy and Barnothy (1968). Substantial theoretical efforts by various other investigators have been devoted to seeking an answer; see for example Avni (1981), Peacock (1982), Ostriker and Vietri (1986), and Schneider (1987). This paper is intended to critically review the proposed answers and the problems which prevent us from arriving at a definite conclusion at present. Since the two most thorough and recent treatments conclude on the one hand that there is probably no important effect for the vast majority of quasars (Ostriker and Vietri 1986) and on the other hand that it is "inescapable" that it is a "dominant effect" given certain plausible conditions (Schneider 1987), it is clear that unanimity has not yet been achieved.

WHAT IS KNOWN

The basic question described above arises from the realization that gravitational lensing events which are not otherwise detectable could substantially magnify (and hence amplify the flux from) the images of distance sources. There are two fundamentally different types of lensing events which could have such an effect. First, there are strong lensing events which produce two or more bright images with angular separations which are not currently observationally resolvable. For point sources of radiation magnification factors are independent of characteristic angular separation scales. The archetypical case of this type is the so called micro-lensing event in which a lensing object of very roughly stellar mass produces a strong flux amplification associated with angular splittings in the general vicinity of 10^{-6} arc seconds (far beyond the reach of current or anticipated observational resolutions). Second there is the possibility of single image, high magnification lensing events caused by extended mass distributions with surface densities near the critical value (at which the lens's focal length is equal to its distance from the observer). The recently discovered giant luminous arcs in rich galaxy clusters are probably examples of such events in which the source is a background galaxy (Lynds and Petrosian 1986, Soucail et. al. 1987, Paczynski 1987, Blandford and Kovner 1987, Grossman and Narayan 1987). If the source were a quasar, the characteristic arc shaped distortion of the image would probably be unresolvable (its size is proportional to the intrinsic size of the source) but the magnification would remain high.

It is important to understand that frequent occurrence of both of these types of problematic lensing events is *a priori* reasonably plausible. Calculation of point mass strong lensing event optical depths (Turner, Ostriker, and Gott 1984) show that even a modest population of compact objects will result in a substantial fraction of the objects at typical quasar redshifts being substantially magnified, particularly when account is taken of the selection bias favoring the inclusion of lensed sources in flux limited samples (Turner 1980). For instance, in a simple toy model in which the baryons deduced to exist from Big Bang nucleosynthesis (contributing 0.14 to Ω_o) are assumed to exist in a uniformly distributed population of compact objects in an $\Omega=1$ universe (with the additional mass supplied by a constant density sea of non-baryonic elementary particles), one fifth of redshift 3 quasars will be strongly micro-lensed with a mean amplification of greater than 2 given a quite modest selection factor of 4. This is by no means an extreme case; much more potent micro-lens populations and selection factors are possible in principle. For single image, high magnification lensing events known classes of objects (e.g., rich galaxy clusters) are probably too rare to give an important effect. However, hypothetical dark objects could well be important; in particular, biased cold dark matter galaxy formation theories (Bardeen et. al. 1986) predict a population of dark "failed" galaxies which will be much more numerous than visible galaxies but which will have substantially lower central surface densities. These may well produce many strong, single image lensing events (Hinshaw and Krauss 1987).

Despite minor modifications by subsequent investigators (Avni 1981, Peacock 1982), the original conclusion (Turner 1980) stands that inclusion of significant numbers of undetected lensing events in observed quasar samples would result in the distortion of derived luminosity functions and evolution. The mechanism of this distortion is most conveniently described as the convolution of some kernel describing the probability distributions of flux amplifications with the true quasar luminosity function. This kernel should include a delta function very nearly at unit amplification to account for the unlensed quasar population and a distribution of larger amplifications for the lensed objects. It appears that the differential probability distribution of amplifications varies as the inverse cube of the amplification for large values for point sources quite genericly (Blandford and Kochanek 1987). This implies that the luminosity function distortion will only be important if the intrinsic luminosity function is steeper than L^{-3} and that it will be most evident at high luminosities in such cases. In fact, the best current determinations of the quasar optical luminosity function at low redshift (where it can be assumed to be free of lensing effects) indicate that it is steeper than L^{-3} at the bright end (see papers by Boyle et. al. and Marshall elsewhere in this volume).

All of the points mentioned so far in this section would seem to suggest that undetected lensing events may well be playing havoc with our studies of quasar statistics, that the effect is in fact an important one. There is, however, one strong empirical indication that this is not the case, namely the fact that the optical luminosity function for quasars and its evolution derived from observations do not resemble those which would be expected to result from lensing distortions. Schneider (1987) whose positive conclusion about the probable importance of the effect was quoted above ignores this issue, arguing that the observations are too uncertain to provide useful constraints. Ostriker and Vietri's (1986) negative conclusion is basicly derived from such an argument. Ironicly, the Ostriker and Vietri analysis was based on an analytic approximation of the evolving quasar luminosity function (Schmidt and Green 1983) which is now (with an order of magnitude more data available) known to be a very poor representation, even in a qualitative sense. However, the new data suggest a result which is no more (and probably less) consistent with the expected effects of undetected lensing. In particular, the constant functional form "luminosity evolution" behavior seen at $z<2.2$ (see papers in this volume) and the declining co-moving density of luminous quasars beyond $z=2.5$ (see Schmidt in this volume) seem to be at odds with the expected results of lensing distortions.

WHAT IS NOT KNOWN

How secure is the argument given just above which suggests that undetected lensing events are rare in observed quasar samples? May we now safely ignore the whole bothersome issue? The answer to these questions depends on the reliability both of the observationally derived quasar luminosity function and its evolution and of the calculated predictions for the nature of lens induced distortions of these functions. In the former case, although there are no apparent problems with the current observational results it would be well to note that major infusions of new data concerning both low ($z<2.2$) and high ($z>3$) redshift quasar populations have occurred within the last year or two resulting in considerable modifications in our ideas about these epochs. Indeed systematic sampling of the high redshift population is really only just beginning. The cautious will want to wait for the dust to settle before making strong claims. In the latter case, it is certain that the existing theoretical calculations are based on approximations which could easily be incorrect and could invalidate some of their conclusions. I will focus on these potential weaknesses of the calculations and discuss several in detail below.

In most calculations to date, the quasars have been treated as having zero true angular size. This approximation may well fail significantly if we are dealing with a population of relatively low mass micro-lenses. The most direct effect of a failure of this assumption is to reduce the probability of very large amplifications and thus, for any fixed microlens population, to reduce the importance of micro-lensing amplifications. However, Schneider (1987) has shown that under simple circumstances, the effect of finite source size is to allow heavily micro-lensed populations to show a false (distorted) apparent luminosity function with any slope steeper than L^{-3}, thus erasing one of the primary signatures of micro-lensing. A realistic treatment of finite source sizes might well be very complex particularly if there were a rough coincidence between the true angular sizes of quasars and the angular critical impact parameter of typical micro-lens. Making a good model of such a situation might well require taking into account the spectrum of micro-lens masses and the distribution of quasar emission region physical sizes including the possible dependence of the latter on wavelength, luminosity, redshift, et cetera. Such a situation could lead to very complicated effects in which micro-lensing effects varied in importance for various classes of quasars, for instance lensing events could even be less important for higher redshift quasars due to larger physical sizes and/or their smaller angular diameter distances in some cosmologies. Of course, the fact that we know very little directly about micro-lens masses or quasar physical sizes will, to say the least, be a serious obstacle to building realistic models.

Standard calculations of micro-lensing statistics normally consider only single, isolated, strong lensing events. They neglect both the possibility of multiple strong scatterings and of the perturbation by weak scatterings of strong events. If micro-lensing plays any significant role, these are likely to be very poor approximations in my opinion. The threshold nature of lensing can effectively couple the effect of mass distributions with very different physical scales (Turner, Ostriker, and Gott 1984); thus it may be important whether or not the micro-lenses are themselves embedded in large scale potential wells. If, as argued below, micro-lenses are clustered then even in situations where the *mean* optical depth is fairly low, the typical event in which a strong scattering occurs is likely to involve multiple strong scatterings. Furthermore, even for a uniformly distributed micro-lens population, the isolated scattering approximation becomes invalid at fairly small mean optical depths; for instance, at $\tau=0.1$ there is typically a second micro-lens within about 3 critical radii of any strong scattering event. The treatment of multiple lensing events and of the coupling of lensing on different mass scales is very complex and poorly understood. It is not presently clear what the effects on cross sections and amplification distributions are likely to be but major modifications are not ruled out. If for instance, flatter amplification probability distributions are indicated, it would have major implications for the possible importance of undetected lensing events.

Yet another common but very weak set of approximations is that the population responsible for undetected lensing events is taken to be a uniformly distributed, non-evolving set of point mass objects. Certainly the micro-lens population which we can see (i.e., the stars) are not uniformly distributed nor are any plausible population of dark point masses. The same aggregation process by which such objects must form is likely to occur to some extent on slightly higher mass scales and result in clumping. If a substantial population of dark point like objects exists, it would be surprising if at least a substantial fraction were not concentrated around galaxies and in galaxy clusters. As mentioned in the previous paragraph, such micro-lens clustering would have important statistical consequences since it would imply that few if any lines of sight would actually have the *mean* optical depth, the vast majority being either much higher or much lower. In addition to worries about micro-lens clustering, it is not clear that point mass objects will be the most important source of undetected lensing events. As discussed in the previous section, single image but substantial amplification events due to extended mass distributions with surface densities near the critical lensing value could play a major role. Such a class of undetected lensing events might have properties significantly different than those of micro-lenses including (at least) a much reduced sensitivity to the sizes of the quasar emission regions. Modeling of such single image lenses will also need to take into account clustering, multiple scatterings, and so forth and may also involve substantial cosmic evolution of the lens properties over the redshift ranges of interest. In view of the wide spread interest in theories of biased galaxy formation, it is clear that this class of possible lenses has received less attention than it deserves; see Hinshaw and Krauss (1987).

Finally I would call attention to the modeling of sample selection biases. It is clearly understood that such effects are likely to play a major role in determining the frequency of lensing events, detected or undetected, in quasar samples. Model calculations routinely include the effect of a flux limit bias; however, this is by no means an adequate description of the biases in most modern samples which depend upon such criteria as UV excess, multi-band color anomalies, emission line detection, flux in non-optical bands (radio, x-ray), variability, lack of proper motion, and so on. Taking proper account of these complex selection biases may be critical to understanding quasar population statistics in general and is likely to be even more critical to assessing the impact of undetected lensing events.

It should be clear that the comments in the above paragraphs are not intended as criticisms of those who have attempted calculations (my own foray into the field being the most idealized of all); in fact it seems unlikely that fully realistic models, in the various senses discussed above, will be possible within the foreseeable future. Rather the point is to emphasize that the conclusions based on available calculations must be regarded with some caution; the Universe has no difficulty realizing situations which we cannot adequately model and could well be doing so in this case. Of course, these complications do not necessarily all tend to increase the importance of undetected lensing events; many (e.g., finite source sizes, microlens clustering, complex selection effects) could just as easily act to reduce it. They do, however, introduce some uncertainties into the situation.

CONCLUSIONS

It is reasonable to ask for the best guess at the answer to the questions raised at the beginning of this article. My own view of such a best guess is that undetected lensing events are quite rare in observed quasar samples and that their effect on quasar population statistics is minimal. The fundamental justification for this conclusion is essentially the Ostriker and Vietri (1986) point that the observed quasar populations do not have properties resembling those expected for a heavily lensed sample; recent improvements in the data seem to make this argument even stronger. Nevertheless, I see no way to certainly exclude the opposite conclusion at present. If, as argued by Schneider (1987), the available observations are seriously misleading (unlikely in my view), or if realistic treatment of the various complications discussed in the previous section were to give results very different from the calculations so far available, it is still possible that the effect could turn out to be an important one. We should keep it in mind as a caveat on straightforward interpretations of quasar statistics much as those studying large scale structure once regarded the (then unpleasant now considered attractive) possibility that galaxies do not fairly sample the true total mass distribution.

The path forward is fairly clear. Improved theoretical models are certainly possible and desirable including probably numerical simulations to treat analytically intractable issues. On the observational side continued accumulation of quasar population statistics (desirable for many reasons) is needed. In addition targeted studies of specific quasars designed to search for evidence of micro-lensing events could be very informative. The remarkable lens system 2237+0305 is likely to be uniquely well suited for such studies (Schneider et. al. 1988).

ACKNOWLEDGEMENTS

Useful conversations with J. P. Ostriker and the support of NASA grant NAGW-765 are gratefully acknowledged.

REFERENCES

Avni, Y. 1981, *Ap. J. Lett., 248,* L95.

Barnothy, J. M., and Barnothy, M. F. 1968, *Science, 162,* 348.

Blandford, R. D., and Kochanek, C. S. 1987, preprint, *GRP-138.*

Blandford, R. D., and Kovner, I. 1987, preprint, *GRP-158.*

Grossman, S., and Narayan, R. 1987, preprint.

Hinshaw, G., and Krauss, L. M. 1987, *Ap. J., 320,* 468.

Lynds, R., and Petrosian, V. 1986, *Bull. Am. Ast. Soc., 18,* 1014.

Ostriker, J. P., and Vietri M. 1986, *Ap. J., 300,* 68.

Paczynski, B. 1987, *Nature, 325,* 572.

Peacock, J. A. 1982, *M. N. R. A. S., 199,* 987.

Schmidt, M., and Green, R. F. 1983, *Ap. J., 269,* 352.

Schneider, D. P., Turner, E. L., Gunn, J. E., Hewitt, J. N., Schmidt, M., and Lawrence, C. R. 1988, *A. J.,* in press.

Schneider, P. 1987, *Astron. Astrophy., 183,* 189.

Soucail, G., Fort, B., Mellier, Y., and Picat, J. P. 1987, *Astron. Astrophy., 172,* L14.

Turner, E. L. 1980, *Ap. J. Lett., 242,* L135.

Turner, E. L., Ostriker, J. P., and Gott, J. R. 1984, *Ap. J., 284,* 1.

DISCUSSION

Stocke : I'd like to make a comment about something I was going to present tomorrow. This viewgraph shows the observed redshift distribution of the complete sample of X-ray selected quasars from the Einstein medium sensitivity survey, independent of the color. You can see, as all the other samples have shown, the peak is at quite low redshift, and that's thought to be a bias with respect to the X-ray selection of these objects. The top panel shows, however, the redshift distribution of X-ray selected quasars within 3 optical diameters of a galaxy brighter than 18th mag. These two distributions differ at the more than 97.5% confidence level, with the sense being that the sample near galaxies has higher redshifts. In that regard, 2 of these 3 objects here from the first complete sample we found are near bright galaxies. The highest redshift object in our sample is very close to a bright elliptical galaxy.

Peacock : One comment about the effect of lensing on the high luminosity end of the luminosity function. You've been talking mainly about microlensing, that is, lenses we don't know exist. As far as lensing by galaxies is concerned, we know of at least one example, PG1115 or whatever. The point to be brought out, and Brian Boyle mentioned it, that you can't fit the bright end of the luminosity function with a power law and an exponential cutoff; the data don't allow the cutoff. My feeling is that this conclusion is biased by just a few extra, luminous objects, such as PG1115. I wish to suggest that if we threw out such objects, then our conclusion might change.

Turner : Despite some weakness in the theory, I think we can clearly believe that you should be particularly worried when you're talking about high luminosities, high redshifts, and flux that originates in small regions, like the X-rays. Those are the situations most likely to be affected by either microlensing or galactic scale lensing.

MacAlpine : Ed, if you were given unlimited access to any facilities you wanted and 10 postdocs, what would you do?

Turner : Go on vacation! Much of what I would do is being done anyway, such as studying quasar statistics directly. But the other thing I would push is very careful monitoring of known galactic scale gravitational lenses, where we know there are stars around and where there ought to be microlensing effects. The 2237 object is a particularly good case to monitor, with its four images split within a 2" x 2" field. If we didn't see microlensing in that and similar

systems, I think we could feel fairly safe about it not being
an important effect.

Schmidt : Ed, I wonder whether you can empirically come up
with an upper limit to the effect of lensing on the apparent
evolution we see. If the density at a given luminosity goes
to a peak at 2 and then drops again at redshift 4, would it
be valid to say that the effect of lensing can be no more
than what you would get out to a redshift of 4, which means
that most of the peak would have to be real?

Turner : Unfortunately I don't think that argument quite
works. In general the lensing effects increase with
redshift, but there are complicating factors that can be more
important. What matters is the angular size of the source
compared to the Einstein ring of the lens. There are
cosmologies in which the angular sizes of the sources get
bigger with redshift, and you might get unpleasant
coincidences with the Einstein ring diameters at $z = 0.5$.

SEARCH FOR GRAVITATIONAL LENSING FROM A SURVEY OF HIGHLY LUMINOUS QUASARS[*]

J. SURDEJ[**], J.-P. SWINGS
Institut d'Astrophysique, Université de Liège, Avenue
de Cointe, 5, B-4200 Cointe-Ougrée, Belgium
P. MAGAIN
European Southern Observatory, Casilla 19001,
Santiago 19, Chile
U. BORGEEST, R. KAYSER, S. REFSDAL
Hamburger Sternwarte, Gojenbergsweg 112, D-2050 Hamburg 80,
Federal Republic of Germany
T.J.-L. COURVOISIER[***]
ST-ECF, European Southern Observatory, Karl-Schwarzschild-
Str. 2, D-8046 Garching bei München, Federal Republic of
Germany
K.I. KELLERMANN
National Radio Astronomy Observatory, Edgemont Road,
Charlottesville, Virginia 22903-2475, USA
H. KÜHR
Max-Planck-Institut für Astronomie, Königstuhl,
D-6900 Heidelberg 1, Federal Republic of Germany

[*]   based on observations collected at the European
      Southern Observatory (La Silla, Chile) and at the
      National Radio Astronomy Observatory (Socorro,
      New Mexico)
[**] also, Chercheur Qualifié au Fonds National de la
      Recherche Scientifique (Belgium)
[***] affiliated to the Astrophysics Division of the Space
      Science Department of ESA

ABSTRACT We recently initiated a high angular resolution
direct imaging survey of a selected sample of Highly Luminous Quasars (hereafter, HLQs) : the observations are being
carried out with the 2.2 m telescope at the European Southern
Observatory (ESO, Chile) and with the Very Large Array (VLA)
at the National Radio Astronomy Observatory (NRAO, New Mexico).
The observing procedures are described in the present paper.
Following the detection of several good candidates for gravitational lensing, we report here on preliminary results concerning the frequency of occurrence of resolved HLQ images.
When more data will become available, this program should contribute to a better understanding of the effects of gravitational lensing on the observed quasar luminosity function, on
the source counts of extragalactic objects and, possibly, on
the QSO phenomenon itself.

1. INTRODUCTION

The first case of a gravitationally lensed distant object
has been identified by Walsh et al. (1979) almost sixty years
after the first observational evidence of light bending by
the Sun (in 1919). There are presently up to eight additional candidates of multiply imaged distant quasars (Weymann
et al. 1980; Weedman et al. 1982; Lawrence et al. 1984;
Djorgovski and Spinrad 1983; Huchra et al. 1985; Hewitt et al.
1985, 1987; Surdej et al. 1987). However, no lensing object
has yet been identified for approximately half of the proposed cases ! In addition, gravitational lensing by an intervening object has recently been suggested for the giant radio
galaxy 3C324 (Le Fèvre et al. 1987), as well as for an extragalactic arc near the galaxy cluster Abell 370 (Soucail et al.
1987).

For the particular case of Q0957+561 A and B, a very
convincing physical model has been proposed by Young et al.
(1980, 1981) : the observed images are produced through the
gravitational lensing of a single distant quasar (z_q = 1.41)
by a giant elliptical galaxy and its associated cluster
(z_L = 0.36). Using the difference in light travel time between
the individual images of Q0957+561, Borgeest and Refsdal
(1984) and Falco et al. (1985) have derived an upper limit on
H_o. It is clear that a statistical evaluation of the occurrence
of gravitational lensing within a well-defined sample of quasars is of prime importance to better understand the QSO luminosity function and possibly the QSO phenomenon itself (cf.
Barnothy and Barnothy, 1968), to test cosmological models
(Refsdal, 1964, 1966) and to probe the luminous and dark matter distribution on various scales in the Universe (see
Canizares, 1981).

The answer to the question "What fraction of QSOs are multiple due to gravitational lensing ?" is very closely related
to that to the question "What is the mass distribution at different scales in the Universe ?". If we consider for instance
scales typical of galaxies and clusters, we know that the answer to the first question is highly dependent on the central
velocity dispersions adopted for the bright elliptical galaxies
and on the maximum surface densities in large galaxy clusters
(Turner et al. 1984). More generally, any prediction made for
the expected number of multiply lensed quasars is very much
dependent on the adopted model (Barnothy and Barnothy 1968;
Tyson 1981; Peacock 1982; Turner et al. 1984; Hinshaw and
Krauss 1987). Once more, we naturally conclude that an observational approach is required to further constrain our under-

standing of gravitational lensing.

Adopting such a viewpoint, we have recently initiated a systematic search for new gravitational lenses. In the following, we consider that the apparently ($m_V \lesssim 18.5$) and intrinsically ($M \lesssim -29.0$) luminous quasars constitute very promising candidates to search for the presence of gravitationally lensed images at arcsec and sub-arcsec angular scales. Indeed :

i)   the HLQs form a particularly high flux limited sample of QSOs for which the probability of detecting multiply lensed images is higher than for a volume limited one (Turner et al. 1984).

ii) The HLQs are the most likely objects for which we may assume that their intrinsic brightness is partially due to gravitational lensing.

iii) The large cosmological distances suggested by the higher redshift values observed for the HLQs imply a high probability for gravitational lenses to be located along their line-of-sights. This is also suggested by the presence of rich absorption systems at redshifts $z_a < z_q$ recorded in the optical spectrum of most HLQs.

iv) We suspect that the paucity of known multiply lensed QSO images with angular separations in the range $\lesssim 2 - 3$ arcsec is mainly caused by observational biases. High resolution imaging of the HLQs with the Hubble Space Telescope, the Very Large Array and ground based optical telescopes under very good seeing conditions ought to bring important clues on the occurrence of lensing effects by galaxies or any other class of unknown massive objects.

A first observing run with the ESO/MPI 2.2 m telescope in November 1986 has led to the discovery of UM673 as a new case of gravitational lensing (Surdej et al. 1987, 1988). This system consists of two lensed QSO images A($m_R = 16.9$) and B($m_R = 19.1$), separated by 2.2 arcsec at a redshift $z_q = 2.719$. The lensing galaxy ($m_R \simeq 19$, $z_L = 0.493$) has also been identified. It lies very near the line connecting the two QSO images, and is about 0.8 arcsec away from the fainter one. Application of gravitational optometry to this system leads to a value $M_0 \simeq 2.4 \; 10^{11} \; M_\odot$ for the mass of the lensing galaxy and to $\Delta t \simeq 7$ weeks for the most likely travel-time difference between the two light paths from the QSO (assuming $H_0 = 75$ km sec^{-1} Mpc^{-1} , $q_0 = 0$). CCD photometric monitoring of UM673 A and B has begun in October 1987 using the Danish 1.5 m telescope

(ESO, La Silla) in order to determine the time delay between the two QSO images.

The techniques used for searching multiply lensed quasars among the HLQs are described in Section 2. Preliminary results on the observed frequency of resolved HLQ images are given in Section 3. The conclusions are presented in the last Section.

2. DESCRIPTION OF OUR OBSERVATIONAL SEARCHES FOR GRAVITATIONAL LENSING

Although our present collaborative search for gravitational lensing was initially set up because of the unique observing capabilities that should be offered by the Space Telescope (exceptional angular resolution, very good dynamic range, etc.), the delay in obtaining access to ST has prompted us to attempt the direct imaging of HLQs from the ground. We describe hereafter the observing techniques followed in our optical and radio searches.

2.1. Ground based optical search

During a total of 18 nights already allocated to our program at the European Southern Observatory, a CCD camera (RCA SID 501 EX chip, 320 x 512 pixels of 30 μm) was used at the Cassegrain focus of the ESO/MPI 2.2 m telescope in order to perform, at high angular resolution, direct imaging of HLQs $m_V \lesssim 18.5$, $M \lesssim -29.0$, Decl. $\lesssim + 20°$). These were selected from the Véron and Véron (1987)'s catalogue of quasars. Under average seeing conditions near FWHM = 1.2 arcsec, we succeeded in observing a total of 111 HLQs through wide- (B, V, R or I), and whenever possible, narrow- band filters chosen to isolate one of their bright redshifted emission lines (Lyα or occasionally CIII]) as well as a nearby portion of their continuum. Whereas a detailed analysis of all recorded CCD frames will provide in the near future information on the dynamic range achieved on each target and on the relevance of lensing effects (Courvoisier et al. 1988), the straight comparison of two such different color CCD frames gives an efficient means of selecting QSOs and/or QSO images with similar redshifts (cf. Djorgovski et al. 1985). Spectroscopic identification of the resolved HLQs becomes necessary at this stage. Until now, only one of the resolved objects has been thoroughly studied spectroscopically; i.e. UM673 (see Surdej et al. 1987, 1988). More information on the other resolved HLQs will be given in Section 3.

2.2. VLA search

A radio detection survey of 19 optically selected HLQs
(Decl. > -40°) has first been carried out with the VLA (C/D
configuration, 6 cm) in January 1987. Among these observed
HLQs, only four have been found to be sufficiently bright
sources in order to be mapped with the VLA at higher angular
resolution. Observations of these were made with the VLA (A
configuration, 6 cm, angular resolution ≃ 0.5 arcsec) in
October 1987. Two of the four sources are resolved, one is
unresolved and the fourth one has insufficient dynamic range
to conclude; we also point out that we did not succeed yet in
optically resolving the two detected multiple radio sources
($\Delta\Theta$ ≃ 1 and 1.4 arcsec).
Furthermore, three of our optically resolved HLQs, known to
be observable at radio wavelenghts, were also mapped with the
VLA in October 1987. Two of these are resolved at 6 cm. Finally, we failed to detect the gravitational lens system UM673
A and B at 3.6 (resolution ≃ 0.4 arcsec) and 6 cm during 30
minute observations with the VLA in the A configuration.

3. PRELIMINARY RESULTS ON THE OPTICAL OBSERVATIONS

In order to delineate as clearly as possible the observational characteristics of the quasars under study, we have
drawn in Figures 1-3 histograms representing the redshift
(Fig. 1), the apparent visual magnitude (Fig. 2) and the absolute magnitude (Fig. 3) of the 111 HLQs imaged with the ESO/
MPI 2.2 m telescope (cf. Section 2.1.). The histogram in
Figure 4 shows the distribution of the seeing conditions
(FWHM) characterizing the best CCD frames that were obtained
for each object.

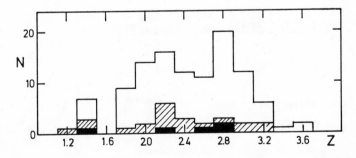

Fig. 1 Histogram representing the redshift of the 111 optically selected HLQs from the Véron and Véron (1987)'s
catalogue of quasars. The dashed and dark areas refer to a
sample of 25 interesting HLQs and to that of 5 very good
candidates for gravitational lensing, respectively, as
described in the text.

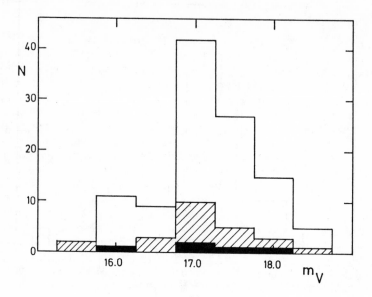

Fig. 2 This histogram refers to the apparent visual magnitude of the 111 optically selected HLQs (see caption in Fig. 1 for additional information).

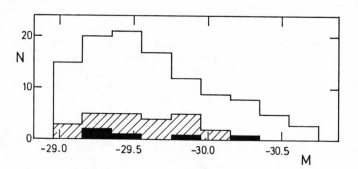

Fig. 3 This histogram refers to the absolute magnitude (H_0 = 50 km sec^{-1} Mpc^{-1}, q_0 = 0, $\bar{\alpha}$ = 0.7) of 110 HLQs. One HLQ with M = -31.6 has not been included (see caption in Fig. 1 for additional information).

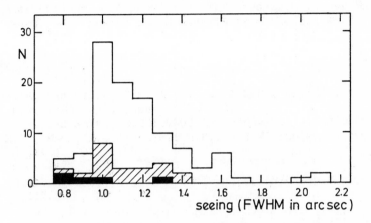

Fig. 4 Histogram representing the distribution of the
seeing disk (FWHM in arcsec) for 106 (out of 111) of the
HLQs that were imaged with the ESO/MPI 2.2 m telescope
(see Text). Data are temporarily missing for 5 HLQs.

    Twenty-five of these 111 luminous quasars were found to
be interesting because of one of the following reasons :
- multiple images,
- elongated image,
- presence of a jet-like feature,
- presence of a faint nearby galaxy,
- the image shows some fuzz,
- etc.
The dashed region of the histograms in Figures 1-4 refers to these
25 interesting HLQs. Furthermore, five of these 25 objects
turn out to be exceptionally good candidates for gravitational
lensing (cf. the several examples shown at the conference).
These are represented in Figures 1-4 by the dark area of the
histograms. It is interesting to note from Figure 4 that four
of the five best potential candidates for gravitational lensing have been identified on CCD frames taken under optimal
seeing conditions (i.e. FWHM $\lesssim$ 1 arcsec). It is clear that
observational biases play a major role in the detection of
gravitational lens effects. An illustration of these is given
in Figures 5 and 6. Figure 5a represents radial intensity
profiles, along one same direction, of the image of a selected
HLQ that has been observed with a CCD behind a Bessel B filter
under various seeing conditions (FWHM = 1.5 and 1.0 arcsec).

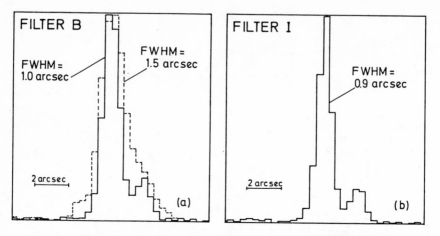

Fig. 5 (a) Radial intensity profiles, along one same direction, of the image of a good lensed HLQ candidate observed with a CCD camera behind a Bessel B filter when the seeing disk was 1.5 and 1.0 arcsec.
       (b) As (a) but for an I filter and a seeing disk of 0.9 arcsec.

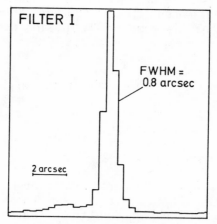

Fig. 6 Radial intensity profile of the image of another good lensed HLQ candidate observed with a high S/N ratio under a seeing of 0.8 arcsec. Note the presence of a very faint secondary image to the left.

This example clearly indicates that the lensed HLQ candidate, appearing barely elongated on the 1.5 arcsec exposure, would have been missed under (slightly) less favourable seeing conditions. Figure 5b was similarly constructed from a CCD frame taken through a I filter and under a seeing (FWHM) ≃ 0.9 arcsec. A more general analysis of the B [, V, R] and I CCD frames available for this HLQ confirms that it is a very good candidate for gravitational lensing. The separation between the two unequally bright, although resolved, images is about 2 arcsec. Figure 6 shows, for the case of another good lensed HLQ candidate, the importance of getting a sufficiently high S/N ratio in order to detect the presence of faint secondary images.

4. CONCLUSIONS

We have shown in this paper that our search for gravitational lensing from a survey of optically selected HLQs appears very promising. Indeed, observational features (multiple images, image elongation, jet-like feature, fuzz, etc.) possibly associated with the HLQ phenomenon have been detected for more than 20 % of the objects under study. While it is not yet known how many of the HLQs are gravitational mirages, at least 5 (out of 111) of the investigated quasars appear to be highly luminous because of amplification of their brightness by gravitational lensing. Since "amplification" does not necessarily mean "multiplicity of images", it could very well be that many more HLQs (and possibly quasars) are gravitationally lensed (cf. Barnothy and Barnothy, 1968). Whereas it is clear that observational biases conspire against the detection of many gravitational lens systems at arcsec and sub-arcsec angular scales, it is likely that the nicest cases ($\Delta\Theta \gtrsim 2$ arcsec) yet to be found will rely more ... on the generosity of the future observing program committees.

Acknowledgments

Part of this research has been supported by NATO grant n° 0161/87 and by the Deutsche Forschungsgemeinschaft (Az Re 439.4). The National Radio Astronomy Observatory is operated by Associated Universities Inc. under contract to the U.S. National Science Foundation.

REFERENCES

Barnothy, J.M., Barnothy, M.F. 1968, Science, **162**, 348.
Borgeest, U., Refsdal, S. 1984, Astron. Astrophys., **141**, 318.
Canizares, C.R. 1981, Nature, **291**, 620; erratum Nature, **293**, 490.
Courvoisier, T.J.-L. et al. 1988, in preparation.

Djorgovski, S., Spinrad, H. 1983, B.A.A.S., **15**, 937.

Djorgovski, S., Spinrad, H., Mc Carthy, P., Strauss, M.A. 1985, Astrophys. J. (Letters), **299**, L1.

Falco, E.E., Gorenstein, M.V., Shapiro, I.I. 1985, Astrophys. J. (Letters), **289**, L1.

Hewitt, J.N., Turner, E.L., Lawrence, C.R., Schneider, D.P., Gunn, J.E., Schmidt, M., Mahoney, J.H., Langston, G.I., Burke, B.F. 1985, B.A.A.S., **17**, 907.

Hewitt, J.N., Turner, E.L., Burke, B.F., Lawrence, C.R., Bennett, C.L., Langston, G.I., Gunn, J.E. 1987, in Observational Cosmology, IAU Symposium n° 124, p. 747, eds. Hewitt, A., Burbidge, G., Fang, L.Z.

Hinshaw, G., Krauss, L.M. 1987, preprint.

Huchra, J., Gorenstein, M., Kent, S., Shapiro, I., Smith, G., Horine, E., Perley, R. 1985, Astron. J., **90**, 691.

Lawrence, C.R., Schneider, D.P., Schmidt, M., Bennett, C.L., Hewitt, J.N., Burke, B.F., Turner, E.L., Gunn, J.E. 1984, Science, **223**, 46.

Le Fèvre, O., Hammer, F., Nottale, L., Mathez, G. 1987, Nature, **326**, 268.

Peacock, J.A. 1982, Mon. Not. R. astr. Soc., **199**, 987.

Refsdal, S. 1964a, Mon. Not. R. astr. Soc., **128**, 295.

Refsdal, S. 1964b, Mon. Not. R. astr. Soc., **128**, 307.

Soucail, G., Mellier, Y., Fort, B., Mathez, G., Cailloux, M. 1987, The Messenger, **50**, 5.

Surdej, J., Magain, P., Swings, J.P., Borgeest, U., Courvoisier, T.J.-L., Kayser, R., Kellermann, K.I., Kühr, H., Refsdal, S. 1987, Nature, **329**, 695.

Surdej, J., Magain, P., Swings, J.P., Borgeest, U., Courvoisier, T.J.-L., Kayser, R., Kellermann, K.I., Kühr, H., Refsdal, S. 1988, Astron. Astrophys., in press.

Turner, E.L., Ostriker, J.P., Gott III, J.R. 1984, Astrophys. J., **284**, 1.

Tyson, J.A. 1981, Astrophys. J. (Letters), **248**, L89.

Véron-Cetty, M.-P., Véron, P. 1987, ESO Scientific Report n° 5.

Walsh, D., Carswell, R.F., Weymann, R.J. 1979, Nature, **279**, 381.

Weedman, D.W., Weymann, R.J., Green, R.F., Heckman, T.M. 1982, Astrophys. J. (Letters), **255**, L5.

Weymann, R.J., Latham, D., Angel, J.R.P., Green, R.F., Liebert, J.W., Turnshek, D.A., Turnshek, D.E., Tyson, J.A. 1980, Nature, **285**, 641.

DISCUSSION

Veron :

What was the estimated amplification factor for the UM object?

Surdej : There is a very recent calculation that shows it could be as high as a factor of several hundred, although that seems almost unbelievable. We had been working with calculations that indicated an amplification factor of a factor of several, less than 10.

MacAlpine : I'm wondering what one should be looking for on Schmidt plates that would lead to finding some of these things. Is there anything else that drew your attention to UM673?

Surdej : No. We just worked from the luminosity of the sources.

Osmer : It seems to me you're doing just the right thing: using a CCD detector, which has a better dynamic range than plates, at a better scale than is used for Schmidt surveys. Consequently, you find these fascinating results.

Turner : I would just mention the well known result that the number of known lenses with separations of a few arc seconds implies that there should be a large number with separations of 1" or less.

RADIO QUASARS AND RADIO GALAXIES –
A COMPARISON OF THEIR EVOLUTION,
ENVIRONMENTS AND CLUSTERING PROPERTIES

J. A. PEACOCK *and* L. MILLER
Royal Observatory, Blackford Hill,
Edinburgh EH9 3HJ, U.K.

ABSTRACT This paper compares the properties of radio-loud quasars with those of elliptical radio galaxies. Apart from well-known differences in radio structure, these two classes of object display no differences in their statistical properties. In particular: (i) the radio luminosity functions appear identical above $P_{2.7GHz} \simeq 10^{25}$ WHz^{-1}sr^{-1} – both quasars & galaxies displaying a 'redshift cutoff' at $z \gtrsim 2$; (ii) both quasars and galaxies may display a trend towards richer environments at the most extreme radio powers – care is therefore needed in distinguishing true epoch dependence of environments from luminosity effects, since objects at the highest redshifts also tend to be those of highest luminosity. We also present a study of the spatial clustering of radio galaxies and discuss the relation of this result to the clustering of radio-quiet quasars.

1. INTRODUCTION

The purpose of this paper is to present a summary of results from several related projects underway at Edinburgh. We shall discuss (i) the relation of radio-loud to radio-quiet quasars; (ii) the cosmological evolution of the quasar radio luminosity function (RLF) and its relationship with the evolution of radio galaxies; (iii) the environments of quasars and radio galaxies and the connection with large-scale clustering of these objects.

Unless otherwise stated, we assume $H_o = 50$ kms^{-1}Mpc^{-1} and $\Omega_o = 1$. Where an explicit dependence on H_o is quoted, we use the standard notation $h \equiv H_o/100$ kms^{-1}Mpc^{-1}.

2. RADIO-LOUD & RADIO-QUIET QUASARS

The first question facing any study of the statistical properties of radio-loud quasars is how these are to be distinguished from radio-quiet quasars. A common approach has been to consider the radio/optical flux ratios – where one normally ends up considering a continuous distribution of the ratio, so that there is a single bivariate luminosity function satisfying a so-called 'colour factorization'.

The alternative approach is to work with two distinct populations. This point of view has been previously advocated by us (Peacock, Miller & Longair 1986), principally on the grounds that optically faint quasars ($B \simeq 20$) are detectable by the VLA in only a few percent of cases. This drastic change in radio detectability argues against simple colour factorization models, and more in favour of two distinct classes, which can be separated in radio luminosity at $P_{2.7GHz} \simeq 10^{24}$ WHz^{-1}sr^{-1}. Our point of view will be that the radio-loud class has properties very similar to those of radio-loud elliptical galaxies: in particular, there is no correlation of radio & optical luminosities, apart from that between polarized compact (beamed?) components (Meisenheimer 1988). Evidence in support of the latter statement comes from VLA observations made by ourselves of a selection of prism-selected quasars with $1.8 \lesssim z \lesssim 2.5$ — i.e. an approximation to a thin redshift shell. Fig. 1 shows the plot of radio against optical powers, which displays clear bimodality and a lack of any radio-optical correlation.

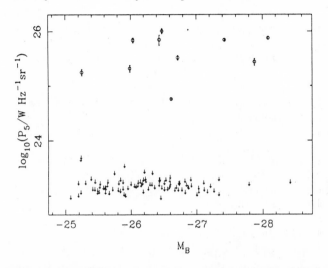

Fig. 1

We shall not be greatly concerned here with the nature of the radio-quiet population, except that they are utterly distinct in properties from elliptical radio galaxies. The natural hypothesis is that these objects are an extension of the Seyfert population, hosted by spiral galaxies. For these, there does appear to be a correlation of radio & optical nuclear luminosities (Edelson 1987) — but this does not conflict with our above statements concerning the more luminous objects.

There are many reasons to expect a close relation between the properties of radio-loud quasars and elliptical galaxies, of which the most direct is the now accumulating evidence that the host galaxies of radio-loud quasars are indeed giant ellipticals (e.g. Malkan 1984). With the exception of the emission from cores and one-sided jets, the radio properties of both classes are very similar (e.g. Owen 1986), and this has naturally generated discussion of models in which the quasar phase appears at intervals in the lifetime of a radio galaxy, or even where the two classes are related via orientation-dependent effects. Whether or not these speculations have any truth, it is of interest to examine the cosmological properties of the two classes to search for further similarities of behaviour.

3. HIGH-REDSHIFT EVOLUTION OF THE RLF

It has been known for some time that the degree of cosmological evolution displayed by radio galaxies & quasars at $z \lesssim 1$ is similar (Laing, Riley & Longair 1983). Evidence is now becoming available that this is also the case at higher redshifts, based on an extensive optical & infrared study of the Parkes Selected Regions (Downes *et al.* 1986; Dunlop 1988). These regions are the deepest in the Parkes survey ($S_{2.7GHz} \geq 0.1$ Jy) and provide a total of 178 radio sources over 216 deg^2. We have taken CCD frames to B $\simeq$ 25, R $\simeq$ 24 and infrared photometry to K $\simeq$ 19, yielding detections of all but 4 of the sources. Spectroscopy is largely complete to R $\lesssim$ 22, providing redshifts for about 50% of the sample, including essentially all the quasar candidates.

For the remaining very faint galaxies, we have relied on redshift estimation using the infrared photometry. Fig. 2 shows the K-z Hubble diagram for 3CR galaxies with some fainter B2 galaxies.

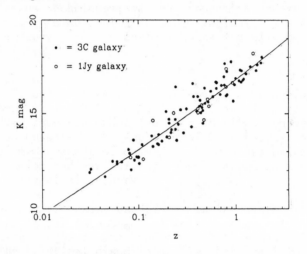

Fig. 2

The 'standard candle' relation is well defined to K $\simeq$ 17.5 (i.e. redshifts well in excess of unity); this is a good deal better than one can do in the optical, as the infrared magnitudes are not corrupted by the extensive star-forming activity known to be common at these redshifts. For galaxies fainter than this, redshift estimates become less certain (although the very recent detection of a radiogalaxy with z=3.395 and K=18.8 by Lilly 1988 may lend some confidence to extrapolation of the K−z relation), but in fact very nearly all the Selected Region galaxies were detected at K $\lesssim$ 17.5. The distribution of K magnitudes is shown in Fig. 3, from which it is clear that very few of the galaxies can lie at z > 2, unless they are implausibly luminous.

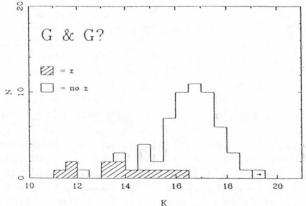

Fig. 3

This result is of great importance in deducing the high-redshift behaviour of the RLF, and parallels the result from the quasars in this sample. The highest redshift in the Selected Regions is z=3.71 for 1351−018 (Dunlop *et al.* 1986), but the next highest is z=2.9 and there are only 3 quasars with z > 2.5 (out of a total of 49). These results confirm a trend noted by Peacock (1985) in the data on flat-spectrum quasars: radio samples at intermediate flux levels contain maximum redshifts no higher than those seen in brighter samples, which tells us that there must be a reduction in the RLF at high redshifts. This situation is illustrated in Figs 4a & 4b, which plot radio power against flux density for the Selected Regions and various brighter Parkes samples. Fig. 4a shows all data, while 4b shows quasars only, with 'guestimated' redshifts indicated by open circles. (sample limits at 0.1, 0.5, 1.5 & 2 Jy are visible)

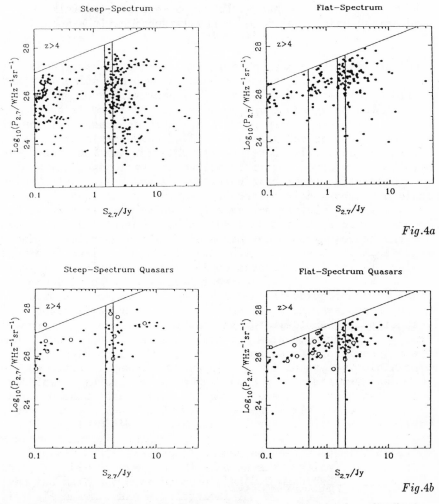

Fig.4a

Fig.4b

Two things are apparent from this diagram: first, the quasar and galaxy distributions are indistinguishable for $P_{2.7GHz} \gtrsim 10^{24.5}$ WHz^{-1}sr^{-1}. This is not an artifact of classification − counting Broad-Line galaxies as quasars will not change the result. The fact that quasar activity has a ~ constant probability above this point (the Fanaroff-Riley division) but is small below it is clearly telling us something important, but no-one really knows what.

Second, it is clear that there exist objects with $P_{2.7GHz} \simeq 10^{27}$ WHz^{-1}sr^{-1} at z $\simeq 1$ in the bright samples which should have high-redshift counterparts in the fainter samples. Since these do not appear, we can deduce that the RLF must have fallen at high redshifts (at least for radio powers in this region); this can be done in a variety of ways.

The data in Fig. 4 can be combined with information on deeper number counts and the local RLF in the free-form expansion method used by Peacock (1985); this yields a set of consistent polynomial RLFs, whose degree of agreement provides an indication of how well the data constrain the RLF. Alternatively, one may simply bin the data in Fig. 4 alone and deduce an RLF directly (with poissonian error bars). The results of both these procedures are shown in Fig. 5, which presents a cut through the RLF at constant radio power ($P_{2.7GHz} = 10^{27}$ WHz^{-1}sr^{-1}).

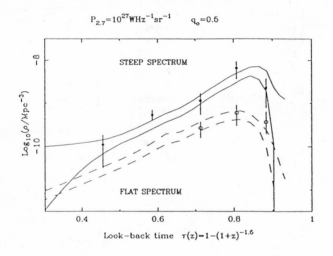

Fig. 5

The evidence for a reduction in the RLF at z $\gtrsim 2$ for both spectral types is quite clear from this figure; although this considers $\Omega_o = 1$ only, a very similar diagram results for $\Omega_o = 0$. The result can be backed up by a 'banded' V/V_{max} analysis as in Peacock (1985). The following table shows the values of $\langle V_e/V_a \rangle$ resulting from a combination of the Parkes Selected Region data with brighter samples, considering z > 2 only

		$\langle V_e/V_a \rangle$	
	n	$\Omega_0 = 1$	$\Omega_0 = 0$
Flat-Spectrum (all)	25	0.328	0.359
Flat-Spectrum Quasars	24	0.342	0.374
Steep-Spectrum (all)	23	0.395	0.423
Steep-Spectrum Quasars	10	0.417	0.431

Could the above result be wrong? For this to be the case, we must have missed some objects at very high redshift; we can see how large this number would have to be by taking the above RLF at $z = 2$ and assuming that it stayed constant at higher redshifts. On this hypothesis, the Selected Regions should contain 18 steep-spectrum sources with $z > 2.5$, which seems highly unlikely in the light of the above data. Nevertheless, caution is advisable; to base extensive cosmological conclusions on ~ 10 missing high-redshift sources would be a risky step. Indeed, this has all along been one of the difficulties in studies of the redshift cutoff; the geometry of the Universe conspires to hide all but a very few objects at $z \gtrsim 3$ (unless unrealistic evolution at high redshift is assumed). One important remaining task is therefore to obtain spectra of some of the objects with K $\simeq 18$, to confirm the prediction that they have $z \lesssim 2.5$.

The next major step in this area will be to determine the high-redshift evolution at lower luminosities. The ideal database for this will probably be the Leiden-Berkeley Deep Survey (Windhorst 1984), which now has optical data of approximately the same degree of completeness as the Parkes Selected Regions. Indeed, Windhorst (1984) claimed that there was evidence for a deficit of low-power galaxies at $z \gtrsim 1$. As we have seen, to establish such a result really requires either spectroscopy or infrared photometry, which are not yet published for the more distant LBDS objects. However, it will soon be possible to test for such behaviour at lower powers. At present, Dunlop (1987) has shown that it is possible to fit all complete-sample and source-count data with a luminosity evolution model: this is very similar to the form which applies for the optical quasar luminosity function at $z \lesssim 2$, but has *negative* luminosity evolution at higher redshifts (provided $\Omega_0 = 1$ and a non-evolving component is added in at the lowest powers to represent spiral & irregular galaxies). Such a model however cannot yet be distinguished from a radically different one which has continuing *positive* luminosity evolution at $z \gtrsim 2$ and some suppression of ρ at high z. It is exciting to think that these fundamental questions about RLF evolution may be settled within the next few years.

4. CLUSTER ENVIRONMENTS

The determination of the local galaxy density about active galaxies is an area of great current interest. In this section, we want to compare the results on quasar environments presented by Green & Yee (1984) and Yee & Green (1987) with those on radio-galaxy environments by Prestage & Peacock (1988) and Yates *et al.* (1988).

The results of Yee & Green (1987) are especially interesting, as they claim to have discovered a change of quasar environment with redshift. If true, such a result would immediately rule out interpretations of quasar luminosity evolution in terms of long-lived quasars. One problem with their result is that it is largely based on radio-selected quasars, for which those at high redshift are also the most radio-luminous. To see whether this matters, consider our knowledge of the correlation of environment with radio power for radio galaxies. In all cases, we will be measuring the density of the environment in terms of the spatial cross-correlation amplitude B_{gr}; for orientation, the values of this parameter corresponding to Abell richnesses 0,1 & 2 are approximately 110, 250 & 380. It was established by Longair & Seldner (1979) that B_{gr} showed a strong dependence on radio power, with $\langle B_{gr} \rangle$ changing by a factor ~ 2 on crossing the Fanaroff-Riley division at $P_{2.7GHz} \simeq 10^{24}$ WHz^{-1}sr^{-1}, in the sense that the weaker sources occupied the denser environments; these results were amplified by Prestage & Peacock (1988). The above investigations nevertheless were unable to study the environments of the most powerful radio galaxies as they were limited to $z \lesssim 0.2$ by the use of photographic material. It then became clear that there was indirect evidence that these more luminous sources (of which the archetypes at lower redshifts are Cygnus A & 3C295) might indeed lie in rich environments (Yates *et al.* 1986). These suggestions have been followed up by deep CCD imaging of very powerful radio galaxies at $0.4 \lesssim z \lesssim 0.7$ (Yates *et al.* 1988). The interpretation of the results is not entirely straightforward, as the visibility of high-z clusters depends strongly on one's assumptions about the evolution of the optical cluster luminosity function. Nevertheless, one can at least compare different classes of object at the *same* redshift. With this caveat in mind, Fig. 6 presents a preliminary plot of B_{gr} against radio power, using a luminosity function which should be close to that of Yee & Green (1987).

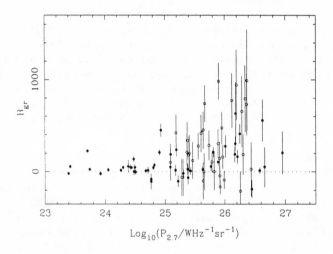

Fig. 6

This shows all Fanaroff-Riley II galaxies from the CCD & photographic studies (filled circles). We also include the radio-loud quasars from Green & Yee (1984) and Yee & Green (1987) (open circles). There is a trend towards higher values of B_{gr} at higher powers for both quasars & galaxies. The quasar points appear to lie somewhat above the galaxy points, but we suspect this may be an artefact produced by remaining differences in luminosity functions; the data may well be consistent with a single B_{gr}–P relation. This is especially interesting as there is now some preliminary evidence (Boyle, private communication) that the environments of radio-quiet quasars at $z \simeq 0.6$ may be rather less dense than for radio-loud objects.

Note that, because most objects in Fig. 6 have bright radio flux densities, one cannot tell statistically from the quasar data whether there is a relation fundamentally with power or with redshift. Indeed, considering the uncertainty in the evolution of the luminosity function, it will probably be difficult to settle this question conclusively. However, taking into account the additional evidence of Yates *et al.* (1986), we suspect that there is indeed a fundamental relation with luminosity. This would make good physical sense: Prestage & Peacock (1988) have argued that the only reason that low-power FRII sources avoid dense environments is that they cannot supply enough ram pressure to overcome the static IGM pressure − a constraint that poses no difficulty for the more luminous galaxies and quasars.

5. LARGE-SCALE CLUSTERING

Finally, we consider the relation of the above environmental results to data on the large-scale clustering of quasars and radio galaxies. The reason this is important is that we know that there is a trend between the richness of a galaxy system and the amplitude of its spatial cross-correlation function (Bahcall & Soneira 1983), so clustering and environmental studies may well have a common interpretation.

It has taken a long time for reliable estimates of quasar clustering to become available. The reason for this is essentially the low number density of quasars. The most reliable determination is probably that produced by Boyle (1988), based on the Durham UVX survey, although similar results have been found using inhomogeneous samples by Shaver (1984) and Kruzsewski (1988), who also claims that the clustering is undetected for $z \gtrsim 2$. At $\langle z \rangle \simeq 1$, the correlation function in redshift space has the form $\xi(s) = (s/s_0)^{-\gamma}$ where γ is consistent with the galaxy value of 1.8 and s_0 lies in the range 5 – 10 h^{-1} Mpc in *comoving* coordinates.

Large-scale structure is defined more easily using radio galaxies, simply because the number density is higher and one can more easily achieve > 1 galaxy in a correlation volume. We are presently working on an all-sky sample of galaxies which approximate the selection criteria $S_{1.4GHz} > 0.5$ Jy; $0.01 < z < 0.1$; $|b| > 15°$. Figure 7 shows the correlation function for the luminous FRI galaxies only (i.e. $10^{23.5} < P_{2.7GHz} < 10^{24.5}$ WHz^{-1}sr^{-1}); there is no detectable clustering from other power ranges.

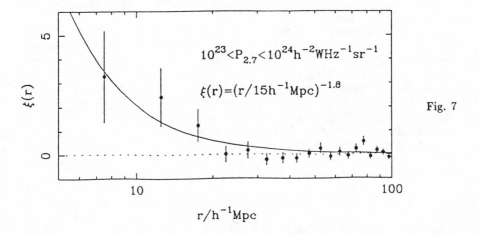

Fig. 7

This result is especially exciting in view of our knowledge both of the environments of radio galaxies and of the spatial correlations of rich clusters. From Prestage & Peacock (1988), we know that the above radio power range should pick out the objects of richest environments at $z < 0.1$ – having average Abell richnesses of about R = 0. The work of Bahcall & Soneira (1983) would suggest a scale-length of $s_0 \simeq 25$ h^{-1} Mpc for R ≥ 1 clusters, with a correlation of s_0 and richness. Sutherland (1988) conversely finds $s_0 \simeq 14$ h^{-1} Mpc with no richness correlation above R = 0. In either case, our results are consistent with the optical determinations for these intermediate-richness systems, but without the problematic selection effects caused by cluster selection. These results should be regarded as rather preliminary, as they are based on only ~ 150 galaxies with redshifts out of a total sample of about 400. We are presently analysing spectra of a further ~ 100 galaxies, which will reduce the error bars in Fig. 7 by a factor ~ 2.

ACKNOWLEDGEMENTS

We are grateful to our colleagues at Edinburgh, on whose work we have drawn in writing this review: Chris Collins, James Dunlop, Malcolm Longair, Andrew Mead, David Nicholson & Mark Yates.

REFERENCES

Bahcall, N.A. & Soneira, R.M. 1983. Astrophys. J., **270**, 20.

Boyle, B.J. 1988. This workshop.

Downes, A.J.B., Peacock, J.A., Savage, A. & Carrie, D.R. 1986. Mon. Not. R. astr. Soc., **218**, 31.

Dunlop, J.S. 1988. PhD thesis, Univ. of Edinburgh.

Dunlop, J.S., Downes, A.J.B., Peacock, J.A., Savage, A., Lilly, S.J., Watson, F.G. & Longair, M.S. 1986. Nature, **319**, 564.

Edelson, R.A. 1987. Astrophys. J., **313**, 651.

Green, R.F. & Yee, H.K.C., 1984. Astrophys. J., **54**, 495.

Kruszewski, A. 1988. Astrophys. J., in press.

Lilly, S.J. 1988. Astrophys. J., in press.

Longair, M.S. & Seldner, M. 1979. Mon. Not. R. astr. Soc., **189**, 433.

Malkan, M.A. 1984. Astrophys. J., **287**, 555.

Meisenheimer, K. 1988. in preparation.

Owen, F.N. 1986. in 'Quasars', proc. IAU symp. no. 119, eds G. Swarup & V.K. Kapahi (D. Reidel), p173.

Peacock, J.A. 1985. Mon. Not. R. astr. Soc., **217**, 601.

Peacock, J.A., Miller, L. & Longair, M.S. 1986. Mon. Not. R. astr. Soc., **218**, 265.

Prestage, R.M. & Peacock, J.A. 1988. Mon. Not. R. astr. Soc., **230**, 131.

Shaver, P.A. 1984. Astr. Astrophys., **136**, L9.

Sutherland, W. 1988. Mon. Not. R. astr. Soc., in press.

Yates, M.G., Miller, L. & Peacock, J.A. 1986. Mon. Not. R. astr. Soc., **221**, 311.

Yates, M.G., Miller, L. & Peacock, J.A. 1988. Mon. Not. R. astr. Soc., submitted.

Yee, H.K.C. & Green, R.F. 1987. Astrophys. J., **319**, 28.

Windhorst, R.A. 1984. PhD thesis, Univ. of Leiden.

DISCUSSION

Heisler : I have a question about the redshift cutoff. In the scatter plot you showed of radio power vs flux, how many quasars with $z > 4$ would you have expected?

Peacock : I didn't intend to show a break at $z = 4$ – that's just a line to guide the eye. The important thing is the absence of very many quasars at redshifts beyond 2.5. This is evidence for a decline setting in very soon after redshift 2.

Schmidt : It used to be important in the early days to understand the relation between the optical and radio luminosity functions, because there was a double selection effect at that time. Now that is less important because you manage to completely identify optically samples of radio selected quasars. On the other hand, you did make a statement early on in which you said that you felt that there was no correlation for quasars between the optical and radio luminosity functions. If that were the case, then the optical magnitude distribution of a full sample of radio quasars should be like that of all quasars, which are mostly radio quiet.

Peacock : I would argue that radio quasars are physically distinct from the main body of quasars and that there is no reason to expect that they have the same optical luminosity function.

A COMPLETE SAMPLE OF FLAT-SPECTRUM RADIO SOURCES FROM THE PARKES 2.7 GHz SURVEY

ANN SAVAGE
Royal Observatory, Edinburgh, Scotland

DAVID L. JAUNCEY and GRAEME L. WHITE
Division of Radiophysics, CSIRO, Australia

BRUCE A. PETERSON and W.L. PETERS
Mount Stromlo and Siding Spring Observatory, Australia

SAMUEL GULKIS
Jet Propulsion Laboratory, California

J.J. CONDON
National Radio Astronomy Observatory, Virginia

ABSTRACT   We describe our complete sample, taken from the Parkes 2.7 GHz catalogue, of flat-spectrum radio sources with fluxes >0.5 Jy. The sample covers all right ascensions and declinations from +10° to -45°, excluding the galactic plane ($|b| < 10°$) and contains some 400 sources. Attention is drawn to the advantages of radio surveys over optical surveys. The survey is used to highlight some selection effects found in optical surveys.

INTRODUCTION

We are investigating a sample of flat-spectrum radio sources drawn from the Parkes 2.7 GHz survey (Bolton et al. 1979, and references therein) complete to 0.5 Jy. The sample covers ⋅.5 sr of sky and comprises 403 sources. Accurate radio positions have been measured for all sources (McEwan et al. 1975; Condon et al. 1977, 1978; Jauncey et al. 1982; Condon et al., unpublished data).   Optical identifications are being made from the SERC/UKST IIIa-J sky survey. Redshifts are being sought with the Anglo-Australian telescope (AAT) (White et al. 1988, and references therein), with results so far for 270 sources. The program is aimed at determining in an unbiased manner the space distribution of quasars over a large area of sky.

THE SURVEY

The basic sample comprises both steep- and flat-spectrum
sources, but we are concentrating initially on the flat-spectrum
sources. The compact nature of these means that accurate radio
positions can be readily measured and thus unique optical
identifications, based on positional coincidence alone, can be
made to the 22.5-mag limit of the SERC J survey.

COMPLETENESS AND SELECTION EFFECTS

Optical identifications have been made for 87% of the sample
and redshifts have been obtained for 67%. We have excellent
positional agreement for 27 stellar objects whose optical
spectra show no features; these we have called "BL Lac types".
Long integrations on three of the objects have eventually
brought up extremely weak emission lines at low redshift. This
content of BL-Lac-type objects is similar to the percentage of
featureless-spectrum blue objects found in optical surveys
(Boyle et al. and Crampton et al. - these Proceedings). This
percentage is however much lower than that of the BL Lacs found
in X-ray surveys (Stocke, Maccacara - these Proceedings). The
total identification content is given in Table I.

TABLE I   Identification content

Type	No red shift	With red shift	Total	Percentage of sample
Galaxies	16	23	39	9.7
BL Lacs	24	3	27	6.7
Quasars	42	244	286	71.0
Empty fields	51	0	51	12.6

Figure 1 gives the differential N(s) distribution. It can
be seen that it is the predominantly weaker radio sources that
have not yet been identified. Figure 2 is a plot of optical
magnitude for the identified sources versus radio flux density.
A weak correlation between optical magnitude and radio flux can
be seen.
Some selection effects in this sample are as follows.

(i)   Our choice of radio-loud quasars may introduce a selec-
      tion effect. Peacock et al. (1986) put forward a hypo-
      thesis for two distinct populations of quasars - "radio-
      quiet" and "radio-loud" - according to the Hubble type

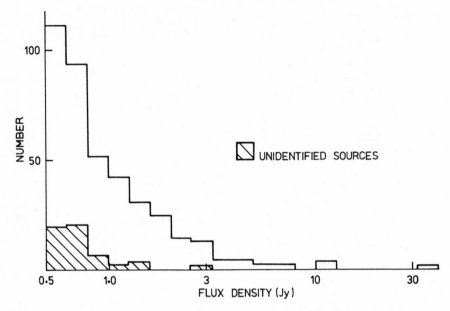

Fig. 1   Number – flux density distribution for all the
          sources in the complete sample.

          of their underlying galaxy. In this scheme radio-quiet
          quasars are analogues of Seyfert galaxies – spirals –
          whilst the radio-selected quasars reside in elliptical
          galaxies.

(ii)   We have restricted our radio sample further to contain
          only the flat-radio-spectrum population. This dis-
          criminates against the steep-spectrum radio population
          that is identified as quasars. If there are any differ-
          ences in the environment or the underlying galaxy for
          these two types of radio-selected quasars we will have
          a further intrinsic effect.

(iii)   The optical identifications are not yet completed, and
          because it is the weakest radio flux sources that are
          missing it may well be that the sample is missing some
          of the highest-redshift objects. Figure 3 shows the
          number-flux distribution for the identified quasars
          divided into redshift ranges. In general the ratio of
          quasars with redshifts >2 to those with redshifts <2
          increases as we go to weaker radio fluxes (although
          there is a peak at fluxes ~1 Jy where nearly half of the
          identified quasars have redshifts >2). The number of

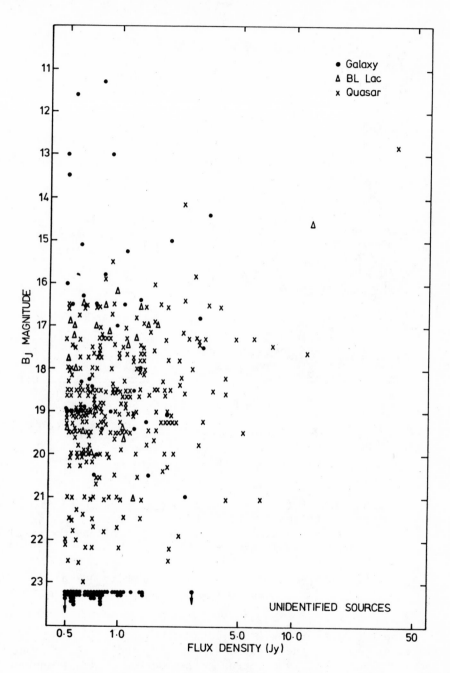

Fig. 2   Optical magnitude (B_J) versus radio flux
density.

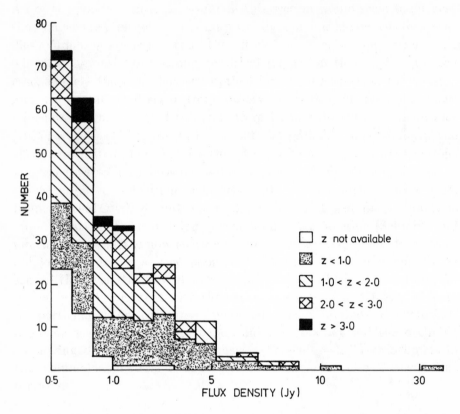

Fig. 3   Number-flux distribution for all the
           sources in the complete sample
           divided into redshift interval.

high-redshift (>3.0) quasars that we have found is consistent with the fraction of sources with z > 3.0 predicted as a function of radio survey limit by the models
of Peacock (1983), namely 3% to 4%.

(iv)   The radio spectra of the highest-redshift quasars are
       not typical of those at lower redshift (Savage and
       Peterson 1983).

The spectra for our 10 objects with z > 3.0 are shown in
Figure 4. Most of the spectra are peaked, indicating compact
components that are synchrotron-self-absorbed at low frequencies and optically thin at high frequencies.

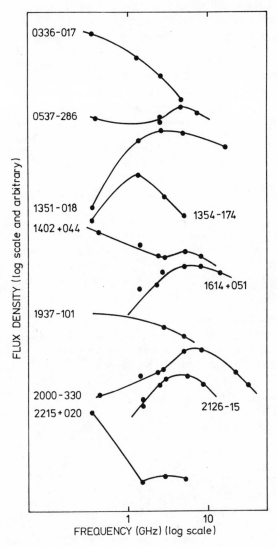

Fig. 4   Radio spectra for quasars with z > 3.0.

ADVANTAGES OF THIS SURVEY

There are a number of advantages to this present survey.

(i)   The survey uniformly covers a large area of sky and
       thus avoids both small-area effects and large-scale
       non-uniform incompleteness found in optical surveys.

(ii)   The large sample size reduces statistical fluctuations
        and provides reliable information on minority popula-
        tions, e.g. the BL-Lac-type objects.

(iii)   Accurate radio positions have been measured for all
        sources, allowing reliable optical identification to
        the 22.5-mag limit of the UKST IIIa-J plates and to
        the 24.0-mag limit of AAT CCD frames, without reliance
        on colour or morphology. This is particularly import-
        ant in avoiding redshift-dependent effects of the type
        found in optically selected samples based on colour
        selection or emission-line selection.

(iv)   The identification rate is high, nearly 90%.

(v)   Redshifts are being sought for all the objects to as
        faint a limit as is possible with the AAT. This in
        turn will provide an unbiased determination of the
        space distribution of quasars to the largest redshift
        yet found for radio-selected objects (Peterson et al.
        1982).

        Given these advantages and bearing in mind the intrinsic
differences and biases which may have crept into our sample,
we now go on to look at the optical properties of our sample
of quasars and compare these with some optically selected
samples of quasars.

QUASAR CONTENT

Of this sample 71% are identified with quasars and we have redshifts for 85% of this subset. We have already excluded some
brighter (BL Lac) objects with no emission lines in their
spectra or very weak emission lines.
        The higher redshift quasars are to be found predominantly
amongst the weaker radio sources and the fainter identifications. It may be that low-redshift objects and also very-high-
redshift objects are missing from our distribution. In Figure 5
we show the number-magnitude distribution for these quasars.
The hatched region indicates those objects for which we have a
measured redshift. The magnitudes for all these objects have
been estimated from the SERC IIIa-J survey following the
method described in Downes et al. (1986). The overall error
in the magnitudes should be < 0.6 mag. It can be seen that we
have obtained redshifts for a fairly representative sample
though the fraction without redshifts increases at the faintest
magnitudes.

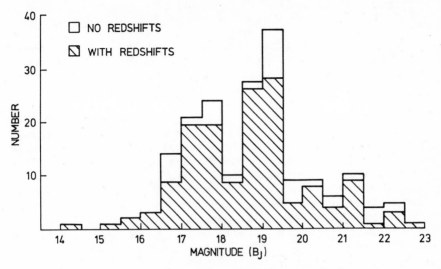

Fig. 5   Number-optical magnitude (B_J)
          for all quasars in the sample.

COMPARISON WITH OTHER SURVEYS

The Redshift Distribution

Figure 6 gives the redshift distribution for this sample. Our
redshift distribution is not inconsistent with a constant co-
moving density of objects with the same absolute radio power.

   A comparison can be made between Figure 6 and Figure 1 of
Wilkes (1986), which is also a redshift distribution for a randomly selected sample of Parkes quasars studied by this group
prior to 1983. Our complete sample confirms the Wilkes finding
that there is no narrow peak at $z \approx 2$ which is found in emission-
line selected samples. In fact our new data further suppress
any such peak, since the main difference between the Wilkes
sample and ours has been an increase in the low-redshift
objects, $z < 1.6$, and a tripling of the number with $z > 2.8$.

Intrinsic Luminosity

The optical Hubble diagram has been plotted using magnitudes from
the SERC IIIa-J survey. No correction has been made to transform
to V since the colours of these quasars are unknown. Our diagram is a scatter diagram, because of the very broad range in
optical luminosity for our identified quasars, and hence is not
reproduced here. We have compared our data with those of
Figure 2b of Wall and Peacock (1985) for a complete sample of
radio-selected quasars with $S_{2.7} > 2.0$. Their figure shows a
very tight correlation with small scatter about a line log z =
2.23 V − 4.2. The scatter in our data is not a function of

redshift and therefore would not appear to be due to strong emission lines contributing to the broad-band colours. We have tried to establish whether this results from the broad intrinsic range in luminosity, or is an effect of poor magnitude estimates, emission lines in the broad-band colours or optical variability, by calculating continuum magnitudes at 1450 Å. We have used the published spectra in Wilkes et al. (1983) and our own unpublished spectra to estimate the continuum absolute magnitudes and have followed the method of Wampler et al. (1984). The broad range in luminosities is still present and is probably intrinsic, although we do need careful absolute continuum magnitudes for all quasars before firm statements can be made.

Our sample contains 10 quasars with $z > 3.0$. We find a large range in optical magnitude, which is an indication of a broad spread in optical luminosity, consistent with the results of Anderson and Margon (1987). We plot our quasars, together with their data, in Figure 7. This is a Hubble diagram of magnitude at 1475 Å versus redshift. They find that 25% of

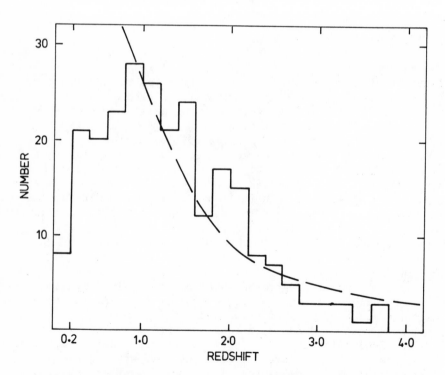

Fig. 6   Number-redshift diagram. The dashed line is normalized for a comoving volume decreasing as $z^{-1.5}$ per unit interval of redshift.

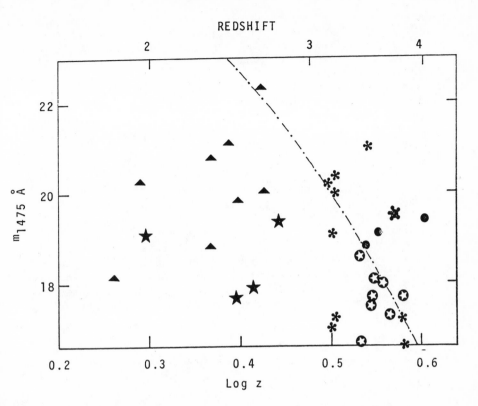

Fig. 7   An adaptation of the Hubble diagram of
          Anderson and Margon (1987) for a variety
          of quasars with z > 3.4. The "snowflakes"
          are our 10 quasars with z > 3.0.

the then-known quasars with z > 3.4 fall beyond the line representing the evolution of maximum redshift with luminosity.
In our sample 50% fall beyond the line. Our data increase the
breadth and scatter in the diagram and are further evidence
that low-luminosity quasars at high redshift are not rare.

Baldwin Effect
Following Wampler et al. (1984), we have plotted in Figure 8
the "Baldwin effect" diagram (Baldwin 1977) in order to compare
objects in common and to use their data for the lower redshift
quasars.
     Our complete sample increases the scatter and broadens the
luminosity range; however, the correlation found by Wampler et
al. (1984) is still apparent.

Emission-line Widths
Figure 9 presents a plot of rest-frame width (FWHM) of Lyα as a

function of redshift for all the quasars in our sample with
z > 1.7. There are 27 sources for which we do not have a
measured FWHM. We can compare this figure with Figure 4 of
MacAlpine and Feldman (1982), which is for optically selected
quasars. Of our quasars 56% have FWHM <3.6 (log scale) whilst
only 42% of their objects fall in this range. Their lowest
velocity width is 3.4 whilst seven (20%) of our quasars fall
below this value. Thus there is a tendency for the radio-
selected quasars to have narrower FWHM. This supports the
hypothesis that optical prism searches are biased towards
stronger-lined quasars. Figure 4 of MacAlpine and Feldman is
consistent with a limiting observed FWHM of 60 Å whilst our
radio-selected quasars are consistent with an observed FWHM
of $\gtrsim 30$ Å over all redshifts.

Figure 10 presents the rest-frame equivalent widths for
Lyα from Wilkes (1986) and from data by this group to date
(White et al. 1988). Although Wilkes used the same data base
she rereduced and remeasured all these data in order that her
own measures would be self-consistent. The Wilkes measures
are generally higher than ours, presumably caused by setting

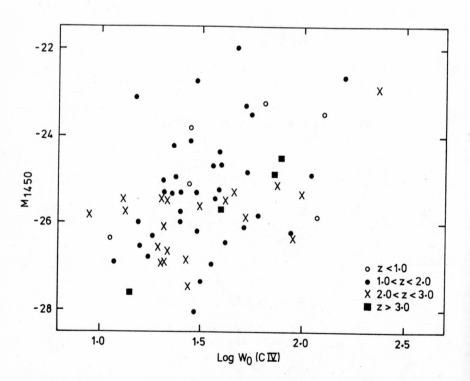

Fig. 8   The "Baldwin effect" for quasars in our sample
with measured C IV 1549 equivalent width.

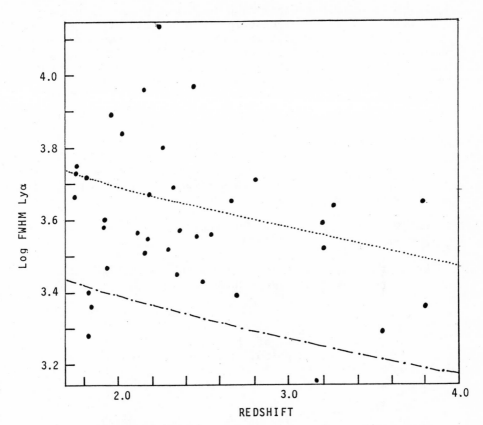

Fig. 9   The FWHM of Lyα as a function of redshift for qua-
          sars in the sample with z > 1.7.   PKS 0529-250 has
          no Lyα emission for z = 2.76, FWHM ≅ 0.   Note that
          an observed equivalent width of 50 Å corresponds to
          a rest frame equivalent width of 12Å at z = 3.

the continuum at different levels.   The mean value of Lyα
equivalent width from 38 quasars is 54 Å ± 37 Å. This is lower
than the value found by Wilkes and quoted by Anderson and
Margon (1987) of 65 Å ± 34 Å, but not significantly so. It is
25% lower than the values adopted by Schmidt et al. (1986) of
75 Å ± 7 Å (the uncertainty being the dispersion of mean
values) determined mainly from measures on optically selected
quasars. Needless to say our distribution is significantly
broader.

Optical Colours
Figure 11 is the B - V, V - R two-colour plot for all stellar
objects in a UKST 6° x 6° field at the south galactic pole (this
plot was kindly provided by Steven Warren and Paul Hewett).
Our measured values for two quasars with z > 3.5 in this region
and two other radio-selected quasars are plotted on this diagram.

Similar colour-colour plots have been made using different combinations of colours but only in the presented plot do all four
quasars stand clear of the main sequence. Our quasars are not
selected by colour-colour techniques, and so a radio-selected
sample is extremely useful to delineate those regions in the
redshift/colour-colour space where multicolour searches may be
incomplete.

DISCUSSION

    (i)   These sources are drawn from a complete radio-selected
               sample which covers a large area of sky and thus avoids
               small-area effects.

   (ii)   Even this radio-selected sample is not free from
               observational selection effects.

 (iii)   Radio samples provide a method of detecting BL-Lac-type
               quasars in significant numbers.

  (iv)   Accurate radio positions are being measured for all
               objects, allowing reliable optical identifications to
               the 22.5-mag limit of the UKST IIIa-J survey without
               any reliance on colour or morphology. The discovery of
               PKS 2000-330 (z=3.78) was instrumental in revealing the

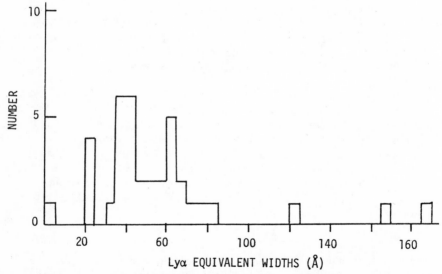

Fig. 10   Distribution of Lyα rest-frame equivalent
widths for the quasars in our sample.

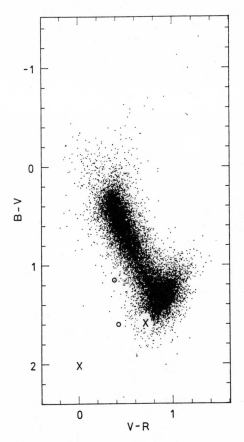

Fig. 11   The B − V, V − R two-colour plot for all
           stellar objects in an SERC/UKST 6° x 6°
           field at the south galactic pole. Two
           optically selected quasars (○), selected
           by techniques other than multicolour
           searches, and two of our radio selected
           quasars (X) are marked.

existence of such high-redshift objects and demonstrat-
ing the properties by which they may be found. Those
sources which remain unidentified to this limit are
being sought on deep RCCD frames taken on the AAT.

(v) We have optical spectral data on a large fraction of
     the quasar identifications. The redshift distribution
     shows no strong evidence for a redshift cutoff. Our
     Hubble diagram shows a large scatter in luminosity.
     We find equal numbers of low-optical-luminosity and

high-optical-luminosity quasars of high redshift, at variance with some optical searches.

(vi)   Our sample contains more objects with narrow Lyα equivalent widths than optically selected samples. The mean rest-frame equivalent width is narrower than that found for optical searches but has a very broad distribution. There is some evidence that the optical luminosity of these quasars also peaks at $z = 2$. Historically, a combination of these properties could spuriously enhance the number of quasars with $z = 2$ found in optical searches, and conversely it could contribute to the difficulty in finding optical quasars with $z > 4$. A combination of the above factors would then serve to produce strong evidence for a steep-redshift cutoff in optically selected samples of quasars. Additionally it would produce spurious predictions for high-redshift numbers in the extrapolations of evolutionary models from lower-redshift quasar data.

(vii) Complete radio samples such as this also serve as an extremely useful tool in revealing incompleteness and bias in optical surveys.

To provide some relevance to this session's title we note that in this sample of 403 radio sources we have one strong gravitational lens candidate - 0.25% of our population!

ACKNOWLEDGEMENTS

This work was supported in part by the National Aeronautics and Space Administration under a grant to the Jet Propulsion Laboratory, California Institute of Technology. We thank the Anglo-Australian Observatory for the use of many of its facilities. GLW acknowledges the receipt of an Australian National Research Fellowship.

REFERENCES

Anderson, S.F. and Margon, B. 1987, Nature, 327, 125.
Baldwin, J.A. 1977, Ap. J., 214, 679.
Bolton, J.G., Savage, A. and Wright, A.E. 1979, Aust. J. Phys.
     Ap. Suppl. No. 46.
Condon, J.J., Hicks, P.D. and Jauncey, D.L. 1977, A. J., 82, 692.
Condon, J.J., Jauncey, D.L. and Wright, A.E. 1978, A. J., 83,
     1036.
Downes, A.J.B., Peacock, J.A., Savage, A. and Carrie, D.R.
     1986, M.N.R.A.S., 218, 31.

Jauncey, D.L., Batty, M.J., Gulkis, S. and Savage, A. 1982, A. J., 87, 763.

MacAlpine, G.M. and Feldman, F.R. 1982, Ap. J., 261, 412.

McEwan, N.J., Browne, I.W.A. and Crowther, J.H. 1975, Mem. R.A.S., 80, 1.

Peacock, J.A. 1983, Proc. 24th Liège Int. Astrophys. Colloquium (held June 1983), p. 272, Univ. Liège, Astrophys. Instit.

Peacock, J.A., Miller, L. and Longair, M.S. 1986, M.N.R.A.S., 218, 265.

Peterson, B.A., Savage, A., Jauncey, D.L. and Wright, A.E. 1982, Ap. J., 260, L27.

Savage, A. and Peterson, B.A. 1983, Proc. IAU Symp. No. 104 (eds. J.O. Abell and G. Chincarini) (held Crete, August— September 1982), p. 57, Reidel, Dordrecht.

Schmidt, M., Schneider, D.P. and Gunn, J.E. 1986, Ap. J., 306, 411.

Wall, J.V., and Peacock, J.A. 1985, M.N.R.A.S., 216, 173.

Wampler, E.J., Gaskell, C.M., Burke, W.L. and Baldwin, J.A. 1984, Ap. J., 276, 403.

White, G.L., Jauncey, D.L., Savage, A., Wright, A.E., Batty, M.J., Peterson, B.A. and Gulkis, S. 1988 – to appear in Ap. J.

Wilkes, B.J., Wright, A.E., Jauncey, D.L. and Peterson, B.A. 1983, Proc. Astr. Soc. Aust. 5, 2.

Wilkes, B.J. 1986, M.N.R.A.S., 218, 331.

DISCUSSION

Foltz : I just wanted to comment on 0528-25, which is the
quasar that you reported as apparently having no Lyman alpha
emission. It does, however, have a very high density damped
Lyman alpha absorption system at nearly the emission line
redshift (which may contain molecular hydrogen). At any
rate, a fair amount of Lyman alpha emission could be removed
by that absorption. Observationally, this is a good test
object for all of us to use to see how well we can find
quasars.

Surdej : Did you find any broad absorption line quasars?

Peterson : The short answer is no. The only quasar that
could be considered a broad line object is 1157 + 014.

AN OPTICAL SEARCH FOR GRAVITATIONAL LENSES

RACHEL L. WEBSTER
Department of Astronomy and Scarborough College,
University of Toronto, Toronto, Ontario, Canada, M5S 1A1.

Paul C. Hewett
Institute of Astronomy,
Cambridge, U.K., CB3 0HA.

ABSTRACT   A new optical search for gravitationally
lensed quasars is described. The search is statistically
controlled and uses UK Schmidt Telescope direct and
objective-prism plates which have been scanned by the
Automated Plate Measuring Facility (APM) in Cambridge,
UK. So far, about 4000 quasars have been examined for
evidence of multiple imaging. A list of candidate lenses
has been obtained, most of which have yet to be observed
spectroscopically. However there are no candidate lenses
in the range 13 - 120". Uses of this survey material
are discussed.

INTRODUCTION

At the present time two groups are undertaking major searches
for gravitationally lensed quasars. The two searches are
essentially complimentary. The VLA search, (see Hewitt et al.
1987), is most sensitive to multiple images of radio quasars
with separations less than 2". The APM search (Webster et al.
1988) is sensitive to multiple images of optical quasars with
separations in the range 2 - 120".

The real strenghts of both these surveys are that the
selection biases can be quantified, and the region of
parameter space searched can be completely described. Since
the density of optical quasars is high, about 4 at $m_B = 19$
and about 20 at $m_B = 20$, the number of quasars which we can
survey is limited only by the rate at which data can be
processed and the candidates observed spectroscopically. If
the probability of gravitational lensing is about 0.001, then
we might expect to find about one multiply imaged quasar in
every 3 fields processed. It is expected, however, that a
higher fraction of bright quasars will be multiply imaged

because of the turnover in the quasar luminosity function at
faint magnitudes.

GENERAL STRATEGY

The gravitational lens search makes use of the APM to scan
direct and objective-prism survey plates of the same area of
sky. For the most part the plates are high quality survey
plates from the UK Schmidt Telescope and form part of the
Southern Sky Survey. The direct plates typically contain
more than 200,000 images in 36 square degrees to a limiting
magnitude of about m_J = 22.5 and an average seeing of 2".
The objective-prism plates have a dispersion of 2400 Å/mm at
4300 Å and about 2000 images per square degree are scanned to
a limit of about m_J = 20. The spectral information on the
objective-prism plates is used to determine quasar candidates,
while the direct plate gives information about the ellipticity
of the images and any close pairs of images. Once the direct
and prism images have been matched, about 10% of the images
are flagged as contaminated or overlapped. Quasars are
selected using the procedure developed by Hewett et al.
(1985). This method has demonstrated a high degree of
success in finding quasars which have been identified using
other techniques. Conservative criteria are used in selecting
the quasar candidates to ensure that the list is as complete
as possible; thus in general it contains 2 - 3 times the
number of quasars expected in the field.

IDENTIFICATION OF GRAVITATIONAL LENS CANDIDATES

The procedure for the identification of gravitational lens
candidates takes account of three different image separation
regimes. The first contains candidates consisting of image
pairs identified as single images by the APM offline analysis.
For images at m_B = 20, this occurs for images with separations
less than 5", while for a pair of images two magnitudes
brighter, the separation is about 7". Quasar candidates
with ellipticities which are twice the standard deviation
form the modal stellar value are inspected on the plates.
Careful simulations enable the detection function, which is
a function of separation, brightest image magnitude and
image magnitude difference, to be calculated. This function
is relatively insensitive to the magnitude of the brightest
image. Figure 1 shows the detection function for two
different values of magnitude difference.
    In the second regime, all images which are separated by
less than 10" on the direct plate are considered. The pairs
are chosen from the direct plate since at these separations,

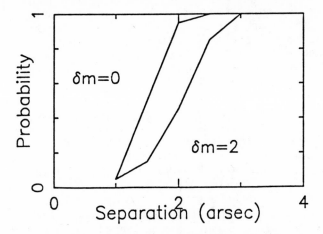

Fig. 1   Plot of the probability of detection of a pair
of images as a function of separation.

more than 10% of spectra on the objective-prism plate are
overlapped. The spectra of the pairs are then inspected for
any possible quasar candidates. Finally, all pairs of quasar
candidates with separations less than 2' are inspected.
There are few pairs with such small separations.
     We estimate the total number of quasars surveyed is
about 4000. More than 100 candidate gravitational lenses
were selected with separations ranging from 1 - 13". Eleven
of these candidates have since been observed spectroscopically,
and three are probably quasars.

DISCUSSION

The optical survey has no candidate lenses with separations
in the range 13 - 120". This result can be used to put
limits on two postulated lensing populations. Narayan et al.
(1984) suggested that clusters with surface densities
greater than critical might account for some of the observed
instances of lensing. One can show that wide separation
doubles with little amplification are at least as likely as
a pair of amplified images near the off-axis caustic. Since
our survey puts a limit on the probability of wide separation
doubles, the likelihood that clusters cause close separation
doubles is constrained. Also the current data can be used
to put an upper limit on the mean mass density of a population
of cosmic strings (Webster and Hewett 1988).
     A wide range of problems, both cosmological and those
specifically related to the nature of the lens can be studied

using gravitational lensing. Canizares (1987) gives an
extensive discussion of these problems. In particular, a
statistical sample of lenses will allow us to directly
calculate the amplification function. This function is
generally described as the differential probability of
amplification by a particular probability A. Theoretical
calculations of this function (Peacock 1986) for simplified
mass distributions have suggested that the function has a
power law form with a slope of -3 for amplifications greater
than a few. The prediction that this form carries over to
more complex mass distributions is being checked numerically
by Dyer and Webster.

     If one takes the available data, however, an interesting
result is obtained. The luminosity function for lensed
quasars can be written as a convolution of the intrinsic
quasar luminosity function and the amplification function.
If we assume that most quasars are unlensed, then the
intrinsic quasar luminosity function can be represented by
a function which has a turnover $m_B = 22$ (Koo 1983). We shall
consider only data in the redshift interval 1.3 - 1.9, since
this is the only interval containing several lensed quasars.
In order to determine the amplification function, both the
total area surveyed and the magnitude limit of the survey
need to be known. The sample from which known lensed quasars
are found is heterogeneous. However by looking at the quasars
in the Veron catalogue (1984), a reasonable estimate of the
magnitude limit is $m_B = 19$, and there are about 350 quasars
in the catalogue in the quoted redshift range. A least
squares fit to the data gives the points and standard
deviations in Figure 2.

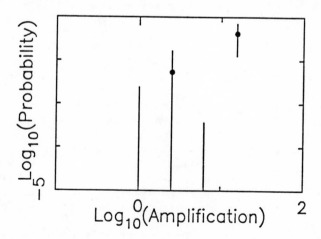

Fig. 2   Plot of the amplification function for known
lensed quasars

This result is clearly not consistent with the predicted slope of -3. The difference arises because the lensed quasars are all relatively bright, and the quasar luminosity function is steep. If the amplification function has a slope of -3, one would expect most of the lensed quasars to have apparent magnitudes near the sample limit. There are three possible explanations for the difference. The analysis is based on three gravitational lenses, which may be a biased sample. The assumed magnitude limit may be too faint. This would mean that there are many more lensed quasars in the range $m_B = 18 - 20$, and that the probability of lensing is higher than has been assumed. The third possibility is that the amplification function is not a smooth -3 power law. A statistical sample of gravitational lenses will help solve this problem.

ACKNOWLEDGMENTS

PCH is supported by Corpus Christi College, Cambridge, and RLW by NSERC grant number A8256.

REFERENCES

Canizares, C.R. 1987,in I.A.U. Symposium 124, ed. G. Burbridge, Reidel.

Hewett, P.C., Irwin, M.J., Bunclark, P.J., Bridgeland, M.J., and Kibblewhite, E.J. 1985, M.N.R.A.S., 213, 971.

Hewitt, J.N., Turner, E.L., Burke, B.F., Lawrence, C.R., and Bennett, C.L. 1987, in Proc. of 13th Texas Symp. on Relativistic Astrophysics, ed. M.P. Ulmer, World Scientific.

Koo, D. 1983, in Proc. 24th Liege Int. Colloq. Astro., ed. J.A. Swings, Universite de Liege.

Narayan, R., Blandford, R., and Nitananda, R. 1984, Nature, 310, 112.

Peacock, J.A. 1986, M.N.R.A.S., 223, 113.

Veron-Cetty, M-P., and Veron, P. 1984, ESO Scietific Report Number 1.

Webster, R.L., and Hewett, P.C. 1988, in Proc. Second Canadian Conf. on General Relativity, ed. A. Coley, B. Tupper, and C. Dyer, World Scientific.

Webster, R.L., Hewett, P.C., and Irwin, M.J. 1988, A.J., 95, 19.

LIMITS ON THE NUMBER OF CLOSE OPTICAL QUASAR PAIRS

D.W. WEEDMAN
Department of Astronomy, The Pennsylvania State
University, University Park, Pennsylvania 16802

S. DJORGOVSKI
Astronomy Department, California Institute of Technology,
Pasadena, California 91125

ABSTRACT   A new search has been conducted for close pairs
of quasars with identical spectra, including both emission
line objects and blue stellar objects. Survey plates
covering 3.9 deg^2 were selected for image quality, full
image widths being 0".8 to 1".2. Although 200 to 400 quasar
spectral images should have been examined, no candidate
pairs with separations < 4" were found. Eight such pairs
from 4" to 10" were found, but none were subsequently
confirmed as lensed quasars. The selection bias is
derived and applied to these limits. It is concluded
that the absence of close pairs expected from gravitational lensing models cannot be explained by observational
selection effects.

INTRODUCTION

Given the pressing mystery of how dark matter is distributed in
the universe, any tool with potential for revealing this matter
is eagerly seized upon. This explains why the theory of gravitational lenses has been considered so rapidly and thoroughly
even though only a few lenses have as yet been observed. The
study of Turner, Ostriker and Gott (1984) demonstrates that
predicting the observable properties of lensed quasars is a
complex effort, incorporating many astronomical assumptions,
but testable predictions are made. The primary parameter that
distinguishes predictions is the relative significance of
isolated lenses, such as individual galaxies, compared to distributed lenses, such as galaxies in clusters. Essentially,
lensing models that rely primarily on individual galaxies
predict smaller separations of lensed quasars than models
dominated by clusters.
     Median separations of lensed quasar images vary from 1"

to 4" among the models of Turner et al. Unfortunately, this
is also a range for which discovery bias, for optically observed quasars, is difficult to evaluate. In comparing the
predictions with existing observations, only one anomaly is
conspicuous. This is the wider separations found for lensed
pairs than expected from the models. It is tempting to attribute this anomaly to selection effects which have entered the
observations. In the present paper, we consider a selected
sample of the highest quality optical survey plates in an attempt to improve the search for close pairs and evaluate the
selection bias. The conclusions emphasize that important
things can be learned even from upper limits, and we encourage
quasar surveyors to undertake similar tests with their own
data.

OBSERVATIONS

Two lensed quasars have already been found serendipitously in
searches of low dispersion spectroscopic survey plates
(Weedman et al. 1982, Djorgovski and Spinrad 1984). These
plates were obtained with large telescopes (the 3.6-m Canada-
France-Hawaii Telescope (CFHT) and the 4-m at Kitt Peak
National Observatory) for the purpose of seeking quasars at
high redshifts by finding objects with conspicuous Ly α emission. The two lensed quasars were of a sample of 151 emission
line quasars published in two surveys (Sramek and Weedman 1978,
Gaston 1983). Only quasars with detected emission lines have
so far been published in such surveys, even though there are
probably many more quasars on the plates showing only continuous spectra. (Primarily, these plates are able to detect
only Ly α and sometimes CIV $\lambda1550$ or CIII]$\lambda1909$ emission.
Quasars at redshifts such that these lines were not present
would not show emission lines). The best of such survey
plates have been obtained with the "grens" on the CFHT. A
plate such as that on which was found the lensed pair 2345+007
typically shows 20 to 25 emission line quasars and over 100
objects with blue continua, or blue stellar objects (BSOs),
many of which are probably quasars. The BSOs are not reported
as candidates because, being faint and without strong emission
lines, follow-up confirmations would be very laborious and so
probably never done.
     Nevertheless, the BSOs on the plates are a potential
source of new lensed pairs. The BSOs taken with the emission
line quasars provide an upper limit to the possible quasars
present, so provide useful limits to the relative fraction of
quasars that show detectable lensed images. Consequently, DWW
selected the seven best plates obtained with the CFHT, all
previously examined for emission line quasars, and searched
them again for close pairs with any kind of similar spectra.

No spectroscopic criterion was applied other than the rejection
of spectra with absorption features classifying them as stars.
In practice, only blue objects show featureless spectra, although many of the BSOs really are blue stars. The plates
examined, listed in Table I, were selected as having images
with full width perpendicular to the dispersion of < 1.2". On

TABLE I   Survey Fields Examined

CFHT Plate No.	UT Date (mo/day/yr)	Field Center RA (1950) Dec.		Spectral Width
$1421\frac{1}{2}$	8/03/81	$16^h 33^m 20^s$	$+33°42'$	$1\rlap{.}''1$
$1424\frac{2}{2}$	8/03/81	$23^h 41^m 48^s$	$+01°03'$	$1\rlap{.}''1$
1425_2	8/03/81	$23^h 46^m 20^s$	$+00°24'$	$1\rlap{.}''2$
1426_2	8/03/81	$00^h 02^m 23^s$	$-00°57'$	$1\rlap{.}''2$
3030_1	1/09/83	$08^h 41^m 40^s$	$+44°44'$	$1\rlap{.}''2$
3044_1	1/10/83	$12^h 57^m 25^s$	$+35°37'$	$1\rlap{.}''0$
3435^1	6/16/83	$13^h 09^m 35^s$	$+28°27'$	$0\rlap{.}''8$

1. Emission-line quasars discovered on this plate are in
    Weedman (1985).
2. Emission-line quasars discovered on this plate have been
    published by Gaston (1983).

the best plate, this width was 0.8". (An image is illustrated
by Weedman et al. 1982). These plates, therefore, represent
those with the best seeing and are about as good as achievable
with existing survey techniques. In this search for pairs, an
upper limit of 10" separation was adopted for listing objects
as possible lensed pairs. Visual inspection was the only
criterion for judging if objects have identical spectra, and
in the spirit of providing upper limits, this criterion was not
overly strict. All possible new candidates for lensed pairs
are given in Table II. The total area surveyed by the seven
plates is 3.9 deg^2.

CCD images and spectra for all candidates except pair #1
were obtained by SD using the 3-m telescope at Lick Observatory. All candidates observed are stars, except that the
fainter of pair 4 is a quasar with a single emission line at
5150 Å, and the fainter of pair 6 may have weak emission lines.
In no case was a lensed pair confirmed. This strengthens the
upper limit to possible pairs in this survey.

TABLE II   Quasar Pair Candidates[1]

No.	Position R.A. (1950) Dec.		Mag. Brighter	Fainter	Companion Location	
1	$00^h02^m03^s.2$	$-00°48'31''$	18	20	9"	N
2	$16^h33^m56^s.7$	$+33°30'27''$	17.5	21	7"	SW
3[2]	$16^h34^m11^s.3$	$+33°25'36''$	19	19	9"	W
4	$16^h34^m21^s.5$	$+33°22'00''$	20	22	7".5	N
5	$23^h41^m08^s.5$	$+00°53'22''$	20	20	4"	NW
6	$23^h44^m52^s.2$	$+00°15'20''$	19.5	20.5	5"	NW
7	$23^h46^m38^s.5$	$+00°22'16''$	19.5	20	7"	SE
8	$23^h47^m44^s.3$	$+00°38'42''$	19	19.5	9".5	N

1. Positions, given for brighter component, were determined from spectra on the survey plates, relative to two SAO stars on a typical plate. Listed positions may be uncertain up to ± 10".
2. All other candidates listed are blue stellar objects. This one is an emission line quasar with what appears to be another image of the emission line, displaced 7".5 N. Continuum of this companion is very weak.

DISCUSSION AND CONCLUSIONS

What we wish to know, for comparison with models, is the fraction of quasars that appear paired as a function of separation. Fortunately, Turner et al. (1984) showed that the magnitude limit of the quasar sample is not particularly important; all that is really necessary is to know the total number of quasars examined. Translating the upper limits for the number of pairs in Table II into percentages of quasars then requires only knowing the total number of quasars that were examined on these plates, regardless of whether we knew they were actually quasars.

Sufficient quasar counts have been made that the number of quasars deg^{-2} brighter than a given magnitude is reasonably well known. A good summary is given by Braccesi (1983). If we know the limiting magnitude of the present survey, the published counts allow a determination of how many quasars are included in the fields examined. Available calibration of the plates indicates that they are complete close to blue continuum magnitude 21, for detecting continuous spectra in addition to emission lines. Gaston (1983) used microdensitometry and the signal to noise ratio of the spectra traced to conclude that the continuum magnitude limit for emission-line quasars is 21.1 ± 0.3 mag on three of the plates used. DWW attempted another calibration based only upon objects with continuous spectra. A CCD image was obtained by W.C. Keel with the 4-m telescope at Cerro Tololo Inter-American Observatory for a small area in the vicinity of quasar 0003-010 listed by Gaston. Fifteen faint objects were identified on the survey plate for which Keel subsequently determined the V magnitudes from the CCD image. We concluded from this comparison that the plate was complete to magnitude 20.5 and 50% to 75% complete at mag. 21. As the objects used for this test were redder than quasars (being stars and galaxies), the B magnitudes that determine detectability on the plates would actually be fainter than the V magnitudes given. It is reasonable to assume, therefore, that the limit in B magnitude for the search described in the present paper is somewhere between 20.5 and 21 mag.

From the tabulation by Braccesi (1983), this range of B magnitude gives from 50 to 80 quasars deg^{-2} expected. As the survey encompassed 3.9 deg^2, from 200 to 300 quasars should have been examined. In fact, 140 emission line quasars had already been noted on the plates. Many of these have continuum magnitudes fainter than the listed limit (Gaston 1983), so can be considered as extra quasars sampled for pairs. A generous range of uncertainty would be that from 200 to 400 quasars have been examined for pairs in the new survey reported in the present paper.

The most important result is that of all these quasars, none are found in pairs with separations less than 4". It is then necessary to decide what is the selection bias for detecting separations of this order or smaller. Even though image sizes are 1.2" or less, even smaller separations could be detected by the blended but widened spectra produced by a pair. When searching such plates, images are continuously compared with nearby spectra of stars so even slightly widened spectra are noticed. Similarly widened spectra could also be produced by barely resolved galaxies, however. For this reason, we have determined a conservative selection bias function for each plate using the criterion that an image pair would be noted if the blended spectrum of the pair had a width twice that of a single spectrum. In this circumstance, two distinct image cores would be seen. This criterion is equivalent to requiring the component of separation perpendicular to dispersion equal or exceed the normal image width. Using widths from Table I folded with a random distribution of position angles for pairs, this selection bias weighted for the total of seven survey plates is as given in Table III.

Limits on the fraction of quasars found as close pairs can now be considered. Using the selection bias applied to from 200 to 400 quasars (the sample uncertainty), a pair with separation of 1".8 or greater, for example, should have been

TABLE III   Selection Bias for this Survey

Pair Separation	Percent Detectable
0".8	0
0".9	4.3
1".0	5.8
1".1	11
1".2	20
1".3	36
1".4	42
1".5	48
1".6	52
1".8	59
2".0	64
2".2	67
2".4	70
3".0	76
4".0	83
5".0	86
6".0	88
8".0	91
10".0	93

found if such pairs occur in 0.4% to 0.8% of quasars. The absence of such a pair is consistent with the quantitative predictions of Turner et al.; their "standard" model for q_o = 0 predicts that 0.5% of quasars should be found as pairs with median separation 1.75", or 0.25% with separations greater than this. They point out that this prediction is actually a lower limit, conceivably by a large factor, because it neglects the basic flux amplification bias and includes only the mass distribution in known objects. Yet the observational limit indicates that this prediction is not greatly underestimated.

With the improved resolution, the absence of close pair candidates remains surprising compared to the two wide pairs (7.3", 3.8") already found in identical surveys. These two were from a total sample of 151 emission line quasars (Sramek and Weedman 1978, Gaston 1983), yielding a literal fraction of 1.3% for separations $\gtrsim$ 4". Using the predictions of Turner et al. (their Figure 11), three times as many pairs are expected for separations > 1".8 compared to > 4". For a sample of 200 to 400 quasars and the selection bias in Table III, 5 to 10 new pairs would have been expected, naively scaling by the wider pairs already found. Meaningful statistical conclusions cannot be drawn from the small sample of pairs currently known, but the fact that disagreement approaches an order of magnitude demonstrates the puzzle that exists. This demonstrates why continued examination of optical quasar survey plates to locate any possible examples of lensed pairs is extremely important, even if results can only be expressed as limits. It also demonstrates the need for further careful studies of existing wide pairs, to confirm that they are indeed lensed objects rather than truely binary quasars (e.g. Djorgovski et al. 1987).

ACKNOWLEDGEMENTS

DWW was a visiting astronomer on the Canada-France-Hawaii Telescope, operated by the National Research Council of Canada, the Centre National de la Recherche Scientifique of France, and the University of Hawaii. We thank the Canadian Allocation Committee for observing time on the CFHT and Rene' Racine for obtaining two of the plates used. SD thanks both Hyron Spinrad and the technical staff for their help in making it possible to obtain observations at the Lick Observatory, University of California. Also, William Keel is thanked for obtaining and processing the CCD image used for magnitude calibration. This research was partially supported by NSF grant AST-8101204.

REFERENCES

Braccesi, A. 1983, in IAU Symp. 104, Early Evolution of the Universe and its Present Structure, eds. G. Abell and G. Chincarini, (Reidel:Dordrecht), p. 23.
Djorgovski, S. and Spinrad, H. 1984, Ap.J. (Letters), 282, L1.
Djorgovski, S., Perley, R., Meylan, G., and McCarthy, P. 1987, Ap.J. (Letters), 321, L17.
Gaston, B. 1983, Ap.J., 272, 411.
Sramek, R.A. and Weedman, D.W. 1978, Ap.J., 221, 468.
Turner, E.L., Ostriker, J.P. and Gott, J.R. 1984, Ap.J., 284, 1.
Weedman, D.W., Weymann, R.J., Green, R.F. and Heckman, T.M. 1982, Ap.J. (Letters), 255, L5.
Weedman, D.W. 1985, Ap.J. (Suppl.), 57, 523.

GALAXY CLUSTER ENVIRONMENT OF QUASARS AND IMPLICATION FOR THE LUMINOSITY FUNCTION

RICHARD F. GREEN
Kitt Peak National Observatory, National Optical
Astronomy Observatories, P.O. Box 26732, Tucson, AZ
85726

H.K.C. YEE
Département de Physique, Université de Montréal,
C.P. 6128, Succ. A,
Montréal, Québec, H3C 3J7   Canada

ABSTRACT   The galaxy group and cluster environment
in which quasars are found is described on the basis
of results from deep systematic imaging surveys. Low-
redshift quasars are rarely found in clusters richer
than Abell richness 0; they are in regions of average
density 2 - 3 times higher than field galaxies and are 6
times more likely to have a companion within 100 kpc
projected distance. This evidence suggests interaction
as the trigger of recent quasar events, requiring a
source function to explain the evolution of the quasar
luminosity function. Radio quasars with redshifts
greater than 0.5 have a high probability of being
located in regions of galaxy density with covariance
amplitudes 8 times greater than that for field
galaxies. The availability of new sites for these
objects at earlier cosmic epochs suggests density
evolution for the ensemble of radio quasars.

INTRODUCTION

The ultimate goal of quasar survey work is to derive the
luminosity function and its changes with cosmic epoch. From
that information, one would like to constrain models of the
formation and luminosity evolution history of individual
objects. A valuable complementary effort is the study of the
environments in which quasars are found. The local galaxy
density, frequency and types of binary encounters, and
condition of the intracluster medium may all play a role in
the formation and fueling of quasars.

The most striking and ubiquitous observational evidence
is the high frequency of close companions near quasars and
Seyfert galaxies (e.g., Stocke 1978, Stockton 1982, Hutchings
et al. 1982, Gehren et al. 1984, Dahari 1984, Malkan 1984,
Heckman et al. 1984, Hintzen 1984, Yee and Green 1984, Smith
et al. 1986, Yee 1987). The frequent presence of a close
companion naturally suggests interaction as a possibility for
triggering or fueling quasar activity. A change in the
interaction rate with cosmic time would have the consequence
of a change in the frequency of quasar activity, reflected in
the observed luminosity function. Models have been
constructed that relate quasar activity to interactions
(Stocke and Perrenod 1981, Norman and Silk 1983, De Robertis
1985, Roos 1985a and b). Whether fueled by stars, gas
donated by the companion, or gas from the host itself, the
prediction is that environmental conditions were more
conducive to the formation of quasars at earlier cosmic
epochs. The more general question to be addressed is that of
the galaxy group and cluster environment in which quasars can
form, and the change of that environment with cosmic time.
Deep, systematic imaging surveys are required to answer that
question. This article will concentrate on results that we
have obtained from a series of systematic surveys.

FAINT IMAGING SURVEYS

Our first direct imaging survey of fields around quasars was
taken with the Palomar 1.5-m telescope and SIT-Vidicon camera
(Green and Yee 1984, Yee and Green 1984). The general
characterization of the result for low-redshift quasars (z <
0.4) is that quasars are found preferentially in regions of
higher than average galaxy density; the average quasar-galaxy
spatial covariance amplitude is 2 to 3 times higher than that
of the galaxy-galaxy covariance amplitude. These conditions
correspond to environments ranging from small groups to poor
clusters; it is extremely rare to find a low-redshift quasar
in a rich cluster. These results were derived by counting
galaxies in the quasar field in excess of the background
density as a function of magnitude obtained from blank-sky
control fields. The excess galaxies were radially
concentrated toward the quasars and had magnitudes in the
range expected for normal galaxies at the quasar redshifts,
giving confidence in the assumption that they are physically
associated host groups and clusters. This first experiment
was redshift-blind; no excess galaxies were detected around
quasars with redshifts such that galaxies more luminous than
the characteristic absolute magnitude, M*, would appear
fainter than the detection completeness magnitude. For this
sample with a range of modest limiting magnitudes, the

contamination by foreground clusters was evidently
negligible.

The second major imaging survey was taken with the
Steward Observatory 2.3-m telescope and thick RCA CCD (Yee
and Green 1987). Filters employed were Gunn r and i; the
typical limiting r magnitude of 22.5 corresponds to M* at
z 0.65. The field of view includes a metric diameter around
quasars with redshifts from 0.4 to 0.6 of 0.5 - 1.0 Mpc.
Thirty-one quasar fields were observed, 21 radio-loud sources
and 10 radio-quiet, with 27 blank sky control fields. The
images were analyzed in a self-consistent way by an image
processing package developed by H.K.Y. Objects were detected
automatically, measured for position and total brightness,
and classified as star or galaxy on the basis of the summed
deviation from a stellar growth curve (Yee, Green and
Stockman 1986). We find first-order agreement with the work
of Koo (1986) and Infante et al. (1986) in deriving a
differential count slope of galaxies as a function of
apparent magnitude of 0.4 in the log. The loss of low
surface brightness objects near the limit make the exact
value slightly uncertain. The nearly constant increase is
not consistent with the K-dimming expected for contributions
from higher redshift galaxies with no evolution. This effect
must be taken into account in interpreting the spatial
association of galaxies with higher redshift quasars.

A method of analysis must be employed that provides a
quantitative measure of the association of excess galaxies
with quasars, one that includes the effects of sampling to a
different depth in the luminosity function at different
redshifts and sampling a different metric area around the
quasar. The two-point spatial covariance function is such a
measure that takes into account the two sampling effects. It
relies on three assumptions: a spherical distribution of
galaxies around the quasar, no change of clustering
properties with redshift, and a "Universal" luminosity
function for each cosmic epoch. Since it was found that the
excess galaxies in the quasar fields were consistent with
being physically associated, it is possible to use that
assumption to construct an observed luminosity distribution
of galaxies as a function of redshift. When properly
normalized, those luminosity functions can be applied to the
observed angular covariance functions to yield the spatial
covariance functions.

The path of analysis then proceeds as follows. Ideally,
one would derive the luminosity function for each associated
cluster; small number statistics make that direct approach
impractical. The quasar fields can be grouped in redshift
bins; however, and an average luminosity distribution of
galaxies derived in that way. Only the radio-loud quasars
show strong excess galaxy counts; however, even these groups

of quasars do not provide enough associated galaxies to
constrain the shape of the luminosity function at the faint
end. The parameterization of Schechter (1976) was therefore
adopted, with an exponential cut-off at bright luminosities
and a power-law toward low luminosities. Two values of the
power-law index were assumed: -1.0 and -1.2, as being
representative of the local form. The former value is from
the Durham/AAT redshift survey reported by King and Ellis
(1985), while the latter is a fit by Sebok (1986). Much
steeper values (Burg 1987) would have some impact on the
exact numbers derived. The luminosity distributions were
derived for epochs with mean redshifts of 0.24, 0.42, and
0.61, and were constructed in the observed r band; the
relative K corrections between the redshift limits for each
average are then very small. The normalization of these
galaxy luminosity distributions is not absolute, because the
sampled clusters do not comprise a volume limited sample.

From these fitted luminosity functions, it is apparent
that changes in the cluster luminosity function have taken
place between $z = 0.24$ and $z = 0.61$. The characteristic
absolute magnitude in the observed r band shows no
significant change in value, while the central rest
wavelength changes from 5200 A at $z = 0.24$ to 4000 A by $z =
0.61$, leading to a dimming of ~0.7 mag for an elliptical
galaxy in the absence of evolution. For any choice of world
model (value of q_o), the measured M* in the observed frame
will lead to predictions about the evolution of the galaxy
luminosity function.

The next step is to get a consistent normalization for
the choice of world model. The procedure is to adopt a zero-
redshift luminosity function of galaxies, and then derive the
evolution in M* that is consistent with the observed r-band
luminosity distributions for the different redshift bins.
The mix of morphological types was based on the local
luminosity functions of King and Ellis (1985) and Sebok
(1986), which had been used to derive the values of the
power-law slope. The brightening in the observed M* is
around 0.9 mag between redshifts 0 and 0.6, at somewhat under
2 sigma significance. This result is based on the assumption
that the luminosity function of the field and the luminosity
function of the galaxies in the quasar host groups and
clusters are identical. This assumption would be invalid if
the elliptical/spiral fraction in the quasar environment were
strongly different from the field, or if the relative
dominance of spirals in the luminosity function changed
sharply with redshift. Neither of these effects is
significant: if the clusters have about 10 times the field
density of galaxies, Dressler (1980) shows that the E + S0
fraction increases from 0.2 to 0.3. In the observed r band,
the distribution of morphological types of type Sb and

earlier contributing to the luminosity function down to M = −20.5 changes from 71% at z = 0 to 64% at z = 0.6. For the purpose of evaluating the richness of quasar cluster environments, we can safely assume that the evolution of the luminosity function of the host clusters with cosmic time represents a contemporaneous change in the field luminosity function.

By parameterizing the change in M* with time in a simple way, the normalization of the luminosity function can be derived for each world model by fitting the observed background counts. The luminosity function is integrated over volume to a limit appropriate for each morphological subgroup, including the change of brightness with z from the K correction and the evolution. The best fits were $q_o \sim 0.5$ for the Sebok function with power-law slope of −1.2, and $q_o \sim 0.02$ for the King and Ellis function with power-law slope of −1.0. In particular, the counts were normalized at r = 20, with the best fitting curve for choice of q_o determined from the faint end. These analysis steps produced a self-consistent combination of luminosity function, evolution, and choice of world model, which were then used for a quantitative measure of the spatial covariance amplitude of the excess galaxies.

RESULTS

The two-point angular covariance function is characterized by an amplitude and a dependence on angle. From counting the excess galaxies in rings around the quasars, a power-law angular dependence was derived identical to that obtained by Groth and Peebles (1977), with a power law index of −0.77 (gamma = 1.77). Normalizing by the luminosity function discussed in the previous section, the spatial covariance amplitude was derived for each field. Values for radio-loud quasars are plotted as a function of redshift, absolute red magnitude, and radio power in the Figure. The results for lower redshifts confirm those of the earlier survey: the average quasar-galaxy covariance amplitude is 2 − 3 times greater than the average galaxy-galaxy covariance amplitude (taken to be 67.5 from Davis and Peebles 1983). For z < 0.50, radio quasars are only rarely found in environments as rich as Abell richness 0 clusters. A sharp contrast is observed for radio quasars in the redshift bin 0.55 − 0.65. The average spatial covariance amplitude is 8 times greater than the average galaxy-galaxy amplitude, an increase of a factor of 3 over the value at lower redshifts. Host cluster galaxy densities are comparable to those of Abell richness 1 clusters. From the Steward 2.3-m imaging sample, some 2/3 of the higher redshift radio quasars fall within these rich

associations. Deeper samples from the CFHT and CTIO 4-meter
telescopes show that fraction to be closer to 1/2, but still
very substantial.

From examination of the three plots in the Figure, it is
clear that the strongest change in average covariance
amplitude is evident as a function of redshift. A trend is
seen with absolute red magnitude, and in the apparent
magnitude-limited sample there is a correlation between
redshift and absolute magnitude. Limiting the sample to a
fixed redshift range, however, no correlation between optical
luminosity and covariance amplitude is seen. Over a very
broad range of radio power, there is a demonstrated trend of
increasing cluster richness with increasing power (Peacock
and collaborators, these proceedings). As is seen from the
lower panel of the Figure, that effect is not operative over
the more limited range of radio power for the objects in the
Steward imaging sample; there is no correlation of the 5 GHz
power with redshift for these objects.

These results have a direct bearing on interpreting the
evolution of the quasar luminosity function. The type of
galaxy cluster supporting a radio quasar at $z = 0.6$ would not
dim below the detectability of this survey by $z = 0.4$. The
fact that no low-redshift radio quasars were found in
substantial clusters suggests that there was a change in the
environment in cluster center between redshifts 0.4 and
0.6. This change in conditions either caused a dimming by
over two magnitudes in the central quasar or caused the
frequency of triggering of short-lived quasar activity to be
drastically reduced. Physical explanations may include a
decrease in binary encounter efficiency in disrupting stellar
or gaseous orbits as clusters of this richness become more
centrally concentrated; or a change in the gas content of all
central galaxies in such clusters, inhibiting the fueling of
the central engine. Both may be correlated with the
development of a hot intracluster medium.

The consequence of a change in physical conditions in
the centers of galaxy clusters conducive to quasar formation
and maintenance is that new sites were available for radio
quasars at redshifts greater than 0.5. In the parlance of
quasar luminosity function evolution, this effect is
equivalent to density evolution - there were more radio
quasars per cubic gigaparsec at earlier epochs. Where are
the modern-day counterparts to these cluster quasars? Work
in progress by De Robertis and H.K.C.Y. on low-redshift radio
galaxies in similar cluster environments has shown that a
substantial fraction of these galaxies, previously classified
as pure absorption-line, in fact have broad, weak H alpha
emission. Since no low-redshift quasars, radio powerful or
quiet, are observed in richer galaxy clusters, it cannot be
the case that the radio source alone is quenched. On the

other hand, it is tempting to draw the connection between the
radio quasars at z = 0.6 and the modern epoch radio galaxies
with a (dust obscured?) remnant broad line region. The
change in environment may be a dynamical one, with a critical
combination of cluster central density and velocity
dispersion, above which quasars cannot be supported. If
richer clusters reached that threshold at earlier cosmic
times, they would not host quasars at lower redshift, but
could provide more sites of increasing richness with higher
redshift.

The increasing availability of sites for quasar
formation could explain the observed increase in the space
density of radio quasars with increasing look-back time.
What is the evidence for the larger population of radio-quiet
quasars? Unfortunately, the BQS sample does not contain any
radio-quiet objects with redshifts between 0.55 and 0.7.
This redshift range represents the maximum contribution of
the 3000 A bump to light in the B band, so the U-B color
excess selection criterion led to systematic incompleteness
in this range. Preliminary evidence from imaging surveys at
the CFHT, the CTIO 4-meter and the INT does not support a
similar abrupt change in preferred environment for radio-
quiet quasars. Although the quantitative analysis is still
in progress, there is little visual impression from the
images that there are rich host clusters comparable to those
for the radio powerful objects.

For radio quiet quasars, the most important factor in
triggering the activity may be the presence of a close
companion galaxy. Yee (1987) investigated images of fields
around a sample of 37 low-redshift radio-quiet objects. For
an average completeness limit corresponding to absolute r
magnitude of -19, some 40% of the quasars had at least one
galaxy within a projected distance of 100 kpc. The sample of
companions was consistent with a random draw from the field
luminosity function, on the assumption that they were
physically associated. Extrapolating the field luminosity
function to fainter absolute mangitudes yields the result
that all low-redshift radio-quiet quasars have companions
brighter than M = -16.5 within a projected distance of 100
kpc. The luminosities of the companions were uncorrelated
with the quasar luminosity; considered together with the
matching of the sample of companions to the field luminosity
function, the implication is that the quasar has little
effect on the internal affairs of the companion. Conversely,
the nature of the companion must not be the primary
determinant of the properties of the quasar event. The
frequency of finding close companions for these quasars is
about 6 times higher than that for random galaxies.

The high incidence of close companions in radio-quiet
quasars strongly suggests that interaction is responsible for

triggering activity. In that case, the interpretation of the
time evolution of the quasar luminosity function must include
a source function, and cannot explain the observations with a
single initial burst of quasar formation.   It is plausible
to interpret the higher volume density of quasars of a given
luminosity at earlier cosmic epochs as an increase in the
frequency of binary encounters; on the other hand, this
interpretation is not unique nor is it demanded by the
data.   The frequency of binary encounters could be constant
or increasing toward the present epoch, gradually becoming
dominant over another primordial mechanism.

It may be possible to use the study of the cluster
environment to constrain quasar lifetimes. If quasars with
redshifts greater than 1 are found in environments which do
not support quasar activity at lower redshifts, and if the
era of quasar formation was largely between z = 2 and 3
(e.g., Schmidt et al. in these proceedings), then the total
luminous lifetime can be meaningfully limited by the
observations. Until then, we must exercise great caution in
deducing the life histories of individual sources from the
ensemble behavior of the quasar luminosity function.

REFERENCES

Burg, R. I. 1987, Ph.D. Thesis, Harvard University.
Dahari, O. 1984, A. J. 89, 966.
Davis, M. and Peebles, P. J. E. 1983, Ap. J., 267, 465.
DeRobertis, M. 1985, A. J., 90, 998.
Dressler, A. 1980, Ap. J., 236, 351.
Gehren, T., Fried, J., Wehinger, P. A., and Wyckoff, S. 1984,
    Ap. J., 278, 11.
Green, R. F. and Yee, H. K. C. 1984, Ap. J. Suppl., 54, 495.
Groth, E. J. and Peebles, P. J. E. 1983, Ap. J., 217, 385.
Heckman, T., Bothun, G., Balick, B., and Smith, E. P. 1984,
    A. J., 89, 958.
Hintzen, P. 1984, Ap. J. Suppl., 55, 533.
Hutchings, J. B., Crampton, D., Campbell, B., Gower, A. C.,
    and Morris, S. C. 1982, Ap. J., 262, 48.
Infante, L., Pritchet, C. J., and Quintana, H. 1986, A. J.,
    91, 217.
King, C. R. and Ellis, R. S. 1985, Ap. J., 288, 456.
Koo, D. C. 1986, Ap. J., 311, 651.
Malkan, M. 1984, Ap. J., 287, 555.
Norman, C. and Silk, J. 1983, Ap. J., 266, 502.
Roos, N. 1985a, Ap. J., 294, 479.
-------- 1985b, Ap. J., 294, 486.
Schechter, P. L. 1976, Ap. J., 203, 297.
Sebok, W. L. 1986, Ap. J. Suppl., 62, 301.
Smith, E. P., Heckman, T. M., Bothun, G. D., Romanishin, W.,
    and Balick, B. 1986, Ap. J., 306, 64.

Stocke, J. T. 1978, A. J., 83, 348.
Stocke, J. T. and Perrenod, S. C. 1981, Ap. J., 245, 375.
Stockton, A. 1982, Ap. J., 257, 33.
Yee, H. K. C. 1987, A. J., 94, 1461.
Yee, H. K. C. and Green, R. F. 1984, Ap. J., 280, 79.
Yee, H. K. C. and Green, R. F. 1987, Ap. J., 319, 28.
Yee, H. K. C., Green, R. F. and Stockman, H. S. 1986, Ap. J.
    Suppl., 62, 681.

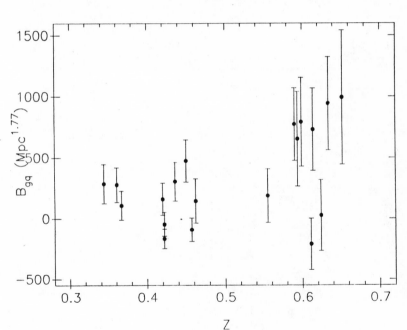

Covariance Amplitude vs. Redshift

Covariance Amplitude vs. Absolute Mag

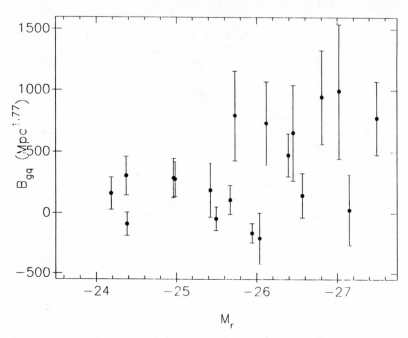

Radio Power vs. Covariance Amplitude

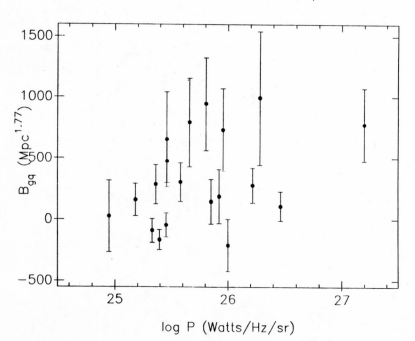

DISCUSSION

Weymann : When you talk about radio power, it's not clear to me what kind of radio sources compose the most part of your sample. Is it extended power, core power, or lobe power?

Green : These were 3C, 4C, and Parkes objects found at low frequencies.

Miller : This point makes sense in the context of the quasar environments. An extended radio source in a dense environment is expected to be luminous for a longer time than in a less dense environment. This provides a physical reason for a radio luminosity effect.

THE SPATIAL CLUSTERING OF QSOS

T. SHANKS
Physics Dept., University of Durham, South Road,
Durham DH1-3LE, England.

B.J. BOYLE
Anglo-Australian Observatory, P.O. Box 296, Epping
NSW 2121, Australia.

B.A. PETERSON
Mount Stromlo & Siding Springs Observatory,
Private Bag, Woden PO, A.C.T 2606, Australia.

ABSTRACT We present the 2-point correlation function of
QSOs in the range $0 \leqslant r \leqslant 2000h^{-1}Mpc$, as obtained from the
UVX survey of 423 $z \leqslant 2.2$, $B \leqslant 21^m$ QSOs of Boyle et al.
(1988a). At small scales ($r \leqslant 10h^{-1}Mpc$, $q_0=0.5$) we find a
clear 4.2σ, detection of QSO clustering. Assuming a
stable evolutionary model the QSO clustering is stronger
than expected if QSOs randomly sample the galaxy distribution and this behaviour is consistent with a group
environmental model for QSOs.
    At larger scales ($10 \leqslant r \leqslant 2000h^{-1}Mpc$) we detect no
significant evidence of QSO clustering for QSOs with
$0.3 \leqslant z \leqslant 2.2$. We test for evolution of the large scale
galaxy correlation function and indicate how the evolu-
tion and position of any weak clustering features detected at large scales may provide powerful cosmological
probes.

INTRODUCTION

The aim of this paper is to describe preliminary results on
the spatial clustering of QSOs from the Anglo-Australian
Telescope (AAT) fibre optic survey of 423 QSOs (Boyle et al.
1988a). The survey consists of 34 40' diameter fibre fields
scattered over eight 5°x5° U.K. Schmidt (UKST) fields, with 4
fields located in each Galactic Hemisphere. U and B magnitudes
for the QSOs were obtained from the COSMOS plate measuring
machine and calibrated using CCD sequences. The sky density
of QSOs with $z \leqslant 2.2$ and $B \leqslant 21.^m9$ is:-

N = 38 ± 9.8 deg^{-2} (22 fields)

with the error being estimated from field-to-field fluctuations. It is this high sky density coupled with the large size of this survey which makes this catalogue uniquely suited to estimating the QSO correlation function at small spatial scales ($r \lesssim 10h^{-1}$Mpc). The catalogue's completeness and large extent in the redshift direction makes it also ideal for probing the structure of the Universe on the largest spatial scales.

The above field-to-field dispersion gives the first basic information on the QSO distribution; the observed dispersion of ±9.8deg^{-2} is in good agreement with the ±10.4deg^{-2} expected for Poisson fluctuations. Thus there is no evidence in this survey for any anisotropy in the distribution of QSOs on the sky.

The overall n(z) relation for our survey is also quite smooth in the range $0 \leqslant z \leqslant 2.2$ and shows no evidence for preferred redshifts etc. However in individual fields the QSO redshift distribution does appear clumpy (see Shanks et al. 1988a); it will be the business of the remainder of this paper to test whether this clumpy appearance is due to noise or to true QSO clustering.

THE 2-POINT QSO CORRELATION FUNCTION

The statistic we use to analyse the QSO distribution is the 2-point spatial correlation function, $\xi_{qq}(r)$. This statistic compares counts of QSO pairs at spatial separation r, with what would be expected for a random distribution. For galaxies, $\xi_{gg}(r)$ is usually represented by the form:-

$$\xi_{gg}(r) \;=\; B_{gg}r^{-1.8} \quad (r \lesssim 7h^{-1}\text{Mpc}).$$

After Osmer (1981) we shall be using a comoving coordinate system to measure QSO separation. For our present purposes we shall assume $q_o=0.5$ and $H_o=100h$km s^{-1}Mpc^{-1}. We use standard methods to estimate $\xi_{qq}(r)$ similar to those used by Shanks et al. (1987). Here the normalisation of ξ_{qq} is made by fitting a smooth polynomial to the observed overall n(z) distribution.

SMALL SCALE RESULTS ($r \lesssim 10h^{-1}$Mpc)

Fig. 1 shows our estimate of $\xi_{qq}(r)$ at small scales. Despite the large size of our QSO survey the figures in brackets show that we are still dealing with relatively small numbers of QSO pairs. Nevertheless we still find 25 QSO pairs with

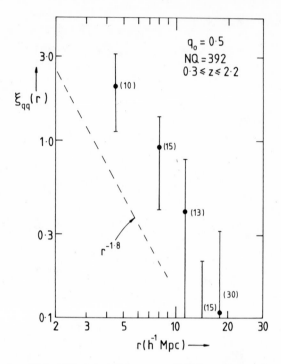

Fig. 1   The QSO 2-point correlation function at small
scales.   Numbers in brackets refer to the QSO pair count.

separation less than the canonical value of $10h^{-1}$Mpc whereas
on a random null hypothesis we would have expected only 11.1
QSO pairs.   Since 23 of the 25 pairs are completely independent, straightforward Poisson statistics suggest that this
represents a 4.2σ result.   The reality of this significance
level is borne out both by simulating random distributions
and by estimated field-to-field errors (see Shanks et al.
1988a).   Although the statistics are poor, Fig. 1 shows that
the <u>form</u> of the QSO correlation function is at least not
inconsistent with the -1.8 power-law expected for galaxies.
The value for ξ_{qq} averaged out over a $10h^{-1}$Mpc radius is

$$\xi_{qq}(r \leq 10h^{-1}\text{Mpc}) = 1.25 \pm 0.45$$

This amplitude is in reasonable agreement with the tentative
result found by Shaver (1984) in his analysis of the inhomo-
genous Veron-Cetty and Veron (1984) catalogue.
    We next compare the observed amplitude of $\xi_{qq}(r)$ with
the amplitude expected for galaxies at our average redshift,
z=1.4.   Based on our initial sample of 172 QSOs we claimed
that the clustering amplitude was higher than expected for

galaxies at z=1.4 (Shanks et al. 1987). This result is still
essentially true but here we emphasise two basic uncertainties.
Firstly the amplitude of ξ_{gg} at scales between 2 and $7h^{-1}$Mpc
is still controversial. Recent results from galaxy redshift
surveys suggest that the amplitude of ξ_{gg} is larger between 2
and $7h^{-1}$Mpc than at smaller scales viz:-

$$B = 16 \pm 2 \; h^{1.8}Mpc^{1.8} \; (0 < r \leqslant 2h^{-1}Mpc)$$
$$B = 33 \pm 3 \; h^{1.8}Mpc^{1.8} \; (2 < r \leqslant 7h^{-1}Mpc)$$

(Shanks et al. 1983, 1988b, Davis and Peebles 1983). Secondly
there is uncertainty over the appropriate evolutionary model
for ξ_{gg}. The simplest assumption is that the galaxy
clustering is stable at scales below $7h^{-1}$Mpc where ξ_{gg} is
large and the clustering is strong. In this case the density
contrast (ie strength) of clustering decreases at higher redshifts where the proper QSO background density is higher by
the factor $(1+z)^3$. However in recent times there has been
increasing interest in 'biased' theories of galaxy clustering
where galaxies cluster more strongly than the mass (eg Davis
et al. 1985). This means that galaxy clusters may still be
expanding almost as fast as the Hubble expansion even at
scales where the galaxy clustering is apparently strong. In
this case a 'comoving' model of galaxy clustering is perhaps
a better approximation rather than the above stable model. In
the comoving case the amplitude of galaxy clustering remains
constant as a function of redshift.
    Fig. 2 shows the predicted B_{qq}:z relations for stable
(solid lines) and comoving (dashed lines) models of various
amplitudes. The above uncertainty in the galaxy correlation
function amplitudes is represented by the range between the
B=16 and the B=33 models. The B=360 models represent the
possible evolutionary behaviours of the rich galaxy cluster
correlation function (Hauser and Peebles 1973, Bahcall and
Soneira 1983). Our observed $\xi_{qq} = 1.25 \pm 0.45$ result averaged
over the redshift range $0.3 \leqslant z \leqslant 2.2$ is shown as the dashed
circle. The fact that the observation lies somewhere between
the stable galaxy (B=16) and stable cluster (B=360) models
was the basis for our previous statement (Shanks et al 1987)
that the QSO clustering is intermediate in strength between
galaxy clustering and cluster clustering. Intuitively this
seems suggestive of a group environment model for QSOs and
this interpretation finds support in the work of Yee and
Green (1987) who found tentative evidence for such a result
from CCD imaging observations of low redshift, radio-quiet
QSOs. As can be seen in Fig. 2 a stable evolutionary model
with a zero redshift amplitude of B=85 gives a reasonable
representation of the QSO clustering amplitude at z=1.4.
    However, it can be seen that if we now allow the higher
B=33 result for the z=0 galaxy clustering amplitude and also

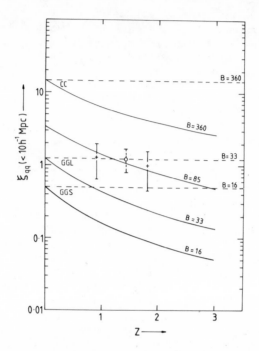

Fig. 2   The amplitude of the QSO correlation function
($q_0 = 0.5$) as a function of redshift.   The solid and dashed
lines represent stable and comoving models respectively.
GGS and GGL represent models based on the range of
amplitudes quoted for the galaxy correlation function.
The CC model is based on the rich cluster correlation
function.

adopt a comoving evolutionary model the agreement with our
observed z=1.4 QSO amplitude is excellent.   Thus a model
where QSOs randomly sample an entirely comoving galaxy distribution is also consistent with our observations of QSO
clustering at z=1.4.

We can try to discriminate between the stable B=85 group
environment model and the comoving B=33 galaxy clustering
model by making separate correlation analyses in the redshift
ranges $0.3 \leqslant z \leqslant 1.4$ and $1.4 < z \leqslant 2.2$.   The 2 crosses in Fig. 2
represent the results of this experiment.   The ratios of
observed to expected QSO pairs was 12/5.2 and 13/6.5 in the
low and high redshift ranges respectively.   As can be seen the
observed change in ξ_{qq} is quite small between these 2 redshift
bins.   But although the comoving model therefore seems to be
in good agreement with the data, the stable model also only
predicts a small amount of evolution in this range and so

cannot be rejected.

We note here that the amount of clustering evolution seen in our data is much smaller than that claimed by either Kruszewski (1988) or Shaver (1988). For example, Kruszewski analysed several slitless QSO surveys and claimed ξ_{qq}(r<16h^{-1} Mpc) dropped from $\xi_{qq} \sim 2.5$ at $z \sim 0.5$ to $\xi_{qq} \sim 0$ at $z \sim 2$. Adopting Kruszewski's bin size we find that this reduces our clustering signal at all redshifts (as might be expected from the fast fall in our ξ_{qq} shown in Fig. 1) and this tends to improve the agreement with Kruszewski's results at higher redshifts. The main difference left between Kruszewski's result and ours is that he sees high correlations at $z \sim 0.5$ whereas we do not. However, because of the small numbers of QSO pairs found at low redshift in both datasets this difference is hardly statistically significant. Shaver's claim of detecting strong evolution also depends on poorly determined QSO pair counts at low redshift. Although by combining all results together there may be some tentative evidence for mild evolution of the QSO clustering, perhaps consistent with the stable model prediction, our overall conclusion is that it may be premature to rule out any model of QSO clustering evolution based on present data.

Direct CCD observations of the environments of low redshift QSOs may also help discriminate between models for ξ_{qq}. Yee and Green (1987) and Boyle et al. (1988b) find that the radio-quiet quasar-galaxy cross-correlation amplitude B_{qg} is about 2.5 times as big as expected on a stable evolutionary model (with $B_{gg}(z=0)=16$ at $r \sim 1h^{-1}$Mpc). Since at z=1.4 our results suggest that $B_{qq}/B_{gg}=2.6$ (assuming $B_{gg}=33$ is appropriate in the range $2 \le r \le 7h^{-1}$Mpc) then there is consistency with a stable group environment model for ξ_{qg} and ξ_{qq} at small and larger scales respectively. The fact that Yee and Green and Boyle et al. observe B_{qg}/B_{gg} remaining constant with redshift over the range $0 \le z \le 0.7$ under stable model assumptions again is consistent with the stable group environment model. On the other hand a comoving model would require B_{qg}/B_{gg} (stable) to increase quickly with redshift and this is not observed; essentially it is very difficult for the comoving galaxy clustering model to explain the high B_{qg}/B_{gg} ratio seen by Yee and Green at $z \le 0.3$ where the dependence on evolutionary model is small.

Now it must be said that the quasar-galaxy clustering could be stable at the $1h^{-1}$Mpc separations where ξ_{qg} is measured and the quasar-quasar clustering at larger scales might still be comoving. However, the natural-looking fit of the comoving galaxy clustering model in Fig. 2 does depend on the assumption that QSOs randomly sample the galaxy distribution; the suggestion from the CCD imaging observations at low redshift that QSOs do occupy special, group, environments leaves the 'comoving' explanation of the strong QSO clustering

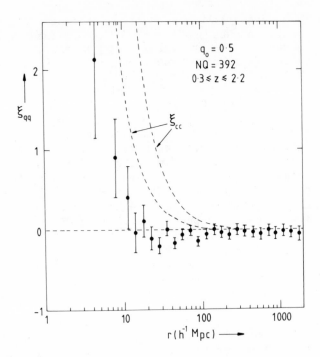

Fig. 3   The QSO 2-point correlation function at large
scales.

looking more implausible. Whether this amounts to a rejection
of biased galaxy formation models is not yet clear.

LARGE SCALE   RESULTS

The results for $\xi_{qq}(r)$ for $0 < r \leqslant 2000h^{-1}$Mpc ($q_o$=0.5) are
shown in Fig. 3. Below r=45h^{-1}Mpc we find that $\sqrt{N}$ errors on
the number of QSO pairs give excellent agreement with the
observed field-to-field dispersion in ξ_{qq}. Beyond
r=45h^{-1}Mpc $\sqrt{N}$ errors increasingly underestimate the field-to-
field errors and in this range the errors shown in Fig. 3 have
been renormalised to take account of this effect. It can be
seen in Fig. 3 that there is no significant detection of
clustering at any scale with r>10h^{-1}Mpc. Thus this result
seems to imply that the Universe appears homogeneous on scales
$10 \leqslant r \leqslant 2000h^{-1}$Mpc. The dashed lines in Fig. 3 represent the
probable range of evolutionary models for the rich cluster
correlation function but the scale-length of the observed QSO
clustering seems much smaller than in either of these models.
In general the QSO correlation function at scales

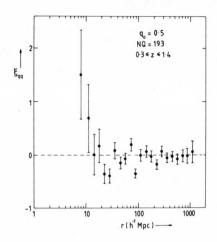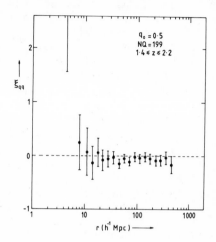

Fig. 4(a) The QSO 2-point correlation function at low redshifts.

Fig. 4(b) The QSO 2-point correlation function at high redshifts.

$10 \lesssim r \lesssim 100 h^{-1} Mpc$ is very similar to the galaxy correlation function derived from galaxy redshift surveys (Shanks et al. 1983, Shanks et al. 1988b). Thus it seems that there is less evidence for strongly positive mass correlations on scales of $100 h^{-1} Mpc$ if galaxies or QSOs prove better tracers of the mass distribution than rich galaxy clusters.

However, it must be remembered that ξ_{qq} is being measured at our average redshift of 1.4 and more significant large-scale clustering may have evolved between then and the present day. Indeed in $\Omega_o^*=1$ gravitational instability models small features in ξ are expected to grow as $1/(1+z)^2$ (eg. Peebles 1980) which implies they will grow by a factor of 6 between z=1.4 and z=0. For low Ω_o models no evolutionary growth is expected. Thus if some development in $\xi_{qq}(r)$ is observed with redshift then this might be taken as a possible indication that $\Omega_o=1$.

We therefore show in Figures 4a,b our estimate of $\xi_{qq}(r)$ in the two redshift ranges $0.3 < z \lesssim 1.4$ and $1.4 < z \lesssim 2.2$. The first impression from these figures is that the correlation function at higher redshift appears in some sense 'smoother' in the range $10 \lesssim r \lesssim 200 h^{-1} Mpc$ than its lower redshift counterpart. Although the lower redshift correlation function shows no tendency to develop systematically positive correlations in

* Ω_o is the cosmological density parameter and $\Omega_o = 2q_o$ in Friedmann cosmological models.

this range, the scatter of the data points around zero seems to have increased in Fig. 4a in a way which is not always wholly explained by the errors (eg the point at $r=85h^{-1}$Mpc is a 3σ deviation from zero). However, it is too early to say that these features are real and have evolved with redshift, particularly since the galaxy correlation function shows no evidence for very strong features in this range (Shanks et al. 1988b). But with their important potential as probes of Ω_0 it will be interesting to test for the reality of these features in bigger QSO redshift surveys.

Finally, Shanks et al. (1987) have pointed out that if a low signal-to-noise feature is picked out in the high redshift QSO and low redshift galaxy correlation functions then a further test of q_0 might be possible. Between $z=0$ and $z=1.4$ features in ξ move by 40% in separation depending on whether $q_0=0.1$ or $q_0=0.5$ is assumed in the computation of ξ_{qq}. Thus if such a feature were found then a comparison of its position at low and high redshift should provide a sensitive q_0 test. This q_0 test would be less subject to the usual evolutionary uncertainties since at least on the simplest gravitational instability models the positions of features in ξ_{qq} are not expected to evolve very quickly. Again this q_0 test may motivate increasing the size of QSO (and galaxy) redshift surveys to determine if any such weak features actually exist.

CONCLUSIONS

We summarise the conclusions as follows:-
(a) At small scales ($r \lesssim 10h^{-1}$Mpc) we have made a clear, 4.2σ detection of QSO clustering. The strength of clustering is stronger than expected on stable evolutionary models of galaxy clustering and suggest a stable group environment model for QSOs, in agreement with the results of Yee and Green (1987). A model where QSOs randomly sample a comoving galaxy distribution is also consistent with our data but fits the low redshift Yee and Green results less well. We find that our data does not show any evidence for the strong evolution of QSO clustering claimed by Kruszewski (1988) or Shaver (1988) but because of statistical uncertainties we cannot rule out the existence of mild evolution in QSO clustering properties.
(b) At large scales ($10 \lesssim r \lesssim 2000h^{-1}$Mpc) we see no evidence of QSO clustering in our $0.3 \lesssim z \lesssim 2.2$ sample. This is consistent with the idea that the Universe is homogeneous at the 10% level at our average redshift of $z=1.4$ We have looked for evolution in ξ_{qq} at large scales and find that for $z<1.4$ ξ_{qq} is less smooth in the range $10 \lesssim r \lesssim 200h^{-1}$Mpc than at higher redshift. However, the lower redshift points still scatter around zero and a bigger survey will be required before this result can be confirmed. If large-scale evolution in $\xi_{qq}(r)$

were detected then it could provide an indication for a high
Ω_0 Universe. Since   another cosmological test may also be
possible based on the position of any weak large-scale feature
that may be detected, it is important to increase the size of
QSO redshift surveys to place further constraints on the
existence of any such features.

ACKNOWLEDGEMENTS

We thank the staff of the Royal Observatory, Edinburgh for
providing UK Schmidt photographs and COSMOS machine measurements. We also thank the staff of the Anglo-Australian
Observatory for support of the fibre optic observations. We
also thank Dr. R. Fong for useful discussions. TS is
supported by a Royal Society 1983 University fellowship. BJB
was supported by an SERC studentship and an SERC fellowship
during the course of this work.

REFERENCES

Bahcall,N.A. and Soneira, R.M. 1983, Ap.J., 270, 20.
Boyle, B.J., Fong, R., Shanks, T. and Peterson, B.A. 1988a,
    in preparation.
Boyle, B.J., Shanks, T. and Yee, H.K.C. 1988b, in preparation.
Davis, M. and Peebles, P.J.E. 1983, Ap.J., 267, 465.
Davis, M., Efstathiou, G., Frenk, C.S. and White, S,D,M, 1985,
    Ap.J., 221, 371.
Hauser, M.G. and Peebles, P.J.E. 1973, Ap.J., 185, 757.
Kruszewski, A. 1988 preprint.
Osmer, P.S. 1981, Ap.J., 247, 762.
Peebles, P.J.E. 1980, "The Large Scale Structure of the
    Universe", Princeton: Wiley.
Shanks, T., Bean, A.J., Efstathiou, G., Ellis, R.S., Fong, R.
    and Peterson, B.A. 1983, Ap.J., 274, 529.
Shanks, T., Fong, R., Boyle, B.J. and Peterson, B.A. 1987,
    M.N.R.A.S., 227, 739.
Shanks, T., Boyle, B.J. and Peterson, B.A. 1988a in
    preparation.
Shanks, T., Hale-Sutton, D., Metcalfe, N., and Fong, R. 1988b
    M.N.R.A.S. submitted.
Shaver, P.A. 1984, Astr. Astrophys., 136, L9.
Shaver, P.A. 1988 preprint.
Veron-Cetty, M.P. and Veron, P. 1984, ESO Sci. Report No.1.
Yee, H.K.C. and Green, R.F. 1987, Ap.J., 319, 28.

QUASAR CLUSTERING IN THE CFHT/MMT SURVEY

DAVID CRAMPTON
Dominion Astrophysical Observatory, National Research
Council, 5071 W. Saanich Rd., Victoria, B.C. V8X 4M6

A.P. COWLEY
Department of Physics, Arizona State University, Tempe,
AZ 85281

F.D.A. HARTWICK
Department of Physics, University of Victoria,
P.O. Box 1700, Victoria, B.C. V8W 2Y2

ABSTRACT The space distribution of quasars selected
from CFHT grens plates is analysed for indications of
clustering among the quasars. Although a few isolated
groups and close pairs of quasars have been recognized,
clustering does not appear to be common. In the
direction 1338+27 a group, containing at least 16
members with a mean z = 1.113 and a dispersion of 0.04
was found, as well as a pair with z = 2.030 and
separation 9.5 arcsec, and a pair with z ~ 1.32 and a
separation of 95 arcsec.

INTRODUCTION

We are conducting an extensive survey of quasars complete to
m < 20.5 with the CFHT blue "grens" and the MMT. (Crampton,
Schade and Cowley 1985; paper I, Crampton, Cowley and
Hartwick 1987, hereafter paper II). The grens survey,
searches, and reduction techniques have been described in
detail in papers I and II. Briefly, the grens plates are all
visually scanned several times to select quasar 'candidates'.
The candidate objects display emission lines or very blue
continua and are generally not spatially extended. The
candidates are assigned classes 1 - 4 depending on the
certainty with which they can be classified. Class 1 objects
display strong emission features, the strongest of which can
usually be identified as Ly α and a redshift estimated.
Class 4 candidates display a blue continuum with weak or
absent emission features. The spectra of the candidates are
then scanned with a PDS microdensitometer and a final

classification of the objects made through examination of the
resulting spectral scans and a final visual inspection of the
plates. Allowance for plate flaws, overlapping orders, etc.
is made in this procedure. Statistics from our previous
surveys indicate that 100% of class 1 objects are quasars,
94% of class 2 objects, 82% of class 3 objects and 47% of
class 4 objects are quasars. Our experience shows that the
main contaminants in our candidate lists are white dwarfs and
blue subdwarfs.

Quasars detected in this way appear to be relatively
free of any selection effects dependent on redshift. The
distribution of redshifts as determined from slit spectra is
shown in Figure 1. The redshift ranges for which the strong
Ly α, C IV, C III], and Mg II lines are in the grens bandpass
are shown at the bottom of the figure. Note that Ly α does
not appear on the grens plates for about two-thirds of these
quasars. The Mg lines at $\lambda 2800\text{Å}$ are not in the spectral
region covered by the grens spectra for $z < 0.25$, and objects
which appear to be extended are not included either, so most
very low redshift 'quasars' are omitted from the survey.
Ly α is redshifted beyond the IIIaJ plate limit at $z \sim 3.4$ so
quasars with higher redshifts are difficult to detect using
our technique, although not impossible since Ly β + OVI
emission is sometimes visible. The higher spectral
resolution provided by the blue grens in good seeing
conditions allows weaker emission lines to be detected than

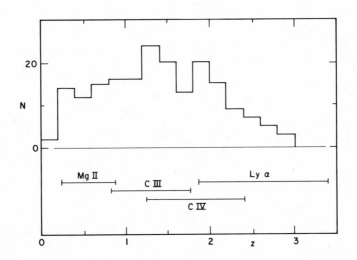

Fig. 1   The redshift distribution for all quasars
discovered with the blue grens and confirmed by slit
spectra. The bars at the bottom indicate the redshift ranges for which the strong emission lines
indicated are visible on the discovery plates.

was the case in previous slitless spectral surveys. Quasars
with weak emission lines were detected at all magnitudes to
the limit of our MMT observations.

An analysis (paper II) of the space distribution of 117
quasars in the first eight 0.52 deg^2 fields completed showed
no evidence for clustering in general on any preferred scale.
One group of seven quasars in the 1338+27 field with z = 1.16
represented an exception, however, since statistical tests
indicate that such a grouping is unlikely to occur by chance.
Consequently, several more grens plates of the surrounding
region were obtained to search for additional group members.
A comparison region near 1639+40 was also observed. Although
hampered by poor weather, the 1338+27 survey region
containing the quasar group was extended to an area of ~2x3
square degrees where more than 500 quasar candidates have now
been identified (Crampton, Cowley, Schmidtke, Janson and
Durrell 1988). The approximate boundaries of the two areas
included in the new extended survey are shown in Figure 2.

Slit spectroscopy of all candidates brighter than m =
20.5, approximately one magnitude brighter than the faintest
quasar candidates detectable on our plates, has been carried
out with the photon-counting spectrograph on the MMT. The
spectral resolution is ~7Å and the spectra have low
signal-to-noise so the accuracy of the redshift
determinations is only ~0.01. So far, only regions of
~2.5 deg^2 surrounding the original field at 1338+27, and a

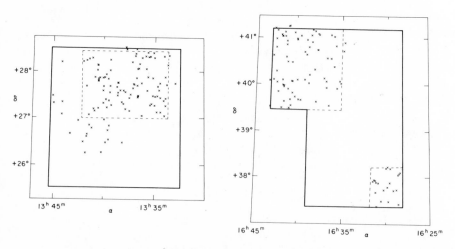

Fig. 2   The areas of sky covered in the surveys of the
1338+27 and 1639+40 fields. Spectroscopic observations
of the quasars are complete to m = 20.5. The crosses
show the positions of verified quasars.

comparable field centered on 1639+40 have been completed.
Combined with results from our previous surveys, redshifts
are now available for 119 objects in the 1338+27 field and 97
in the 1639+40 field. The positions of these confirmed
quasars are marked by the crosses in Figure 2. For
computational ease, the analysis of quasar clustering was
restricted to the 92 quasars in a rectangular 2.7 deg^2 area
of the 1338+27 field for which observations are relatively
complete, and to 60 in a similar 2.5 deg2 area of the 1639+40
field.

CLOSE PAIRS OF QUASARS AND THE 1338+27 GROUP

1343.4+2640 Pair
One very close pair (separation 9.5 arcsec) of quasars with
similar redshift in the 1338+27 field was immediately obvious
during examination of the grens plates. Subsequent
observations (Crampton et al. 1987) showed that the quasars
have identical redshift, z = 2.030, but their spectra are
sufficiently different that they are more likely to be just a
physically related pair rather than the result of a
gravitational lens. The measured velocity difference between
the two quasars is only ~120 km s^{-1}, and the projected linear
separation is only 39h^{-1} kpc (H$_0$ = 100h km s^{-1} Mpc^{-1}). They
could be located in a bound group of galaxies, but such
galaxies are unlikely to be detectable at this redshift
(Bahcall, Bahcall and Schneider 1986).

1336.5+2804 Pair
The next closest pair of quasars with similar redshift (z ~
1.32) has a separation of 95 arcsec. The rather poor spectra
we have of these quasars do not rule out identical redshifts,
although the measured difference is Δz ~ 0.01. If at the
same distance, the linear separation of these quasars in
comoving coordinates is ~400 h^{-1} kpc.

1338+27 Group(s)
One principal aim of our recent survey was to investigate
whether the grouping of quasars with z ~ 1.16 in the 1338+27
direction extended beyond the boundaries of the initial
survey. Apparently it does, although we do not yet know its
full extent. The redshift distributions of the quasars in
1338+27 and the 1639+40 field are shown in Figure 3. The
excess at z = 1.1 in the 1338+27 field is readily apparent.
As seen projected on the sky, the concentration of the
quasars of similar redshift is even more striking (Figure 4).
The quasars at z ~ 1.13 appear to be concentrated into two
subgroups; the five in the NW subgroup have a mean z = 1.113
with a dispersion of 0.016, and the 11 in the other subgroup

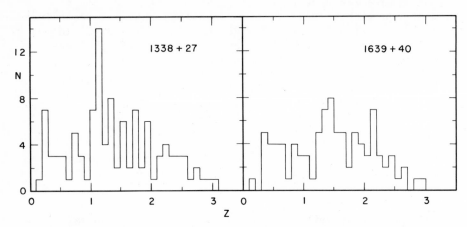

Fig. 3   The redshift distribution for quasars in the
1338+27 and 1639+40 fields.

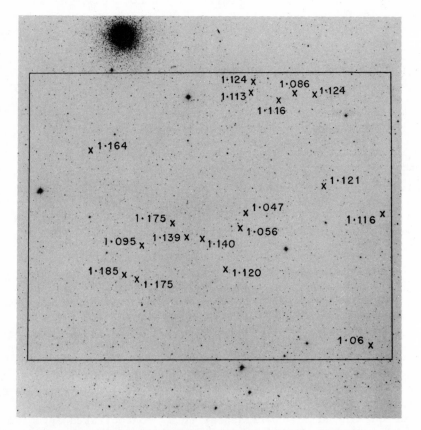

Fig. 4   The positions of quasars with redshifts near
z = 1.13 shown projected on the sky. M3 is at the top.

have a mean z = 1.124 and a dispersion of 0.046. The angular
extents of these subgroups correspond to linear dimensions in
comoving coordinates of ~2 x 10h^{-1} Mpc and 10 x 50 h^{-1} Mpc
respectively. The coordinates and magnitudes of the 16
quasars in these two subgroups are listed in Table 1. No
other similar concentrations of quasars have been found,
either in any of our other 8 complete fields (see paper II),
or by other surveys.

Current ideas suggest that a large number of associated
galaxies can be expected near these quasars, a sizeable
fraction of which are likely to be interacting. No extensive
deep optical imagery of the region has so far been carried
out, but Djorgovski (private communication) has obtained CCD
images showing several field galaxies of the expected
brightness and size in one small part of this field.
Additional observations are underway.

TABLE I   Members of the z = 1.13 Group

Name	m	z	
		Subgroup 1	
1335.3+2820	20.2	1.124	
1335.8+2820	20.4	1.086	
1336.2+2818	19.8	1.116	
1336.8+2823	19.7	1.124	
1336.8+2820	19.0	1.113	
Mean	19.8	1.113	disp 0.016
		Subgroup 2	
1333.7+2743	19.4	1.116	
1335.1+2752	19.2	1.121	
1336.9+2743	19.1	1.047	
1337.1+2738	20.2	1.056	
1337.4+2726	17.5	1.12	
1338.0+2735	19.0	1.140	
1338.4+2736	20.0	1.139	
1338.7+2740	19.2	1.175	
1339.4+2733	20.4	1.095	
1339.5+2723	20.1	1.175	
1339.8+2725	19.0	1.185	
Mean (11 quasars)	19.4	1.124	disp 0.046

CLUSTERING

Our previous analysis (paper II) of separations between
quasars in eight relatively small fields showed no evidence
for large scale clustering when compared to random
distributions (in agreement with previous results, cf. Osmer
1981). We have recently (Crampton, Cowley and Hartwick 1988)
analysed the data on the quasars in the two extended fields
in a similar manner, although the results are presented
slightly differently. A representative redshift distribution
(a smoothed version of the distribution shown in figure 1)
was first constructed from the data in all our complete
fields. Next, redshifts were chosen at random from this
distribution and assigned randomly chosen angular positions
within each survey area. Three-dimensional separations were
then computed in comoving coordinates from each member to all
others in these artificial surveys assuming an $\Omega_0 = 1$,
$H_0 = 100h$ km s^{-1} Mpc^{-1} cosmology. The results from 100 such
artificial samples for each field were averaged to form the
two-point correlation function, $X_i(r)$. Since no significant
differences were found between the results for the two
fields, the results were averaged. The correlation function
is shown as a function of the separation between the quasars
in Figure 5 for the combined sample. There is no evidence
for significant clustering at any scale, in agreement with
the results found previously for the eight smaller fields.
Shaver (1987) has recently summarized evidence for clustering
on the smallest scales considered here, 10 h^{-1} Mpc and has
suggested that such clustering is more pronounced for z <
1.5. Although the errors are large, the two-point

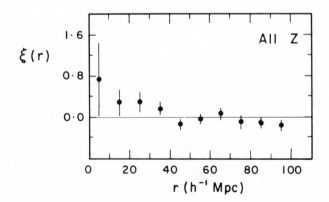

Fig. 5   The two-point correlation function for quasars
in the 1338+27 and 1639+40 fields plotted as a function
of separation.

correlation function for a separation of 10 h^{-1} Mpc computed
from our data is not significantly different from zero, and
is considerably smaller than that found by Shaver. Shaver
(1987) and Kruszewski (1987) also give evidence that the
amplitude of the correlation function for small separations
is a function of redshift. In Figure 6 our results are shown
as a function of three equal intervals of "growth time", i.e.
since perturbations should grow as $(1 + z)^{-1}$, the redshift
intervals have been divided in this manner. Although some
points (namely at 25 and 65 Mpc) in the intermediate range
(middle panel) deviate from zero by about 2σ, there are no
obvious trends as a function of redshift. For r > 10 h^{-1} Mpc
our conclusions are similar to those of Shanks et al. (1987)
who found no significant evidence for clustering but
tentatively point out a possible small amplitude "trough –

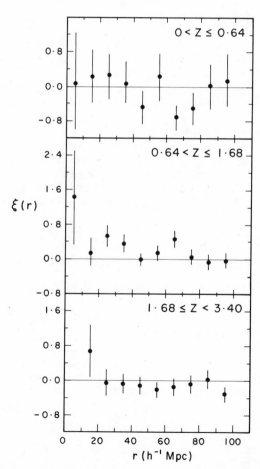

Fig. 6   The two-point correlation function for
different redshift ranges.

peak - trough - peak" form for the correlation function with the peaks occurring at 28 and 56 h^{-1} Mpc. Examination of the pairs contributing to the r ~ 25 h^{-1} Mpc feature in our data indicates that most belong to the group at z ~ 1.13 in the 1338+27 field which we drew attention to above. On the other hand most of the excess pairs at r ~ 65 h^{-1} Mpc can apparently be attributed to quasars with 1.35 < z < 1.65 in the 1639+40 field. The fact that the features in the correlation function are only marginally significant, that they can be identified with a single grouping in each case, and that there is no sign of similar features in the adjacent redshift ranges seems consistent with the idea that significant isolated groupings of quasars may exist, but that clustering on large scales is not a general phenomenon.

SUMMARY

In agreement with our analysis of the clustering of quasars in eight small fields, analysis of a comparable number in two larger fields indicates that clustering is not common. Isolated pairs and groupings do exist however, and our recent observations have confirmed and increased the number of members of the z = 1.13 group at 1338+27. This group and a complex of quasars in the 1639+40 field have separations with characteristic sizes similar to those observed by other workers. Although it is tempting to correlate these peaks with predictions by some theoretical models of the early universe, an order of magnitude more data is required to obtain reliable statistical evidence of quasar clustering.

REFERENCES

Bahcall, J.N., Bahcall, N.A., and Schneider, D.P. 1986, Nature, 323, 515.
Crampton, D., Schade, D., and Cowley, A.P. 1985, A.J., 90, 987.(paper I)
Crampton, D., Cowley, A.P., and Hartwick, F.D.A. 1987, Ap.J., 314, 129. (paper II)
Crampton, D., Cowley, A.P., and Hartwick, F.D.A. 1988, in preparation.
Crampton, D., Cowley, A.P., Schmidtke, P., Janson, T., and Durrell, P. 1988, in preparation.
Crampton, D., Cowley, A.P., Hickson, P., Kindl, E., Wagner, M., Tyson, J.A., and Gullixson, C. 1987, preprint.
Kruszewski, A. 1987, preprint
Shanks, T., Fong, R., Boyle, B.J., and Peterson, B.A. 1987, M.N., 227, 739.
Shaver, P.A. 1987, in NATO Symposium on The Post-Recombination Universe.

DISCUSSION

Surdej : Is there a significant difference in your pair of quasars with a separation of 9.5"?

Crampton : The velocity difference of 120 km/sec is not significant.

Surdej : So why do you claim that it's not a possible case of gravitational lensing?

Crampton : Because the spectra look so different.

Surdej : What about microlensing? I understand that it can produce different line strengths in the different images.

Turner : Microlensing is more likely to affect the line strengths than the widths. I wanted to raise a different question on the error bars you and the previous speaker have used for the covariance function. Two potential problems are that the arrangement of groups can affect the calculations (a large single group vs a collection of smaller groups) and that if clustering does occur, then each point is no longer independent.

Osmer : I agree that the group David showed is very tantalizing. I am surprised (and a little disturbed) that it doesn't produce a significant signal in the analyses.

Shanks : Just to reassure Ed Turner, of the 25 pairs I talked about with separations less than 10 Mpc, 23 were completely independent. That's why we used root n statistics as an approximation.

Surdej : When you select optical candidates, there is a level at which doubt sets in about whether or not to include an object. What do you do about marginal candidates?

Crampton : We select candidates down to the faint limit, about mag 21.3, but the data to which I referred only go down to 20.5. We just don't have time to observe all the faint objects.

Cowley : We rank candidates in four groups and find that the success rate varies from 100% in the best group to less than half in the lowest group.

Crampton : I can also point out that our surface density is as high as anyone's down to mag 20.5, 29 per sq deg, so that if we are missing objects, so is everyone else.

Boyle : To reiterate a point Pat made, I think it's sobering to see what looks like such a wonderful cluster of quasars on the sky turn up at only 2 sigma in your correlation function. I think it should warn people against overinterpreting peaks in n(z) distributions. Tom made the point that we see a number of fields where we have 5 or more quasars separated by a small interval in redshift, 20 Mpc or so, but they don't produce a significant signal in the correlation function. Now, this may be a problem with the sensitivity of the correlation function, but it may also be a signal to beware of apparent groups on the sky.

Green : What's the status of optical magnitudes for these objects?

Crampton : Optical magnitudes have been the worst part of our data. We have had to base them on Palomar Schmidt plates, which only go down to about mag 21.

Turner : As a veteran of the galaxy covariance function business, I think it is important to understand what a complicated and in some respects dull tool the covariance function is for studying these problems. To give an example, imagine that the real quasar clustering function has a lot of power, but it is manifest in an occasional, rare cluster with a lot of members. In addition, there is a population of loose binary quasars. Then, if the observational sample does not happen to contain one of the rich clusters, the analysis will not show any evidence for clustering and indicate that the limits are quite tight. It all depends on what's in your sample.

Shanks : That's true. On the other hand, if you are only concerned with rejecting the null hypothesis of no clustering, then the error estimates we derive are in general correct.

Turner : I agree that the correlation function is a good test against the hypothesis of a smooth distribution.

QUASAR CLUSTERING IN PERSPECTIVE

PETER A. SHAVER
European Southern Observatory, Karl-Schwarzschild-Str. 2,
D-8046 Garching bei München, Federal Republic of Germany

ABSTRACT   A review is given of measurements of quasar
clustering to date, in the context of other indications
of large-scale structure at low and high redshifts. It is
suggested that existing quasar samples be combined to
provide the most accurate possible measure of quasar
clustering and its evolution.

INTRODUCTION

The evolution of large-scale structure in the universe is of
great interest for a variety of reasons. It has a bearing on
fundamental properties of the early universe, the presence and
composition of dark matter, the mass density of the universe,
and the formation of galaxies, clusters, and superclusters.
Until recently, our knowledge of large-scale structure was
limited to very low redshifts (galaxy clustering and super-
clustering at $z \sim 0$), and very high redshifts (the microwave
background at $z \sim 1000$), and nothing was known about structure
in the intervening 99%. Quasars have long offered the promise
of exploring the evolution of structure over most of the
history of the universe, and quasar samples are now large and
deep enough that this promise can be fulfilled.

DETECTIONS OF QUASAR CLUSTERING

Rapid progress has been achieved in the measurement of quasar
clustering, as summarized in Table 1. Variations on two basic
methods have been used. In one, a random sample subject to the
same selection effects as the observed sample is generated by
Monte Carlo techniques; comparison of these two samples as a
function of linear separation of pairs of quasars then yields
the correlation function. In the other it is assumed that the
clustering is predominantly on small scales, so that it can be
detected by comparing the incidence of quasar pairs of small
projected and/or radial separation with those of large

TABLE I, Measurements of the Quasar Correlation Function

Sample	Method	Reference
Véron Catalogue *	2a	Shaver (1984)
UVX QSOs (172)	1a	Boyle (1986); Shanks et al. (1987)
Véron Catalogue *	2b	Kruszewski (1988a); Shaver (1988a)
3 Homogeneous Samples (376)	1a	Iovino & Shaver (1988)
9 Homogeneous Samples (629)	1c	Kruszewski (1988b)
UVX QSOs (398)	1a	Shanks et al. (1988)
Hewitt-Burbidge Catalogue **	2a	Zhu & Chu (1988); Chu & Zhu (1988)
Véron Catalogue *	1b	Anderson et al. (1988)

* Véron-Cetty & Véron (1987)
** Hewitt & Burbidge (1987)
(Numbers in parentheses indicate sample size)

Methods

1. Random Comparison Sample
        (a) 3D randomization
        (b) z randomization
        (c) z scrambling

2. Normalization to Large Scales
        (a) radial and tangential coordinates separately
        (b) radial and tangential coordinates combined

separation; both groups are subject to the same selection effects, which therefore cancel in the comparison, so the method can be applied to large but inhomogeneous catalogues. Small-scale clustering in such catalogues can also be studied through the use of random comparison samples generated by randomly "jiggling" the redshifts, enough to destroy any real clustering, but not so much as to significantly violate the selection envelopes.

Quasar clustering has now been confirmed at the 5σ level. Several independent studies are listed in Table 1. There are also indications of possible evolution from several of these

studies (Kruszewski 1988a,b; Shaver 1988a,b; Iovino & Shaver
1988; Anderson et al. 1988), as shown in fig. 1, although this
is disputed by Shanks et al. (1988) on the basis of their UVX

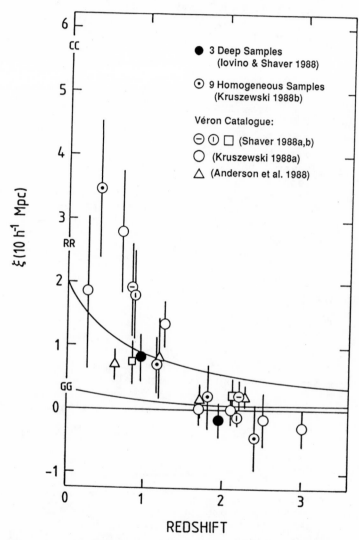

Fig. 1   Amplitude of the quasar correlation function at
10 h^{-1} Mpc (comoving, q$_0$ = 0.5), as a function of redshift, from several different studies as indicated. On
the left axis are marked the amplitudes of the correlation function at z $\lesssim$ 0.1 for galaxies (GG), clusters (CC
- Bahcall & Soneira 1983), and radio galaxies (RR -
Peacock et al. 1988). The two curves show the expected
evolution for stable clustering.

sample. Certainly the magnitude of any such evolution is not yet well determined, and larger samples are clearly required.

Another question of considerable interest is whether radio-loud and radio-quiet quasars have the same clustering properties. Chu and Zhu (1988) have suggested that radio quasars are more strongly clustered, although because they considered all redshifts together, and radio quasars are concentrated to lower redshifts where clustering appears to be stronger, it is not clear whether their result is due to the redshifts or the radio properties. An analysis has therefore been made using only quasars with confirmed redshifts in the range $0.1 < z < 1.5$ from the Véron catalogue. At linear separations < 5 h^{-1} Mpc (comoving, $H_0 = 100h$ km s^{-1} Mpc^{-1}, $q_0 = 0.5$) there should be 0.6 quasar pairs with at least one member radio-loud ($S_{6Ghz} > 0.1$ Jy) if they are clustered like radio-quiet quasars. Instead, 4 such pairs (excluding gravitational lens candidates) are found. The probability of such an excess occurring is $< 10^{-3}$, and the 4 pairs comprising it do not have unusual properties or redshifts. One would obviously feel more comfortable with a larger statistical margin, but radio quasars are rare, and it will be some time before this tentative result can be placed on a more secure foundation.

For most of this work gravitational lensing is probably unimportant, but it is a potential source of contamination on small scales ($\lesssim 0.1$ h^{-1} Mpc). The reverse problem is far more severe, and it is very difficult to distinguish bona fide cases of gravitational lensing from close physical pairs of quasars. However, if the clustering of quasars on all scales increases towards low redshifts, then at least a statistical distinction can be made, as gravitational lensing should be more prevalent amongst high redshift quasars. Preliminary studies (Shaver 1988b) suggest that such a distinction can already be discerned.

COMPARISON WITH OTHER MEASURES OF STRUCTURE

There is a rapidly growing body of information about large-scale structure at low and high redshifts. At the lowest redshifts an apparent structural hierarchy has revealed itself. Redshift surveys (e.g. de Lapparent, Geller & Huchra 1988) show that galaxies appear to be distributed on the surfaces of large (20-25 h^{-1} Mpc diameter) voids which are underdense by at least an order of magnitude. This structure implies a minimum correlation amplitude. Gas-rich dwarf galaxies which manifest themselves as "HII-galaxies" have a correlation amplitude one-third that of normal spirals (Iovino, Melnick & Shaver 1988). On the other hand, elliptical galaxies, located preferentially in clusters, are more

strongly clustered than spirals on scales < 10 h^{-1} Mpc (Davis
& Geller 1976). Still more strongly clustered are radio
galaxies of Fanaroff-Riley class I, which are found in the
cores of Abell richness class 0 clusters (Peacock et al.
1988). And finally, clusters and superclusters of galaxies
appear to be most strongly clustered (Bahcall & Soneira 1983;
Bahcall & Burgett 1986).

The environments of quasars at modest redshifts ($z \sim 0.5$)
have been examined, and it is found that quasars generally
have companions, with the quasar-galaxy correlation amplitude
intermediate between those of galaxies and clusters. Radio-
loud quasars have the richest environments, comparable in some
cases to Abell clusters (Yee & Green 1987, and references
therein). This is consistent with the observed clustering of
quasars amongst themselves. Given their gregarious nature it
is not unexpected that they cluster more strongly than
galaxies generally (fig. 1). And as radio quasars appear to
have the richest environments, they may be expected to be
associated with the strongest clustering, as is observed.

A parallel situation may apply at high redshifts,
although here the evidence is of course less direct. The Lyα
forest absorption lines in quasar spectra exhibit weak
clustering (Webb et al. 1984) relative to the extrapolated
galaxy correlation (assuming stable clustering), consistent,
for example, with their being associated with dwarf galaxies.
Heavy-element absorption lines, thought to be associated with
normal galaxies, are more strongly clustered than the
extrapolated galaxy correlation (Sargent et al. 1980); this
difference could be due to the presence of multiple discrete
absorbing clouds in galactic halos. There is a strong excess
of these heavy-element lines, and a paucity of Lyα lines, near
the redshifts of radio quasars (Foltz et al. 1986; Bechtold
1988), presumably due to the dense surrounding clusters. Other
evidence for rich environments of high redshift quasars
includes their distorted radio structures (Barthel 1986), and
the high incidence of associated absorption in quasar pairs
(Shaver & Robertson 1984). The parallels with the low redshift
universe are tantalizing, and it will be of great interest to
firmly establish detailed relationships between low and high
redshifts.

Quasars, spanning a large range in redshift, offer the
most immediate prospect of elucidating the evolution of
structure over the history of the universe. However, other
evidence on the evolution of structure can also be gleaned
from galaxy clustering at moderate redshifts (out to
$z \sim 0.5-1.0$), and Lyα clustering at high redshifts ($z \sim 2-4$),
and some comparisons can already be made. In both of these
cases there is tentative evidence for rapid evolution of
structure (Jones et al. 1988; Webb & Carswell 1988), as in the
case of quasars. By so examining the clustering of different

types of objects over a large range of redshifts, it should eventually be possible to understand the evolution of structure on a variety of scales.

PROSPECTS

Our knowledge of quasar clustering can immediately be improved simply by combining the different samples that already exist. Beyond confirming the mere existence of evolution, however, we will clearly want to examine the detailed behavior of this evolution at all redshifts and as a function of many parameters, and to pursue the study of structure to the highest possible redshifts. Fortunately the database for this work is rapidly expanding, and it may be expected that by the mid-1990's, just a decade after the first detections of quasar clustering, the subject will already be well developed.

REFERENCES

Anderson, N., Kunth, D., Sargent, W.L.W. 1988, preprint.
Bahcall, N.A., Burgett, W.S. 1986, Astrophys. J. 300, L35.
Bahcall, N.A., Soneira, R.M. 1983, Astrophys. J. 270, 20.
Barthel, P.D. 1986, in Quasars (ed. G. Swarup & V.K. Kapahi; Reidel, Dordrecht), p. 181.
Bechtold, J. 1988, in High Redshift and Primeval Galaxies, in press.
Boyle, B.J. 1986, thesis, University of Durham.
Chu, Y., Zhu, X. 1988, preprint.
Davis, M., Geller, M.J. 1976, Astrophys. J. 208, 13.
Foltz, C.B., Weymann, R.J., Peterson, B.M., Sun, L., Malkan, M.A., Chaffee, F.H. 1986, Astrophys. J. 307, 504.
Hewitt, A., Burbidge, G. 1987, Astrophys. J. Suppl. 63, 1.
Iovino, A., Melnick, J., Shaver, P.A. 1988, Astrophys. J. (submitted).
Iovino, A., Shaver, P.A. 1988, Astrophys. J. (submitted).
Jones, L.R., Shanks, T., Fong, R. 1988, in Evolution of Large Scale Structures in the Universe (ed. J. Audouze & A. Szalay; Reidel, Dordrecht), in press.
Kruszewski, A. 1988a, preprint.
Kruszewski, A. 1988b, in Proceedings of 2nd Ringberg Workshop on High Energy Astrophysics, in press.
de Lapparent, V., Geller, M.J., Huchra, J.P. 1988, Astrophys. J. (in press).
Peacock, J.A., Miller, L., Collins, C.A., Nicholson, D., Lilly, S.J. 1988, in Evolution of Large Scale Structures in the Universe (ed. J. Audouze & A. Szalay; Reidel, Dordrecht), in press.

Sargent, W.L.W., Young, P.J., Boksenberg, A., Tytler, D. 1980,
    Astrophys. J. Suppl. 42, 41.
Shanks, T., Fong, R., Boyle, B.J., Peterson, B.A. 1987,
    M.N.R.A.S. 227, 739.
Shanks, T., Hale-Sutton, D., Boyle, B.J. 1988, in Evolution of
    Large Scale Structures in the Universe (ed. J. Audouze &
    A. Szalay; Reidel, Dordrecht), in press.
Shaver, P.A. 1984, Astron. Astrophys. 136, L9.
Shaver, P.A. 1988a, in Evolution of Large Scale Structures in
    the Universe (ed. J. Audouze & A. Szalay; Reidel,
    Dordrecht), in press.
Shaver, P.A. 1988b, in The Post-Recombination Universe (NATO
    ASI), in press.
Véron-Cetty, M.-P., Véron, P. 1987, ESO Scientific Report
    No. 5.
Webb, J.K., Carswell, R.F., Irwin, M.J. 1984, BAAS 16, 733.
Webb, J.K., Carswell, R.F. 1988, in preparation.
Yee, H.K.C., Green, R.F. 1987, Astrophys. J. 319, 28.
Zhu, X., Chu, Y. 1988, in Proceedings of 2nd Ringberg Workshop
    on High Energy Astrophysics, in press.

DISCUSSION

McCarthy : I think you left an important close pair out of
your table. Djorgovski has one with a separation of 3.5",
redshift about 1.2, spectra that look as identical as the
best confirmed lenses, and yet only one member is a radio
source. It looks as good as any lens, but it's almost
certainly a true close
pair.

Shaver : Was it published?

McCarthy : Yes, it was in the June Astrophysical Journal.

Shaver : Thanks.

Boyle : I find it a wee bit disturbing that the confirmation
of the result we obtained with the correlation function comes
from the Iovino and Shaver compilation, which overlaps
considerable with our work. Have you analyzed the data
separately to see if there really is the confirmation?

Shaver : Yes, we do obtain confirmation even when the AAT
data are excluded.

Shanks : We did look for gravitational lenses in our sample
of 400 quasars. While we cannot pick up lenses with
separations less than 5", we do have sensitivity at 5-10",
and we found no lenses.

RESULTS FROM A NEW GRISM SURVEY FOR QUASAR CLUSTERING

PATRICK S. OSMER
Kitt Peak National Observatory, National Optical
Astronomy Observatories, P. O. Box 26732, Tucson,
Arizona 85726

PAUL C. HEWETT
Institute of Astronomy, Madingley Road, Cambridge CB3
OHA England

ABSTRACT   We present results from a blue grism survey
carried out with the CTIO 4m over a 0.5 x 12 deg strip
of sky. The plates were searched visually and then
scanned with the APM. As a result of follow-up
spectroscopy, a uniformly selected sample of 127
quasars with confirmed redshifts in the interval
$0.3 < z < 3.4$ has been assembled. A first analysis
shows no evidence for clustering, with the greatest
sensitivity being on the 50 - 600 Mpc scales.

INTRODUCTION

Quasars continue to offer the only information on the large
scale distribution of matter for redshifts between 1 and 4.
Their own distribution in space, which can be derived from
their redshifts and positions on the sky, is of interest in
itself and, to the extent quasars indicate the distribution
of galaxies, for studies of how galaxy clustering may have
evolved. In addition, brighter quasars allow the
distribution of intervening absorbers along the line of sight
to be probed. Here we report on preliminary results from a
new quasar survey designed specifically to investigate their
distribution in space. Because the sample is based on
machine scans of grism plates, calibrated photometry, and
slit spectroscopy, it will also be of use for other problems
in quasar research.

PROPERTIES OF SURVEY

One objective of the survey was to establish a new quasar
sample independent of all previous ones, which in some cases
have overlaps among themselves and therefore have statistical
interdependences which are difficult to evaluate. Thus, a
field was chosen arbitrarily in the Northern Galactic
Hemisphere at galactic latitude 50 deg and away from known
quasar surveys. The shape of the survey was set to be a thin
connected strip on the sky, as numerical simulations (Osmer
1981) showed such a shape to be sensitive to possible
clustering on a wide range of scales. The survey was carried
out with the CTIO 4m telescope and prime-focus grism, which
has a 30 minute square field of view and yields a nearly
linear dispersion of 1500 A/mm. Sensitized IIIa-J plates
were used. Quasars as faint as magnitude 21.5 are detectable
on the plates. However, because the detection probability
depends on both magnitude and line strength, the detection
limit cannot be described by a single limiting magnitude
(e.g. Osmer and Gratton 1987). The survey field covers 6.1
sq deg in a strip centered at 12^h10^m, $-11°$ (1950). The
extent in declination is 30 arc min, and the range in right
ascension is from 11^h47^m to 12^h36^m.

Searches for clustering require good uniformity in the
survey material and detection process along with enough
information on calibrations to allow corrections for residual
non-uniformity to be derived. In this survey we have made
use of both visual scans of the plates to detect quasar
candidates and scans by the Cambridge APM to produce a
catalog of all measurable spectra. The machine scans also
include UK Schmidt direct plates of the fields; these were
combined with CCD photometry from the CTIO 1.5m telescope to
yield a uniform magnitude system for all the objects. As a
result, the sample is effectively complete to a uniform,
broad-band magnitude limit. The APM, of course, has been
used extensively for the analysis of UK Schmidt objective
prism plates (Hewett et al 1985, Foltz et al 1987), and the
same software applies in large part to the grism plates. The
main difference is that the grism plates have a higher and
more linear dispersion than the UK Schmidt plates.

Prior to carrying out the analysis with the APM,
approximately 75 quasars and emission line galaxies had been
confirmed with slit spectroscopy of candidates selected from
the visual scans of the plate material. These quasars were
used to develop selection criteria for the APM data which
would permit the impartial detection of a larger and
homogeneous set of candidates. The final criteria resulted
in a candidate list that included virtually all the
previously found quasars plus additional objects. The two
adopted criteria were both based on the presence of an

emission line in the APM spectra. In the first pass through
the data, all objects with emission lines stronger than an
equivalent width of 50 Å and a signal to noise in the line
greater than 6 were chosen. In the second pass, the
combination of a strong ultraviolet flux and a detectable
emission line with a signal to noise stronger than 8 was
required. These limits were judged to provide a good balance
of finding a high percentage of quasars without including too
many spurious objects. The first criterion picked all
objects with strong emission lines, in accord with the visual
samples originally produced with the slitless spectrum
technique. The second allowed the inclusion of objects with
weaker lines and what amount to ultraviolet excesses. It
takes advantage of the high information content of the grism
plates and significantly extends the range of quasars that
can be detected, especially for $z < 1.8$, where Lyman alpha
emission is not visible. It should also be noted that this
approach was aimed primarily at quasars with $L\alpha$ and C IV
lines of normal strength and with criteria that were
relatively easy to quantify. Unusual or weak-lined quasars
were not sought for this program, although they can
undoubtedly still be found from the plate material.

The objects obtained from the APM scans were then
observed with the CTIO 4m telescope and two dimensional
photon counting spectrograph. This led to the compilation of
the final sample, which consists of 127 quasars with
confirmed redshifts in the interval $0.3 < z < 3.4$. The
uncorrected surface density of the sample is 21 per sq deg.
If we allow for the fact that 15% of the spectra are not
accessible because of overlaps, the true surface density of
the sample is close to 25 per sq deg. Of the 127 quasars, 78
are from the APM scans and 49 are from the visual searches of
the plates. The spectroscopic observations showed that 40
candidates from the APM sample were stars or objects without
emission lines whose nature could not be determined from the
low signal-to-noise ratio of the data.

ANALYSIS FOR CLUSTERING

We used both qualitative and quantitative approaches to look
for possible clustering in the data. The former is useful
because, in effective displays of the data, the eye is very
good at picking out features to which numerical tests may be
insensitive. On the other hand, quantitative tests are
essential to characterizing any clustering that may be
present and can be very sensitive to particular types of
clustering.

In the case of the present survey, we inspected the so-
called pie diagram, in which we plotted comoving r as the

radial coordinate and right ascension as the angular
coordinate. This is the indicated choice because of the
narrowness of the strip of sky in declination. Although some
intriguing arcs and voids were noted in the data, we doubted
their significance on the basis of simulated catalogs we
generated and plotted in the same way. We used the original
data to obtain a smoothed version of the redshift
distribution and then produced random samples of data that
followed the redshift distribution and that had a uniform
probability in right ascension. We found that the simulated
catalogs showed similar structures often enough to make us
skeptical of the reality of those seen in the real data.

We next used three quantitative tests to investigate the
data in detail: the nearest neighbor test, the correlation
function, and a test for voids. The nearest neighbor test is
most sensitive to an excess of clustering over random
fluctuations at the smallest scales, mainly because it is the
distance over which no other objects are seen. The
correlation function is good at small and intermediate scales
and has the advantage of indicating at which scale the
clustering is occurring. The test for voids, which amounts
to estimating the number of times no quasars occur in volumes
of specified size, is perhaps the best indicator of structure
at the largest scales.

Application of the nearest neighbor test yields a mean
value of $r = 40.4$ Mpc for the whole sample (comoving
coordinates, $H_0 = 100$, $q_0 = 0.5$). The results of 100
simulations of the data are a mean r of 41.3 Mpc and $\sigma =$
2.6. If we limit the sample to $1.3 < z < 2.6$, where the
number density is highest, we obtain a mean of $r = 32.7$ Mpc
for the data and a mean of $r = 31.3$ for the simulations, with
$\sigma = 1.9$. It is clear that there is no significant difference
between the data and the simulations. Although the problem
of overlapping spectra reduces the sensitivity of detection
of very close pairs, the effect matters only for $r < 0.3$ Mpc
at $z = 2$, which is not a problem in the current context.

The correlation function may be calculated following the
formalism of Osmer (1981). Note that this is a truly three-
dimensional calculation because the redshifts are used
together with the position on the sky for each object, which
in turn is taken to be the center of the coordinate system
from which the comoving distances to the other members of the
catalog are calculated. The selection and edge effects are
estimated by generating sample catalogs with similar global
properties to the real catalog but for which the objects are
otherwise placed at random.

These first simulations allow for the non-uniformity in
redshift, as described above, but not for any plate-to-plate
variations. As a result, the tests will, if anything,
overestimate any clustering that may be present. We have

found a correlation between the plate quality, as measured by
the signal-to-noise ratio in the spectra, and the number of
quasars detected per plate. This information will be
incorporated into the next simulations to be carried out.
However, as can be seen from the following discussion, it is
unlikely that our main results will be affected when the
plate-to-plate variations are taken into account.

     The results for the correlation function, which are
shown in Figs. 1 - 4, provide no evidence for significant
departures from the sample catalogs on any scale, with the
main sensitivity being for $50 < r < 600$ Mpc ($H_0 = 100$, $q_0 =$
0.5), where the one sigma limit is 0.1 to 0.15 for the
correlation function. The graphs show both the unnormalized
function to give an idea of the number of pairs observed and
expected at different scales, and the normalized function, to
permit comparison with other studies. The data from the
actual quasar catalog are shown as crosses. In the graphs
showing the number of pairs, the lower line is the mean of
the simulated catalog while the upper line is the +1 sigma
limit. In the graphs for the normalized correlation
function, the lines are the +1 and -1 sigma limits, with zero
being the expected value for the simulated data.

     Finally, to test for the possible significance of the
large scale voids noted by eye in the pie diagrams, a test
for voids was performed following the approach of White
(1979). Points were put down at random within the survey,
following only the trend in redshift, and then the largest
radius to which no objects in the true catalog were found was
determined. A large number of such determinations for both
the real data and the simulated catalogs indicated no
evidence for voids occurring in the real data at a frequency
above what is expected by chance.

     The conclusion of all the analysis to date, therefore,
is that there is no evidence for clustering or voids in the
new sample.

DISCUSSION

While the lack of detectable clustering is consistent with
many previous studies, it would appear at first sight to be
in conflict with the recent results reported by Shanks et al
(1987 and this meeting) and by Shaver (1987 and references
therein). Note, however, that the present survey has very
poor sensitivity at the $r < 10$ Mpc scale and that the bulk of
the quasars have $z > 1.3$. The clustering detections by the
above authors are only on the $r < 10$ Mpc scale, with a hint
that the clustering is weaker for $z > 1.5$. Specifically, the
present survey has 2 pairs with $r < 10$ Mpc, while 4 might
have been expected according to the results of Shanks et al,

not a significant difference. Therefore, there does not appear to be any direct conflict of the present results with their work.

It is also of interest to compare the present results with those of Crampton, Cowley, and Hartwick (1987) from their major survey with the CFHT grens. They find from a correlation function analysis that there is no evidence in their data for clustering of the sample as a whole, but they do note the existence of a group of six quasars with redshifts near 1.17. They present further data on this group at this conference.

The present data are most useful for setting limits on clustering on the $50 < r < 600$ Mpc scale, where we have seen that no clustering is detectable and the 1 sigma upper limit is 0.1 to 0.15 in the correlation function. This information supports the concept of a uniform distribution of quasars on the large scale for redshifts greater than about 1.3. As Shanks et al (1987) point out, quasars set more stringent limits on the presence or absence of large scale structure for scales of one to a few hundred Mpc ($H_0 = 100$) than do galaxies or the X-ray and microwave backgrounds. Thus, the present data provide additional evidence for the uniformity of the mass distribution of the universe on such scales.

REFERENCES

Crampton, D., Cowley, A. P., and Hartwick, F. D. A. 1987, Ap. J., 314, 129.
Foltz, C. B., Chaffee, Jr, F. H., Hewett, P. C., MacAlpine, G. M., Turnshek, D. A., Weymann, R. J., and Anderson, S. F. 1987, A. J., 94, 1423.
Gratton, R. G., and Osmer, P. S. 1987, Pub.A.S.P., 99, 899.
Hewett, P. C., Irwin, M. J., Bunclark, P., Bridgeland, M. T., Kibblewhite, E. J., He, X.-T., and Smith, M. G. 1985, M.N.R.A.S., 213, 971.
Osmer, P. S. 1981, Ap. J., 247, 762.
Shanks, T., Fong, R., Boyle, B. J., and Peterson, B. A. 1987, M.N.R.A.S., 227, 739.
Shaver, P. 1987, in IAU Symposium 130, "Evolution of Large Scale Structures in the Universe", in press.
White, S. D. M. 1979, M.N.R.A.S., 186, 145.

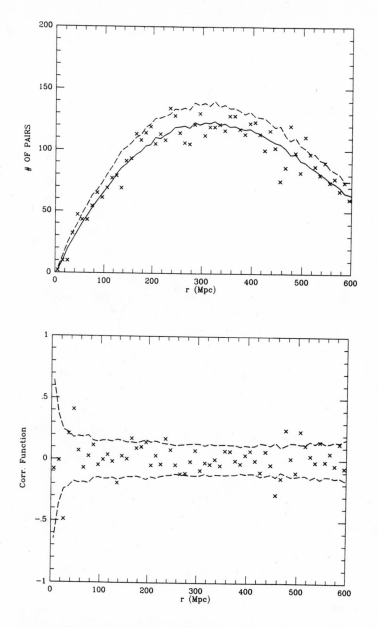

Fig. 1 (upper)   Number of pairs vs. r(Mpc) for 10 Mpc
bins.   Solid line – mean of simulations.   Dashed line –
upper 1σ envelope of simulations.

Fig. 2 (lower)   Correlation function for 10 Mpc bins.
Dashed lines – upper and lower 1 σ envelope of
simulations.

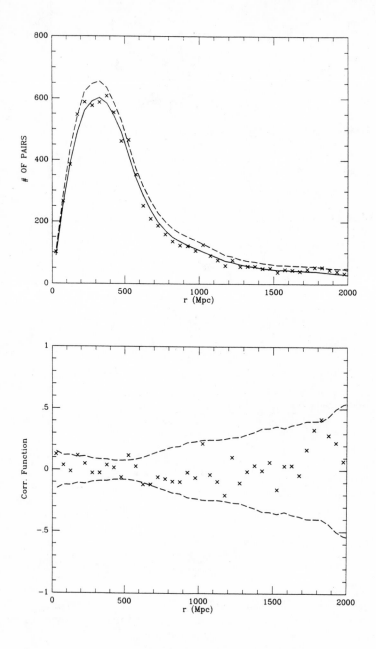

Fig. 3 (upper)   Same as Fig. 1, for 50 Mpc bins.

Fig. 4 (lower)   Same as Fig. 2, for 50 Mpc bins.

Statistical tests applied to the spatial distribution of quasars in several fields

E. GOSSET, J. SURDEJ, J.P. SWINGS
*Institut d'Astrophysique, Université de Liège, Avenue de Cointe, 5
B-4200 Cointe-Ougrée (Belgium).*

Abstract: We discuss the application of different statistical tests to the study of the spatial distribution of quasars. Emphasis is given to the necessity of simultaneously using fundamentally different methods in order to reach a good level in the understanding of the true nature of the distribution. Applications to data sets of optically selected quasars lead to the detection of a clustering at a typical scale of 10 arcminutes. Furthermore, the spatial distribution of quasars in a field around NGC 450 shows a deviation from randomness, towards clustering, at a scale of $10 \, h^{-1}$ Mpc.

INTRODUCTION

Over the last decade, and since the pioneering work of Osmer (1981), many astronomers have tried to analyze the distribution of quasars: the initial results suggested a random distribution. However, Shaver (1984) has pointed out a tendency towards clustering, tendency later confirmed by himself (Shaver, 1987) or by other authors such as Kruszewski (1987), Anderson et al. (1987), etc. Nevertheless, these results were based on a statistical method applied to a catalog and we feel that, if some observational biases can be taken into account, the existence of less well understood or even unknown effects cannot be ruled out. It seems to us complementary to study more homogeneous (although limited) samples of quasars in order to confirm the phenomenon. Both approaches are necessary, and we decided to concentrate on the latter. We have already reported preliminary results from the analysis of our surveys (Gosset et al., 1986); an up-to-date report is given in Swings et al. (1988). The present paper is dedicated to comments on the various statistical tests used for the analysis of the different samples.

THE STATISTICAL TESTS

The question is: *are quasars uniformly distributed at random ?* There is no doubt that, to give an answer to such a question, it is necessary to make use of statistical tests. Whereas a visual inspection may be very powerful in detecting anomalies or deviations from randomness, only a quantitative approach will help to know whether such effects are attributable to the underlying process or are simply due to statistical fluctuations of the sampling. Fortunately, it is now common practice to utilize a statistical method when analyzing data. However, our experience in this matter suggests that this is still not sufficient. Each method has its own advantages, sensitivity, property but also has its defects and weaknesses. Therefore, an exhaustive knowledge or understanding of the data is *not* accessible when one uses just one method. The combined results of several independent methods are, without any doubt, essential to reach a high degree of confidence in the conclusions of the analysis.

We have searched the literature in order to gather several good methods (details are given in Gosset, 1987a). We have retained a combination of five of them for their fundamentally different nature.

The first selected method is the MBA (Multiple Binning Analysis). This is the most ancient test and it has been widely used. Nevertheless, it lost some of its importance with respect to other more recent tests. Essentially, such a weakness is not due to the nature of the method of analysis but to the choice of the statistic. Gosset and Louis (1986) have put the MBA back to its right place among the most useful tests by introducing a statistic based on a randomization process. They have introduced the 4 within 16 randomization and the 8 within 64 randomization tests for the two- and the three-dimensional MBA's, respectively.

The second method is the CFA (Correlation Function Analysis). It consists in making counts in concentric rings sequentially built-up around each object of the sample. This method is extremely sensitive to edge effects and a correction has therefore to be applied. The only way to properly perform this correction is to calculate the exact measure of the domain actually explored when making counts. Such an approach can induce systematic effects, and the best way to estimate these consists nowadays in computing the mean cross-correlation function between simulated populations of uniformly distributed (usually) individuals and the data (as suggested by Sharp, 1979). One of the relevant estimators is unbiased, and its dispersion over the simulations gives a good approximation of the error associated with the autocorrelation function.

The third method is the NNA (Nearest Neighbours Analysis) which is based on the mean distances to neighbours. It is sensitive to edge effects too; no rigorous correction can be applied, although a good approximation can be obtained by using simulated populations mimicking the data. This test is somewhat less powerful but gives some additional information such as, for example, the number of individuals per cluster.

The fourth method, the PSA (Power Spectrum Analysis; Webster, 1976), as well as its generalized version (GPSA; Peacock, 1983), has a good reputation of flexibility and of great sensitivity. However, we found that this characteristic is overrated and we would just like to show a simple example. We have generated a 2-D synthetic population heavily clustered at a scale of 0.05; the clusters were built in such a way that they have a tendency to repel each other, and the characteristic scale of the inhibition was taken to be 0.12. So, the clusters are not randomly distributed but, rather, exhibit a deviation towards regularity. Figure 1 illustrates the variation of the statistic Q' of the PSA as a function of the spatial frequency explored. A peak at $1/\lambda = 7.5$ is clearly visible; it corresponds to the inter-cluster regularity ($7.5 \sim 1/0.12$). It is nevertheless worth noticing that, besides this, there is no visible trace of clustering ($20 \sim 1/0.05$). To be sure that the clustering is present, we have applied the MBA in the configuration of the 4 within 16 randomization test. The run of the relevant normal statistic Z is shown in Figure 2 as a function of the investigated characteristic scale. We obtain a deviation greater than 4σ towards clustering, in good

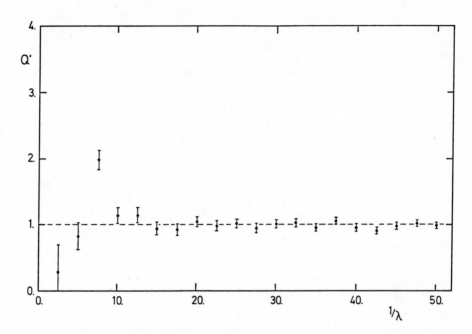

Figure 1

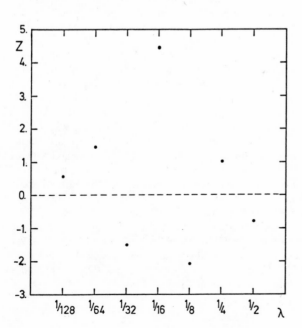

Figure 2

agreement with the known characteristic of the synthetic population. The MBA also reveals a deviation towards regularity at a scale of 1/8, in good agreement with the PSA. Three things are to be retained as conclusions at this particular point: (1) the PSA is sometimes difficult or impossible to interpret and can lead to false conclusions; (2) this is another proof of the interest of the randomization tests; (3) again, the use of several methods is most useful.

Finally, a fifth method has been taken into account. Effectively, the first four are not well suited for analyzing data at a typical scale of the order of that of the whole field. For the one-dimensional space, such a test, only sensitive to large scale deviations, exists: it is the well-known Kolmogorov-Smirnov test. Peacock (1983) has proposed an Extended Kolmogorov-Smirnov test for the two-dimensional case while, recently, Gosset (1987b) has introduced the three-dimensional EKS test. The EKS tests are also helpful in adjusting a probability density function to the distribution of the individuals. They can be used jointly with the NNA in order to correct for edge effects or as a preliminary step to the application of the GPSA.

We conceived a set of computer codes in order to test those five selected methods and we will use them in the analysis of the data.

THE DATA ANALYSIS

Data from different fields have been analyzed. A brief description of the three main surveys is given in Swings et al. (1988). In a preceding report (Gosset et al., 1986), we concluded that the quasars in the NGC 450 field were clustered on the celestial sphere with a typical scale of 10 arcminutes. We have subsequently analyzed the distribution of quasars in the NGC 520 field and in the ESO field #300. Both of these samples show the same deviation towards clustering at a similar scale of about 10 arcminutes.

The NGC 450 sample has been analyzed in a three dimensional space (α, δ, z). Because the number of identified quasars ($\sim$ 60) is quite small, it results a very low volume density. For this reason, some difficulties may arise. The GPSA is difficult to apply and no convincing result in any direction can be drawn. Provided we authorize some mixing of the scales, it is possible to apply the MBA. The 8 within 64 randomization test detects a deviation towards clustering with a significance level of 0.04 and at a typical scale somewhere between 6 h^{-1} Mpc and 70 h^{-1} Mpc. The CFA detects a deviation, greater than 4σ, towards clustering and with a characteristic scale of about 12 h^{-1} Mpc. In order to take into account selection effects on the redshift, we also computed the standard deviation on the basis of non-uniform simulated populations obtained by randomizing the redshifts. This attitude leads to a highly conservative test. The relevant significance level for the 12 h^{-1} Mpc clustering is 0.04 and is to be considered as an upper limit. The agreement with the results from the MBA is interesting. The NNA test indicates that we have to deal with a tendency to form pairs. Finally, we would like to note that the deviation in this particular field is mainly caused by the presence of the quasars Q0107-025A and Q0107-025B, whose spectra are described in Surdej et al. (1986).

CONCLUDING REMARKS

We believe that a good and thorough analysis of the distribution of a sample of quasars for the detection of clustering, contagion or regularity, requires the simultaneous use of several methods (at least MBA, CFA, NNA, PSA and EKS). We have suggested what appear to presently be the best configurations for these tests. The analysis of our samples leads to the general detection of a clustering on the celestial sphere at a typical scale of 10 arcminutes. A deviation towards spatial clustering with a highly conservative significance level of 0.04 and a scale of approximately $10\ h^{-1}$ Mpc is also reported for the field of optically selected quasars around NGC 450.

REFERENCES

Anderson, N., Kunth, D., Sargent, W. L. W.: 1987, *Astron. J.*, in press.
Gosset, E.: 1987a, *Analyse de nuages de points. Applications astronomiques et étude de la distribution des quasars*, Ph.D. Dissertation, Université de Liège.
Gosset, E.: 1987b, *Astron. Astrophys.*, **188**, 258.
Gosset, E., Louis, B.: 1986, *Astrophys. Space Sci.*, **120**, 263.
Gosset, E., Surdej, J., Swings, J. P.: 1986, in *Quasars*, IAU Symposium no.119, eds.: G. Swarup, V. K. Kapahi, Reidel, Dordrecht, 45.
Kruszewski, A.: 1987, *Preprint*.
Osmer, P. S.: 1981, *Astrophys. J.*, **247**, 762.
Peacock, J. A.: 1983, *Monthly Notices Roy. Astron. Soc.*, **202**, 615.
Sharp, N. A.: 1979, *Astron. Astrophys.*, **74**, 308.
Shaver, P. A.: 1984, *Astron. Astrophys.*, **136**, L9.
Shaver, P. A.: 1987, *ESO Preprint*, no.534.
Surdej, J., Arp, H., Gosset, E., Kruszewski, A., Robertson, J. G., Shaver, P. A., Swings, J. P.: 1986, *Astron. Astrophys.*, **161**, 209.
Swings, J. P., Surdej, J., Gosset, E.: 1988, in *Optical surveys for quasars*, these proceedings.
Webster, A. S.: 1976, *Monthly Notices Roy. Astron. Soc.*, **175**, 61.

CAPTION FOR FIGURES

Figure 1: Run of the statistic Q' of the bi-dimensional PSA as a function of the spatial frequency $1/\lambda$ (see text for details).

Figure 2: Run of the normal statistic Z of the MBA in the configuration of the 4 within 16 randomization test as a function of the investigated characteristic scale (see text for details).

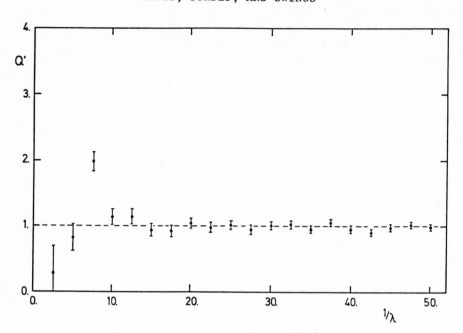

Figure 1

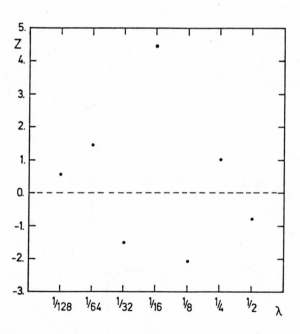

Figure 2

EVOLUTION OF QUASAR CLUSTERING

A. IOVINO and P. SHAVER
European Southern Observatory, Karl-Schwarzschild-Str. 2,
D-8046 Garching bei Munchen, Federal Republic of Germany

ABSTRACT   We have combined three different samples,
totalling 376 quasars, to study the evolution of the
clustering properties with cosmic time. Clustering is
detected at low redshifts ($z < 1.5$) on scales less than
$10\ h^{-1}$ Mpc at the 4.7 σ level, but not at high redshifts
($z > 1.5$). The difference between these two results is
significant at the 2 σ level.

SAMPLES AND METHOD USED

The three samples used in this study are summarized in Table
I. They have different magnitude limits and surface densities,
and like virtually all quasar surveys they are incomplete. But
they are homogeneous, subject to relatively simple selection
effects, and so can be reliably used in clustering analyses.
The redshift distribution of the three samples is quite smooth
(Fig. 1), and the surface density distribution is essentially
constant over the fields, except for the Barbieri et al.
sample, where there is some indication of mild ($\lesssim 25\%$)

TABLE I   The Three Samples

Search Technique	Area (sq. deg.)	Magnitude Limit	Number of Quasars	Reference
UVX	4.0	20.9	171	Boyle (1986)
grens	5.2	20.5	125	Crampton et al. (1987)
UVX	10.0	19.5	80	Barbieri et al. (1987)

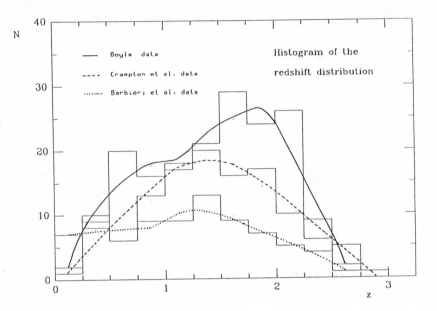

Fig. 1   Redshift distribution of the three samples used.
The curves indicate the smoothing adopted.

vignetting. The correlation function method has been used to
analyze the samples. Many random simulations were made for
each of them, exactly reproducing both the redshift
distribution and the angular distribution of the real data.
These samples have then been used to evaluate ξ according to
the formulae

$$\xi = \frac{N_{obs}}{N_{RAND}} - 1 \quad ,$$

$$\Delta\xi = \sqrt{\frac{\xi + 1}{N_{RAND}}}$$

(Peebles 1980).

RESULTS

The results of our analysis are illustrated in Figs. 2 and 3
and in Table II. We have found evidence for clustering of
quasars at low redshifts, up to comoving scales of 10 h^{-1} Mpc

TABLE II   Number of Quasar Pairs with Separations
< 10 h^{-1} Mpc

	z < 1.5	z > 1.5
expected	4.7	4.0
observed	15	5

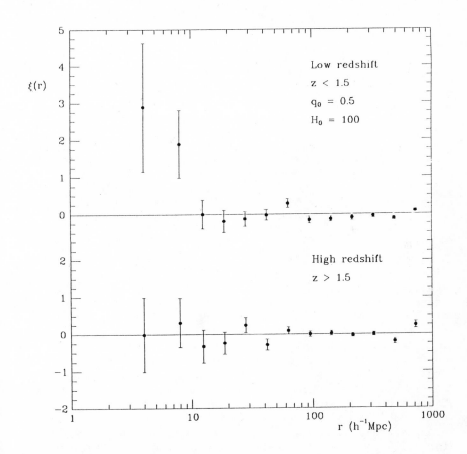

Fig. 2   Quasar two-point correlation function at low
(z < 1.5) and high (z > 1.5) redshifts, from the three
samples used.

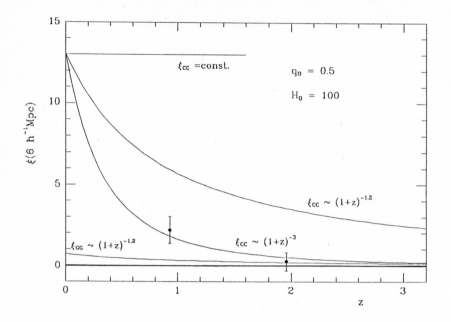

Fig. 3   Amplitude of the quasar correlation function at 6 h^{-1} Mpc as a function of redshift, from the three samples used. The smooth curves correspond to possible forms of evolution of the correlation function for galaxies (GG) and clusters (CC): comoving (ξ = constant), stable clustering ($\xi \sim (1+z)^{-1.2}$), and collapsing ($\xi \sim (1+z)^{-3}$).

(15 pairs found instead of the 4.7 expected, at z < 1.5). By contrast there is no indication of clustering in the high redshift bin (5 pairs found instead of the 4 expected at z > 1.5). Thus, there is evidence for the evolution of clustering, at a formal significance level of 2σ.

   At low redshifts the amplitude of the quasar-quasar correlation function is higher than that for galaxies, whatever evolutionary model is assumed, but lower than that for rich clusters except for the most extreme collapse models. There are indications, however, that clustering scales with richness (Peacock et al. (1987) have recently shown, for example, that the clustering scale length is about 13 h^{-1} Mpc for radio galaxies of Fanaroff-Riley class I which are known to lie preferentially in clusters of Abell richness class 0). Quasars are also located preferentially in groups or clusters of galaxies (Yee and Green 1986); radio-quiet quasars (the majority of objects in this study) have poorer environments than radio-loud quasars. The present result, in which quasar

clustering is intermediate between that of galaxies and rich
clusters, is therefore consistent with what is known about the
environments of quasars.

The evolution of the quasar-quasar correlation function
appears to be relatively rapid, implying gravitational
collapse on cluster scales, although the present uncertainties
do not exclude stable or even comoving clustering. It is
unlikely that this evolution is due to a selection effect;
clustering appears to be stronger for more luminous objects,
at least in the case of galaxies. Gravitational lensing would
produce more pairs at higher redshifts, and the likely angular
separations are smaller than those found in the present study.
The evolution most likely reflects a genuinely increasing
concentration of objects in clusters. If so, it provides a
direct measure of the evolution of structure on Megaparsec
scales.

REFERENCES

Barbieri, C., Cristiani, S., Iovino, A., Nota, A. 1987, A&A,
    to be submitted.
Boyle, B.J. 1986, Ph.D. Thesis, University of Durham.
Crampton, D., Cowley, A.P., Hartwick, F.D.A. 1987, Ap.J., 314,
    129.
Peacock, J.A., Miller, L., Collins, C.A., Nicholson, D.,
    Lilly, S.J. 1987, in IAU Symp. 130 on Evolution of Large
    Scale Structure in the Universe, eds. Audouze, J.,
    Szalay, A., Reidel, Dordrecht, in press.
Peebles, P.J.E. 1980, The Large Scale Structure of the
    Universe, Princeton Series in Physics.
Yee, H.K.C., Green, R.F. 1986, in IAU Symp. 119 on Quasars,
    eds. Swarup, G., Kapahi, V.K., Reidel, Dordrecht.

THE ROE/ESO LARGE SCALE AQD-SURVEY FOR QUASARS

A. IOVINO and P. SHAVER
European Southern Observatory, Karl-Schwarzschild-Str. 2,
D-8046 Garching bei München, Federal Republic of Germany

R. CLOWES
Royal Observatory, Blackford Hill, Edinburgh EH9 3HJ,
Scotland, U.K.

ABSTRACT   We present a report on the progress and aims
of the ROE/ESO large-scale AQD (Automated Quasar
Detection, Clowes et al. 1984, Clowes 1986) survey for
quasars.

Seven connected fields near the south galactic pole have been
processed with the aim of extracting a large and complete
sample of quasars. The photographic material consists of
IIIa-J objective-prism and associated direct plates from the
1.2 m. UK Schmidt Telescope in Australia. These plates were
measured by COSMOS (MacGillivray and Stobie 1984), the fast
plate-measuring machine at the Royal Observatory in Edinburgh.
  The direct plates were used to obtain for each image in
5.35×5.35 deg^2 areas of each field:
i)   the celestial coordinates, with accuracy better than one
     arcsecond,
ii) a classification as star-like or galaxy-like
     (MacGillivray and Stobie 1984),
iii) a classification for overlapping for the corresponding
     image on the prism plate (Clowes 1986); an image was
     classified as a geometrical overlap if a second image
     occurred within ± 70 pixels in the dispersion direction,
     and ± 3 pixels perpendicular to the dispersion (1 pixel
     is 16 μm).
  Coordinate transformation from the direct plates to the
corresponding prism plates was established, and all spectra
corresponding to images with B ~ 21 on the direct plates were
extracted in spectrum blocks of 128 × 8 pixels.
  Approximately 430,000 spectra were extracted from the
seven plates, and it is these that are processed by AQD.
  Compared with visual searches of prism plates, AQD has
the great advantage that the selection criteria and,

consequently, the physical selection effects, are known, predefined, and rigidly mantained. It can select quasar candidates by emission lines, absorption lines, spectral discontinuities, ultraviolet excess and red excess, but for this survey we have used only the two most productive criteria: emission lines and ultraviolet excess. Note that the selection by ultraviolet excess uses the continuum and so avoids biases from spectral lines. All the spectra that were classified as geometric overlaps were excluded. In this way we have obtained a total number of ~ 9500 quasar candidates from the five best fields for a limiting magnitude of B ~ 20.5. Naturally, by no means all of the objects that satisfy the selection criteria are quasars, and, ideally, multi-object spectroscopy should be used to state exactly wich objects are quasars. In the meantime a probability based approach can be highly successful, permitting progress with the high probability subsets.

For this probability based approach we devised a grading scheme, in which points are awarded according to quasar-like properties, and the sum of the points for a given candidate should be directly related to the probability of its being a quasar.

Using this grading scheme we have: ~ 5500 good candidates, that have ~ 40% probability of being quasars, ~ 1500 very good candidates, that have ~ 70% probability of being quasars (these are a subset of the 5500 'good candidates').

We have therefore obtained with the ~ 1500 high-probability candidates from the five best plates approximately half of the total number of quasars listed in the Véron & Véron catalogue (Véron-Cetty and Véron 1987). The redshift range covered by the AQD survey extends up to z ~ 3.3 (the redshift for which the Lyα line is within the sensitivity range of the IIIa-J emulsion).

We should also stress that some of the objects rejected by us as non-quasars are of great interest for other fields of research (e.g. intergalactic HII regions, emission lines galaxies, white dwarfs).

The large database from our survey is tremendously valuable for many areas of quasar reseach. Some examples follow:

a. Clustering of Quasars
Quasars are the only objects at large redshifts that can be readily detected with present techniques and they are, therefore, the primary indicators of the structure of the universe at such early epochs. Quasar clustering has been detected on scales of $\lesssim 10$ h^{-1} Mpc (H$_o$ = 100 h km s^{-1} Mpc^{-1}), and there is tentative evidence that it decreases with increasing redshift (cf. Shaver, this volume, and references

therein). Our sample is well placed to investigate any
evolution in clustering with cosmic time.

b. Close Pairs

Close pairs of quasars (angular separation < 5 arcmin.),
usually arising from positional coincidence on the sky, are of
great astronomical interest. Common absorption in the spectra
of the two quasars yields information on the properties of the
absorbing matter along the line of sight, while associated
absorption yields information on the environment of the
foreground quasar. At the same time close pairs can also
provide a direct test of the cosmological hypothesis of
redshifts.
      From our high probability subset we expect to obtain
~ 100 quasar pairs with angular separation < 5 arcmin. In just
one of our fields we have 26 such pairs, some of which have
already been successfully observed at La Silla and the
Anglo-Australian Observatory.

c. Gravitational Lens Candidates

Candidates for gravitational lenses are close pairs of quasars
of small angular separation (< 10 arcsec.).The interest in
gravitational lenses arises from the possibility of obtaining,
from the position and luminosities of the images, an estimate
of the total mass distribution (both luminous and dark) that
acts as the lens. Our data can be used for a systematic search
for candidates for gravitational lenses - namely, quasars that
have a companion on the direct plate within a few arcseconds.

d. High Redshift Quasars

The existence and location of a cut-off in the redshift
distribution of quasars would point to a preferred epoch of
formation. Our samples of quasars, being complete within the
criteria adopted to select them, are well suited to studies of
the redshift distribution in the range 1.8-3.3.

REFERENCES

Clowes, R.G., Cooke, J.A., Beard, S.M. 1984, MNRAS, 207, 99.
Clowes, R.G. 1986, MNRAS, 218, 139.
MacGillivray, H.T., Stobie, R.S. 1984, Vistas Astr., 27, 433.
Véron-Cetty, M.P., Véron, P. 1987, ESO scient. rep. no. 5.

THE STRUCTURE OF THE HOST GALAXIES OF RADIO-LOUD QUASARS- COOLING FLOWS , STAR FORMATION AND QUASARS

W. Romanishin
Physics Department
Arizona State University
Tempe, AZ 85287-1504

ABSTRACT We discuss evidence linking cooling (accretion) flows, star formation, and quasar activity. In a number of galaxy clusters with x-ray detected cooling flows, there is unmistakable evidence for ongoing star formation in the central 10-20 kpc of the cluster dominant galaxy. When fit in the blue with a standard de Vaucolueurs "$r^{1/4}$" profile, these galaxies have smaller values of the effective radius parameter (r_e), compared to "normal" ellipticals of similar total absolute magnitude, due to the central concentration of the blue light from young stars. A sample of ~two dozen radio-loud quasar host galaxies have been analyzed with an image modelling program to derive de Vaucouleurs parameters for the host galaxies. The hosts of these radio quasars tend to show the same behavior in the $M_B - r_e$ diagram as the star- forming cooling flow galaxies, implying there is significant centrally concentrated star formation in these host galaxies. This ongoing star formation could provide fuel for the nucleus, as well as produce large numbers of compact stellar remnants which may be connected with the nuclear activity.

INTRODUCTION

What physical processes "triggers" activity in galactic nuclei (including quasar activity)? Recently, a number of studies have stressed the role of galaxy interactions/ and or mergers as possible triggers (see for example Dahari 1985). Although the interaction picture may explain many AGNs, other physical mechanisms for AGN triggering must be explored. One such mechanism is that of cooling or accretion flows, which can occur on both galaxy cluster and individual galaxy scales. Recent work has shown that there is star formation involving significant numbers of massive stars in at least some dominant galaxies in cooling flow clusters. This paper will outline the possible links between star formation in elliptical galaxies, such as the cooling flow cluster dominant galaxies *and* the hosts of radio-loud quasars, and AGN activity.

COOLING FLOW GALAXY PHOTOMETRY

At the present time, the central galaxies in at least five cooling flow clusters show evidence, from two-color surface photometry showing blue central regions, of a significant young stellar population in their inner 10–20 kpc . These clusters are: Perseus (NGC 1275) and PKS 0745–191 (Romanishin 1987), 2A 0335+096 (Romanishin and Hintzen 1988), A 1795 (Romanishin 1988), and A 2597 (Romanishin, in preparation). Color profiles, in a broadband optical color approximating (B–I), are shown in the references given. Besides producing the color gradient, the young starlight also affects the *structure* of the galaxy, particularly in blue light, as the young stars are not distributed uniformly throughout the galaxy. The central concentration of young stars causes the galaxies to have a smaller value of r_e , the effective radius, than the same galaxy without young stars. The young stars also brighten the galaxy somewhat, but since cluster central galaxies are very luminous to start with, the young star light is only a minor perturbation on the total blue luminosity. For example, in both NGC 1275 and PKS 0745–191 , the r_e value in the blue is only one half that measured in I , where the effects of young stars should be much diminished, while the total increment of blue light is only about 15%. Figure 1 shows the B and I surface brightness profiles of NGC 1275, and the best fit $r^{1/4}$ lines. Although the central star formation does noticeably distort the profile in B from a pure $r^{1/4}$ profile, the distortion is rather small, and could not be detected for more distant galaxies.

QUASAR HOST GALAXY STRUCTURE

An image modelling program, developed by the author, has been used for the past several years to study the host galaxies of quasars. For details of the program, and its application to a large sample of quasars, see Smith *et al.* 1986 (hereafter SHBRB) and references therein. The program gives "best-fit" parameters for quasar host galaxies for an assumed galaxy type (elliptical with $r^{1/4}$ profile or spiral with exponential disk plus $r^{1/4}$ bulge). SHBRB show that, where it is possible to reliably determine the type of the underlying galaxy, the radio- quiet quasars prefer spiral hosts, and the radio- loud quasars prefer elliptical ($r^{1/4}$) hosts. A more detailed analysis of the photometric properties of the radio quasar host galaxies from SHBRB, plus new data on another dozen radio- loud quasars (Romanishin and Hintzen, in preparation) strongly suggest that the hosts of radio-loud quasars are *not* normal ellipticals, but rather tend to have smaller values of r_e for their luminosity. This trend is, of course, exactly what we see in the cooling flow cluster central galaxies which show ongoing star formation. The r_e , M_B points for the quasars from these two studies is shown in Figure 2, along with the line for

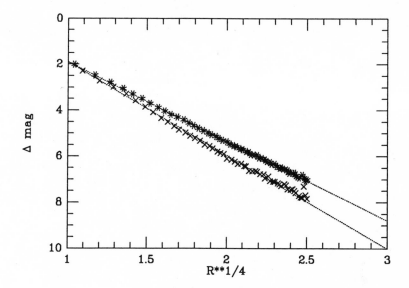

Figure 1. The surface brightness profiles of NGC 1275 (Perseus cluster) in I band (upper points) and B band (lower points) are plotted, with arbitrary vertical offset. The horizontal axis is $r^{1/4}$, where r is in kpc. The dotted lines are straight line fits to the data. Note that the B profile is steeper than the I profile, so that r_e in B is *less* than in I. In B, $r_e = 18$ kpc, while in I, $r_e = 33$ kpc. NOTE: Points within $4''$ radius have not been plotted, as they are contaminated by the nuclear point source.

"normal" ellipticals (Romanishin 1986). The two stars are the blue passband points for the dominant galaxies in the Perseus and PKS 0745−191 clusters, showing the effect discussed above.

DISCUSSION

The structural similarities of host galaxies of radio- loud quasars and cooling flow cluster central galaxies provides evidence that star formation, perhaps induced by cooling flows, might be related to quasar activity. Of course, the structural properties of the quasar hosts may also be due to some other cause, such as galaxy interactions and/or mergers. Other evidence possibly implicating cooling flows in at least some quasars comes from the large [O II] cloud seen around 3C 275.1 (Hintzen and Romanishin 1986), a quasar which is apparently in the brightest galaxy in a cluster, and analysis of the emission line gas around 3C 48 (Fabian *et al.* 1987).

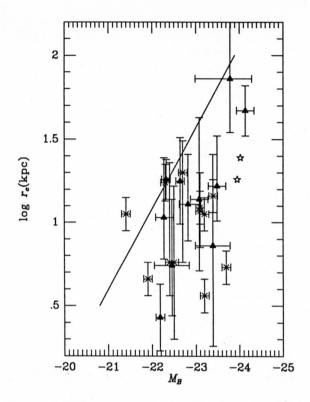

Figue 2. Host galaxy total blue absolute magnitudes and effective radii (measured in the blue in the quasar's rest frame). The x's are from SHBRB, and the triangles from Romanishin and Hintzen (in preparation). The two stars are the blue passband points for NGC 1275 and PKS 0745−191 . $H_0 = 50$ km sec^{-1} Mpc^{-1} assumed. The straight line is for "normal" ellipticals from Romanishin (1986).

Even if cooling flows are only important for a small subset of AGNs they are still worth careful study. The most obvious such subset would be quasars in the dominant galaxies in clusters, of which 3C 275.1 is the best example. Because imaging surveys of low- redshift quasars have shown little tendency for quasars to lie in rich clusters, it might be thought that such objects are very rare and perhaps not relavent to quasars in general. However, while this may be true at low redshifts, Yee and Green (1987) report evidence that quasars at z greater than 0.5 or so are more likely than lower redshift quasars to lie in Abell richness clusters. This shows that the cluster environment (cooling flows?) may be important for high redshift quasars. Although we have stressed the possible importance of cooling flows in dominant cluster galaxies, x-ray evidence exists for cooling flows in poor clusters (Canizares, Stewart, and

Fabian 1983) and in individual elliptical galaxies (Thomas *et al.* 1986). Further study is needed to determine if the cooling flows in these types of objects result in significant star formation.

How might cooling flow induced star formation and AGNs be related? Obviously, mass loss from young stars can provide large amounts of gas to fuel an active nucleus. In addition, such star formation activity will produce large numbers of compact stellar remnants (neutron stars and perhaps black holes). Models for AGN activity involving clusters of accreting compact objects can account for much of the known behavior of certain types of AGNs (Pacholczyk and Stoeger 1986).

ACKNOWLEDGEMENTS

Most of this work has been done in collaboration with Paul Hintzen. This work has been partially supported by the HEAO Guest Investigator Program under grant NAG8-576 awarded to W.R.

REFERENCES

Canizares, C. R., Stewart, G. C., and Fabian, A. C. 1983, Ap.J. , 272, 449.
Dahari, O. 1985 Ap. J. Suppl. , 57, 643.
Fabian, A.C., Crawford, C.S., Johnstone, R.M., and Thomas, P.A. 1987, M.N.R.A.S. , 228, 963.
Hintzen, P., and Romanishin, W. 1986, Ap.J. , 311, L1.
Pacholczyk, A. G., and Stoeger, W. R., 1986. Ap.J. , 303, 76.
Romanishin, W. 1986, A.J. , 91, 76.
Romanishin, W. 1987. Ap.J. (Letters) 323, L113.
Romanishin, W. 1988 in Cooling Flows in Clusters and Galaxies
      ed. A. C. Fabian, Dordrecht:Reidel (in press).
Romanishin, W., and Hintzen, P. 1988. Ap.J. (Letters) , 324, L17.
Smith, E.P., Heckman,T.M., Bothun,G.D.,Romanishin, W., and Balick,B. 1986, Ap.J. , 306, 64 (SHBRB).
Thomas, P. A., Fabian, A. C., Arnaud, K. A., Forman, W., and Jones, C. 1986, M.N.R.A.S. , 222, 655.
Yee, H. K. C., and Green, R. F. 1987, Ap.J. , 319, 28.

THE EINSTEIN OBSERVATORY EXTENDED MEDIUM SENSITIVITY SURVEY AND THE X-RAY SELECTED QUASAR SAMPLE

T. Maccacaro[1-2], I.M. Gioia[1-2], S.L. Morris[3], R. Schild[1], J.T. Stocke[4], and A. Wolter[1].

1) Harvard-Smithsonian Center for Astrophysics, Cambridge, MA
2) Istituto di Radioastronomia, Bologna, Italy
3) Mount Wilson and Las Campanas Observatory, Pasadena, CA
4) Center for Astrophysics and Space Astronomy, Boulder, CO

ABSTRACT. We present here a status report on the Extended Medium Sensitivity Survey and on the x-ray selected quasar sample. We briefly review some of the most important results obtained in the past from the analysis of a smaller but fully identified sample of x-ray selected quasars as well as some of the newest results obtained from the present, much larger, but still incomplete collection of x-ray selected Active Galactic Nuclei.

1. Introduction

So far at this workshop, and quite correctly, the emphasis has been on optical surveys. Optically selected quasars still outnumber quasars selected by all other means combined and the optical band is by far the region of the electromagnetic spectrum which provides us with the largest body of information on quasars. There are however other means to survey for quasars, and, as these surveys become more efficient and sensitive, the number or radio, infrared, and x-ray selected quasars increases to become a significant fraction of the total number of quasars known. Previous speakers have expressed concern about the understanding of the selection effects which are inherent to each method used to search for quasars. In this respect surveys of quasars conducted by radically different techniques (e.g. infrared surveys, x-ray surveys, etc.) are of extreme importance since they complement optical surveys and have the potential to sample a different region of the underlying quasar population. A comparative study of the properties of quasars selected by very different techniques is also a check on the relative efficiency of each technique.

The substantial improvement in sensitivity which has been achieved in x-ray astronomy thanks to the use of imaging techniques has already allowed the discovery of a large number of x-ray selected quasars and Active Galactic Nuclei. X-ray selection, as we shall see later on, favors the detection of low-redshift and low luminosity objects so that many of the x-ray selected objects come from the boundary region between "classical" quasars and Seyfert galaxies. Figure 1 shows the Hubble diagram for UVX and x-ray selected AGN, the latter being represented by filled circles. It is evident that x-ray selected AGN occupy a rather unpopulated region of the diagram. At any given redshift they are in fact characterized, on the average, by a fainter magnitude (i.e. lower luminosity) than that of UVX selected quasars. The *Einstein Observatory*

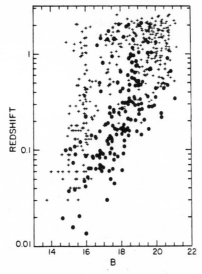

Figure 1: Hubble diagram for UVX quasars (crosses; from Boyle et al., 1987), and x-ray selected AGN (filled circles).

has thus contributed to further crumbling the distinction between quasars and Seyfert galaxies and to promote the use of the term AGN to refer to these objects in general. Our working definition of AGN includes quasars and Seyfert galaxies but excludes BL Lac objects. This exclusion is based on the fact that the number counts for BL Lacs is significantly different from the number counts of AGN, thus suggesting that their evolutionary properties are quite different (Maccacaro et al., 1984; Stocke et al., 1982). A detailed description of the properties of x-ray selected BL Lacs is given in the following paper (Stocke et al. 1988).

But let me now get to the point and describe the x-ray survey we are conducting and the results that we have so far obtained with respect to x-ray selected AGN.

2. The Medium Sensitivity Survey (MSS) and the AGN sample

The *Einstein Observatory* Medium Sensitivity Survey (MSS) was initiated several years ago (Maccacaro et al., 1979; 1982) with the aim of collecting a large homogeneous and statistically complete sample of x-ray selected objects suitable for statistical studies. The published sample consists of 112 sources with flux in the range 1×10^{-13} to 1×10^{-11} erg cm^{-2} s^{-1} in the 0.3-3.5 keV band. This sample was the result of the analysis of about 350 IPC images of the high galactic latitude sky (corresponding to $\sim$90 deg^2 of sky). The uniqueness of the MSS sample stands on the fact that it is a flux-limited x-ray catalog of fully identified sources, thus allowing unbiased statistical studies of their properties (Stocke et al., 1983 Gioia et al., 1984). Table 1 summarizes the optical content of the MSS and shows how AGN constitute the most numerous class of sources (50% of the total and $\sim$70% of the extragalactic population). At a flux level of 1×10^{-13} erg cm^{-2} s^{-1} there are approximately 3 AGN/deg^2, a negligible fraction of the 100 quasars/deg^2 which are found in optical surveys of 22 mag objects. The surface density of x-ray selected AGN is sufficient however to allow the definition of large samples if one can survey even a few hundreds deg^2 of sky to this limiting sensitivity.

Table 1

Medium Survey Optical Content

Active Galactic Nuclei	57
Clusters of galaxies	18
BL Lacs	4
Normal galaxies	5
Stars	28
Total	112

ROSAT, (Truemper 1984), an x-ray telescope now scheduled for launch in early 1990, will survey the whole sky with an average sensitivity not too far from 10^{-13}erg cm^{-2} s^{-1}. It has been estimated that ROSAT will find of the order of 100,000 x-ray sources. A very conservative estimate indicates that the number of AGN hidden among these sources will largely exceed 20,000. Optical spectroscopy will be of course needed in order to find these AGN and determine their redshift. The task of identifying all of them is however prohibitive, given their expected number and the resources presently available. We are thus concentrating on the *Einstein* MSS and on its extension since it represents our best chance to gather a large, complete and fully identified sample of x-ray selected objects.

Using the small sample of 56 AGN extracted from the MSS we have been able to derive their x-ray LogN-LogS and to study their luminosity function and cosmological evolution. Figure 2 shows the derived number-counts for x-ray selected AGN and BL Lac objects. It is worth noting how different the two curves are. AGN are characterized by a rather steep LogN-LogS (integral slope = 1.71), while the BL Lac number-counts are consistent with the Euclidean slope at high fluxes and exhibit a significant flattening at fluxes below 10^{-12} erg cm^{-2} s^{-1} (Stocke et al., 1982, Maccacaro et al., 1984). As discussed by John Stocke in the following paper, the data from the extension of the survey seem to confirm this behavior.

Since for the AGN sample complete redshift information is available, it is possible, and more instructive, to study their luminosity function. Within the framework of pure luminosity evolution Maccacaro, Gioia and Stocke (1984) have determined that less evolution is required by the x-ray data than by the optical data. When the evolution of the x-ray luminosity with cosmic time is expressed by $L_x(z) = L_x(z=0) \exp[C \times z/(1+z)]$, the data are best fitted by the value C = 4.85. For comparison, values of C of about 7 are obtained for the evolution of the optical luminosity of optically selected quasars (e.g. Marshall et al., 1983). The MSS sample is too small to allow the study of the luminosity function in different redshift shells. Maccacaro, Gioia and Stocke (1984) have thus used all the objects in the sample and, by de-evolving their luminosity (according to the derived evolution function), have determined the present epoch luminosity function (see Figure 3). It is interesting to note that the low luminosity flattening seen in the x-ray luminosity function at $L_x \sim 2 \times 10^{25}$ erg s^{-1} Hz^{-1} (@ 2 keV) is consistent with the flattening which has been more recently detected (Boyle et al. 1987; Weedman 1986) in the optical luminosity function at about $M_b \sim -22$.

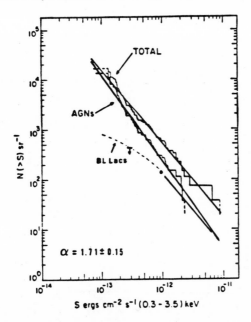

Figure 2: The x-ray LogN-LogS for the extragalactic sample, for the AGN, and BL Lac objects subsamples (adapted from Gioia et al., 1984 and Maccacaro et al., 1984).

The relatively small size of the MSS sample constitutes an intrinsic limit in these and other studies. It is evident that the analysis of larger samples would give us a more satisfactory understanding of the behavior of x-ray selected AGN and BL Lacs. The low luminosity end of the AGN x-ray luminosity function where the flattening occurs is poorly determined. A more accurate determination of the AGN cosmological evolution is needed, with the capability to discriminate between different evolution models and to constrain within small uncertainties the relevant parameters. The study of the AGN x-ray luminosity function and its cosmological evolution will then provide the best approach (short of launching AXAF!) to the determination of the contribution of quasars and Seyfert galaxies to the diffuse x-ray background.

We have thus undertaken to expand the survey, extending the search for sources to additional IPC images.

3. The Extended Medium Sensitivity Survey (EMSS)

Since 1984 we have been working at this extension by analyzing all the IPC observations available in the *Einstein* Data Bank that meet well defined selection criteria. The

resulting Extended Medium Sensitivity Survey (EMSS) consists of about eight times as
many sources as were in the original MSS. The quality of the data has also improved
since all the IPC images used have been reprocessed with the REV.1 processing system.
The optical program, undertaken to spectroscopically identify all the sources, is still in
progress.

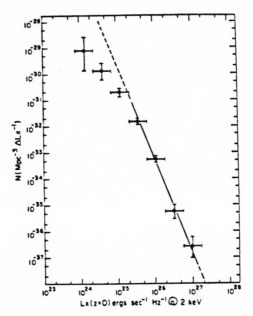

Figure 3: The x-ray luminosity function for AGN in the MSS sample (adapted from
Maccacaro, Gioia, and Stocke (1984)).

A description of the selection criteria for inclusion of IPC images and x-ray sources in
the survey is given in Gioia et al., 1988. In the interest of clarity we repeat them here.
We have searched the entire *Einstein* Data Bank for IPC fields with centers located
outside the galactic plane ($|b^{II}| > 20^0$) to avoid regions of high galactic absorption as
well as high stellar density. The exposure times of the IPC observations used range
from $\geq$ 800 s to 40000 s, the average value being 3600 s. We have excluded from the
analysis those fields centered on : (i) very bright (> 0.6 cnt s^{-1}) and/or extended
targets, (e.g. supernova remnants, very rich and nearby clusters of galaxies); (ii) groups
and/or associations of targets, (e.g. groups of nearby galaxies, star clusters or stellar
associations as Hyades, Pleiades etc.); (iii) all the Abell clusters with a distance class D
$\leq$ 3. This selection was necessary to prevent including in the survey sources physically
related to the target and thus not truly serendipitous.

Sources are accepted if the significance of the detection equals or exceeds 4 times
the r.m.s. level, and if they are physically unrelated to the target. To ascertain the
serendipity of the x-ray sources, each IPC image has been manually inspected. The
sources found were evaluated one by one as to the likelihood of association with the
target source and consequently excluded or retained in the catalog. Given the improved
detection algorithm and IPC background determination, we have been able to extend

the search for sources to the entire IPC field of view instead of restricting it to the central $32' \times 32'$ region, as in the MSS. Taking into account that the area under the detector window supporting structure ("ribs") was omitted from the survey (together with the central $5'$ radius region around the target), the useful area per IPC field searched for sources is about 0.54 deg^2. In the MSS each field contributed for about 0.27 deg^2.

We have analyzed a total of 1435 IPC fields, for a total area of 780 deg^2 of high galactic latitude sky. Eight hundred and thirty-six serendipitous sources have been detected. The relevant parameters of the EMSS are listed in Table 2.

Table 2

Einstein Extended Medium Sensitivity Survey

IPC images analyzed	1435
Lim. sensitivity[1]	$6 \times 10^{-14} - 3 \times 10^{-12}$
Area for IPC field	0.54
Total area of sky	780
Sources detected	836
Flux range[1]	$7 \times 10^{-14} - 10^{-11}$
Energy range	$0.3 - 3.5$ keV
Significance of detection	$\geq 4\sigma$
Sources identified[2]	605
Sources with radio data	549

[1] erg cm^{-2}s^{-1}
[2] Identifications as of January 1988.

4. Identification and Classification of the x-ray sources

As with the MSS sample we have undertaken to spectroscopically identify all the sources in the EMSS. At the time of this writing (January 1988) 605 sources out of 836 have been positively identified. Figure 4 shows the flux distribution of the EMSS sources and of the subset of identified sources. Sources which are still unidentified have been classified as either "extragalactic" or "galactic" on the basis of their x-ray to visual flux ratio f_x/f_v (Maccacaro et al., 1988). The f_x/f_v is in fact a reliable method to separate galactic from extragalactic sources prior to spectroscopy (Stocke et al., 1983, Gioia et al., 1984).

Three color (B, V, R) photometry with the CfA CCD camera on the 61 cm telescope of the Whipple Observatory is being obtained for all the AGN and BL Lacs north of -20^0 declination. About 80% of the AGN and BL Lacs have already been observed. Moreover all the sources identified with an extragalactic object, or those classified as such using the f_x/f_v method, north of -20^0 declination, have been observed at 6 cm with the NRAO Very Large Array in the C configuration (snapshot mode). We are still in the process of analyzing these data.

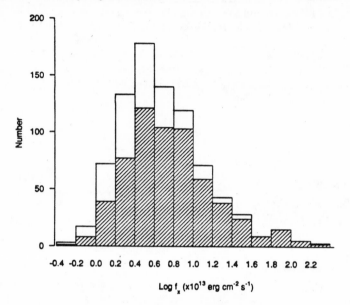

Figure 4: X-ray flux distribution of all the sources in the EMSS sample. Shaded area corresponds to spectroscopically identified sources.

Figure 5, a) through d) show the redshift, apparent magnitude, absolute magnitude and x-ray luminosity distribution respectively, for the AGN which have been so far identified and for which the relevant measurements are available. We stress that the subset of identified AGN is not necessarily representative of the final sample since the identification process has not proceeded at random. In particular priority has been given to the brightest x-ray sources (see Figure 4) and to those sources for which a radio detection was also available.

The substantial fraction of AGN already available has allowed us to obtain some preliminary results on studies for which completeness of the identification is not critical. In particular we have been able to study the x-ray energy distribution of x-ray selected AGN, showing that these objects are characterized, in the soft x-ray band, by a variety of spectral indices. We have also presented evidence that a number of the high redshift quasars found may have been gravitationally "microlensed" by a foreground galaxy. These results are discussed in detail by Maccacaro et al., (1988) and by Stocke et al., (1987) respectively. It is clear however that for most studies (e.g. determination of luminosity function and cosmological evolution) we need to complete the identification program.

5. Summary

The EMSS is an x-ray survey containing 836 sources, 605 of which have been already identified. 280 x-ray selected AGN and 28 BL Lacs or BL Lac candidates have been

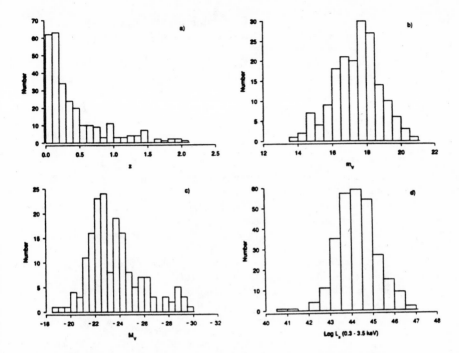

Figure 5: The redshift (a); apparent magnitude (b); absolute magnitude (c); and x-ray luminosity (erg s^{-1}, in the 0.3-3.5 keV band) (d) for the EMSS AGN.

thus far discovered. Upon completion of the optical identification program we estimate that the x-ray selected AGN sample will count about 400 objects and the x-ray selected BL Lac sample some 30 objects. Three color photometry (B, V, R) and a VLA map (5 GHz) will be available for all the AGN and BL Lacs north of -20^{0} declination.

It is a pleasure to thank the organizers of the NOAO workshop "Optical Surveys for Quasars" for inviting us to present the results from the *Einstein Observatory* Medium Sensitivity x-ray survey. This project would not have been possible without the continuous support of the MMT and FLWO time allocation committees. We also thank K. Modestino for assistance in preparing this manuscript for publication. The EMSS is supported at the Center for Astrophysics by NASA contract NAS8 30751 and by the Scholarly Study program SS88-03-87 and at the University of Colorado by NASA contract NAG8-575.

References

Boyle, B.J., Fong, R., Shanks, T., Peterson, B.A., M.N.R.A.S., 227, 717, 1987

Gioia, I.M., Maccacaro, T., Schild, R., Stocke, J., Liebert, J., Danziger, J., Kunth, D., and Lub, J., Ap. J., 283, 495, 1984.

Gioia, I.M., Maccacaro, T., Morris, S.L., Schild, R.E., Stocke, J.T., and Wolter, A., in "Large-Scale Surveys of the Sky", J.J. Condon and F.J. Lockman eds.

in press, 1988.

Maccacaro, T., Feigelson, E., Fener, M., Giacconi, R., Griffiths, R.E., Murray, S., Schwartz, D.A., and Zamorani, G., [Abstract], Bulletin of the American Astronomical Society, 11, 771, 1979.

Maccacaro, T., Feigelson, E., Fener, M., Giacconi, R., Gioia, I.M., Griffiths, R., Murray, S., Zamorani, G., Stocke, J., and Liebert, J., Ap. J., 253, 504, 1982.

Maccacaro, T., Gioia, I.M., and Stocke, J., Ap. J., 283, 486, 1984.

Maccacaro, T., Gioia, I.M., Maccagni, D., and Stocke, J., Ap. J. (Letters), 284, L23, 1984.

Maccacaro, T., Gioia, I.M., Wolter, A., Zamorani, G., and Stocke, J., Ap. J., 326, in press, 1988.

Marshall, H.L., Avni, Y., Tananbaum, H., and Zamorani, G., Ap.J., 269, 35, 1983.

Stocke, J., Liebert, J., Stockman, J., Danziger, J., Lub, J., Maccacaro, T., Griffiths, R., and Giommi, P., Mon. Not. Roy. Astr. Soc., 200, 27P, 1982.

Stocke, J., Liebert, J., Gioia, I.M., Griffiths, R.E., Maccacaro, T., Danziger, J., Kunth, D., and Lub, J., Ap. J. 273, 458, 1983.

Stocke, J.T., Morris, S.L., Gioia, I.M., Maccacaro, T., Schild, R. and Wolter, A., 1988, this conference.

Truemper, J., in "X-ray and UV emission from Active Galactic Nuclei", W. Brinkmann and J. Truemper eds. MPE report 184, p. 254, 1984.

Weedman, D.W., in "The Structure and Evolution of Active Galactic Nuclei", p. 215, Giuricin, G., Mardirossian, F., Mezzetti, M., and Ramella M., eds., 1986.

DISCUSSION

Windhorst : Is the fraction of X-ray sources that are also radio sources consistent with the results of deep surveys; i.e, about 10%?

Maccacaro : It's probably slightly higher, although there are all the usual uncertainties and questions relating to the detection limits of the different surveys.

Green : Do the new counts of sources lie roughly on the extrapolation of the count slopes that you adopted in your first published results?

Maccacaro : So far everything agrees with the results obtained from the first sample. We cannot use the quasars yet because their observations are still incomplete. But, we have done it for the cluster galaxies, and it stayed flat. We did it for the entire extragalactic population, which we could do because we could use the X-ray to optical flux ratio to discriminate between galactic and extragalactic sources, and it gives a very consistent slope.

Schmidt : Are most of the new BL Lac's again rather brighter than the X-ray limit?

Maccacaro : Yes, and you will hear more about this from John Stocke in a few minutes.

Unidentified speaker : Are there any unusual properties of the normal galaxies?

Maccacaro : Well, sometimes they are overluminous in X-rays.

Stocke : If I could amplify briefly, there are two types which appear to be normal galaxies according to spectra in the blue, but which are not. One type covers highly reddened AGN's, which may have H alpha to H beta ratios greater than 20 or evidence from IR work for B band extinctions greater than 2-3 magnitudes. The second type, of which we have only one example so far, looks like a normal elliptical galaxy but has extended X-ray emission of luminosity about 2 x 10 to the 43 erg/sec.

Cowley : How do you decide when you have made an identification? From my own experience, I worry that you stop when you get an AGN, which is why we may tend to keep finding only more and more of the same objects.

Maccacaro : First of all, our error circle is not that large, only 45" or so. It will contain from 1 to a maximum of 5 objects.

Cowley : Down to some limit, but there may be things a lot fainter.

Stocke : I'll just elaborate in two ways. One is that we work outside the galactic plane, which reduces the problem considerably compared to yours, and second, for the original survey, we looked at everything down to the Sky Survey limit, regardless of color.

Cowley : OK, that's what I wanted to hear. You don't just stop when you find an AGN.

Stocke : Now we do if we find an AGN with a reasonable X-ray to optical flux ratio. It was just too exhausting to keep on as we did originally.

Cowley : It's self-fulfilling, isn't it?

Stocke : We think in a sample of 800 sources we only have a couple that are in there by chance.

Maccacaro : I must add that for a number of identified sources we were able to obtain a better X-ray position from EXOSAT, for example, and in every case, to our satisfaction, the new position was consistent with the proposed identification.

Schmidt : Ed Turner?

Turner : I was just waving.

THE EINSTEIN OBSERVATORY EXTENDED MEDIUM SENSITIVITY SURVEY SAMPLE OF BL LACERTAE OBJECTS

John T. Stocke[1], Simon L. Morris[2],
I.M. Gioia[3,4], Tommaso Maccacaro[3,4],
R.E. Schild[4], and A. Wolter[4]

1) Center for Astrophysics and Space Astronomy
   University of Colorado
2) Mount Wilson and Las Campanas Observatories
3) Istituto di Radioastronomia, Bologna, Italy
4) Harvard-Smithsonian Center for Astrophysics

ABSTRACT: We present new results on BL Lacertae Objects discovered in an extensive x-ray survey of the high galactic latitude sky performed by the Einstein satellite. While BL Lacs have been found primarily in high frequency radio surveys, we show that x-ray emission is an efficient new method for finding such objects and will offer an outstanding opportunity to find many of these objects in the near future using the ROSAT all-sky survey. It is also shown that x-ray selected BL Lac objects have somewhat different optical and radio properties than their radio selected counterparts as would be expected from their different selection technique.
        Using a complete, flux limited sample of 16 x-ray selected BL Lacs, we show that their number counts differ decisively from those of x-ray selected QSOs. Moreover, using the few redshifts available for these objects and limits on redshift as determined from resolving the optical images of most of the remainder of the sample, a preliminary $<V/V_{max}>=0.37\pm0.06$. This value suggests an evolution in luminosity and/or density opposite in sense to that of QSOs.

INTRODUCTION

While the number of known QSOs has sky rocketed over
the last several years, and will continue to do so
based upon the on-going surveys described at this
conference, the number of known BL Lacertae Objects
has not enjoyed a similar trend. For example, while
the number of QSOs increased from ~600 to ~3600 in
the decade between the first and third Burbidge
Catalogs (Burbidge, Crowne and Smith, 1977; Hewitt
and Burbidge, 1987), the number of BL Lacs increased
from ~30 to ~80 for a discovery rate of only 5 per
year. No wonder we know so little about these
objects!

Moreover, recent surveys for QSOs have not
proven extremely efficient in finding BL Lacs. The
x-ray surveys and the high frequency radio surveys
are the most efficient methods although the new
optical surveys using several colors should also
find BL Lacs (e.g. APM color surveys; although an
admixture of elliptical galaxy starlight will cause
problems for these methods). Proving that the
spectrum of an object is truly "featureless" adds to
the difficulty in identifying large numbers of BL
Lacs.

The success rates of various techniques for
finding BL Lacs are shown in Table I including
results from the survey discussed herein based upon
the Extended Medium Sensitivity Survey (EMSS). The
basic properties of the EMSS are described in detail
in the previous article by Tommaso Maccacaro et al
in this volume.

The complete optical identification of the
original 112 MSS sources yielded a disappointing 4
BL Lacs but all 4 had x-ray fluxes $>10^{-12}$ erg/s/cm^2
in the soft x-ray band; a factor of 10 above the
faintest sources in that sample and thus a very
unusual distribution of source fluxes to say the
least (see Stocke et al, 1985; Maccacaro et al,
1984). In the continuing optical identifications of
the EMSS, 97% of all sources with $f_x \geq 8 \times 10^{-13}$
ergs/s/cm^2 have now been identified north of
declination $-20°$ and 21 new BL Lacs (16 definite and
5 marginal; see criteria below) have been found
(~ 0.03/deg^2), all with $f_x > 10^{-12}$! Indeed, in all of
the EMSS identifications ($f_x \geq 0.8 \times 10^{-13}$) made thus
far only 7 additional BL Lacs and BL Lac candidates
have been discovered with $f_x < 8 \times 10^{-13}$ so that the
unusual distribution of x-ray fluxes among the BL
Lacs found earlier is confirmed.

Since centimeter radio emission is detected in 20 out of 21 of the high flux BL Lacs (fr $\geq$1 mJy), a combined x-ray/radio selection process yields the identification of BL Lacertae Objects at the extremely high rate of 2 in 5 (see Table I). Therefore, a very efficient method for finding several hundred new BL Lacs will be available soon using the **ROSAT** all-sky survey. A VLA survey of bright (fx $\geq$10-¹²) **ROSAT** point-like sources will give arcsecond positions for hundreds of BL Lac Objects with a 40% success rate; the remaining 60% of these sources will be almost exclusively low redshift (Z $\leq$0.3) AGN.

TABLE I   Survey Efficiencies for Finding BL
            Lacertae Objects

Survey	Flux Limit @ Survey Frequency	# of BL Lacs	Total # of Sources	Percentage Efficiency	Reference
1 Jy "all sky"	1 Jy @ 5 GHz	44	518	8.5%	1
S5 radio survey	0.25 Jy @ 5 GHz	14	185	7.5%	2
PG sample	B=16.5 @ B-band	4	1715	0.4%	3
HEAO-1 A-2	1 X 10⁻¹¹ @ 2-10 keV	4	61	6.5%	4
Original MSS	1 X 10⁻¹³ @ 0.3-3.5 keV	4	112	3.6%	5
EMSS "bright"	8 X 10⁻¹³ @ 0.3-3.5 keV	21	178	11.8%	6
EMSS "bright" + VLA radio	8 X 10⁻¹³ @ 0.3-3.5 keV 1 mJy @ 5 GHz	20	52	38.4%	6

¹ Kuhr et al, 1981.
² Kuhr, private communication.
³ Schmidt and Green, 1985.
⁴ Piccinotti et al, 1983.
⁵ Maccacaro et al, 1985.
⁶ This paper.

THE EMSS BL LAC SAMPLE

In order to compile a complete, x-ray flux limited sample of BL Lacs from the EMSS sample, the following **observational** criteria are used:

1. A member of an EMSS sub-sample which has been completely identified optically. At present this means: X-ray flux $f_x \geq$ 8 X 10^{-13} ergs/s/cm^2 in the 0.3-3.5 keV band and Declination $\geq$-20°.
2. In addition, the x-ray emission must be point-like in the IPC to avoid confusion with clusters of galaxies. This criterion currently eliminates 4 sources whose x-ray IPC images **may** show extendedness.
3. Radio flux $f_r \geq$ 1 mJy (this eliminates one object which is not detected at $f_r <$0.17 mJy; 3 sigma limit)
4. Optical spectrum "featureless" in the following senses: (a). no emission lines with W >5A and (b) if a CaII "break" is visible due to the underlying galaxy it must have a contrast <25% ensuring the presence of a significant power law component. These criteria are currently limited to the spectral region 3400-6000A.

As has been shown previously (Stocke et al, 1985), these criteria seem sufficient to define an object as a BL Lac Object. That is, there are enough similarities between the x-ray selected objects defined using the above criteria and traditional BL Lacs found by radio means to warrant the use of this label. The alternative is to admit a new class of "active" extragalactic object similar to the BL Lacs that we might term "x-ray galaxies". Sixteen EMSS sources meet the 4 criteria above and so constitute the complete, x-ray and radio flux-limited sample used to derive number counts in the next section.

At present we are requiring no optical polarization or variability data to define the sample although such observations are currently underway. Broad-band optical photometry for at least 2 epochs is now available for 19 of the 21 high flux BL Lacs (including the radio quiet object and the objects which are imbedded in x-ray emission which may be extended). Variability is seen in all observed so far at a level of at least a few tenths of a magnitude. An optical polarization survey is also underway and polarization has been positively detected at the few percent level in 5 of 9 objects observed at this writing (B. Jannuzi, private communication). Moderate resolution (~8A) spectroscopy in the 6000-8500A region and increased SNR spectra in the blue are also being obtained. Almost all of the EMSS BL Lacs easily meet the "featureless" criterion above [typically W (emission lines)<1A and CaII break contrast $\leq$10%] so that criterion #4 is based upon the SNR of the faintest EMSS BL Lacs (V~20).

Reanalysis of the IPC x-ray data on the EMSS BL
Lacs is also underway and includes both a careful
test for extended x-ray emission in the four cases
mentioned above and a measurement of the
distribution of power law spectral indices for all
of the BL Lacs (see Maccacaro et al, 1988; for first
spectral index results from the EMSS).

Based upon the optical, radio and x-ray data
for the EMSS BL Lacs gathered thus far, the
following generalizations can be made concerning the
basic properties of x-ray selected BL Lacs (XBLs)
compared to their radio selected counterparts
(RBLs):

1. The optical spectra of XBLs reveal a higher
starlight fraction than do RBLs (see also Halpern et
al, 1986);

2. The optical polarization percentage of XBLs is
systematically of a lower amount than in RBLs;

3. Although there are some notable exceptions (e.g.
H0323+022; Doxsey et al, 1983) XBLs are
systematically less variable in total optical flux
and polarization than RBLs; and

4. XBLs have systematically different overall energy
distributions as shown in Figure 1 below.

Also note the following in Figure 1:

1. The radio-to-optical flux ratio is very similar
for all of the XBLs **except for** the radio-quiet
candidate 1E1603.6+2600 which lies well outside the
distribution. Because this object is not detected in
the radio additional optical observations are needed
before we can safely conclude that it is the first
radio quiet BL Lac. This object could still prove to
be a DC or C2 white dwarf (see e.g. Wegner et al,
1985). We also note that the previous radio quiet BL
Lac candidate 1E1704.9+6046 (Chanan et al, 1982) has
been detected by us at the VLA at 1.8 mJy and lies
within the distribution of radio-to-optical flux
ratios shown in Figure 1. So to date it is premature
to suggest that any radio quiet BL Lacs exist.

2. The x-ray to optical flux ratio of x-ray selected
BL Lacs exceeds that of any previously known class
of extragalactic object.

3. The first three observational properties listed
above are consistent with XBLs being the same type
of object as RBLs but with a viewing angle somewhat
offset from the radio and optical beaming axis.
Figure 1 supports this idea in that if the radio and
optical emission in the x-ray selected objects were
boosted by relativistic beaming (radio boosted
somewhat more than optical), the XBL boxes would

move into the area of the RBL ellipses.
4. The FR I type radio galaxies (small x's in Figure
1), which are thought to be the parent population
for the BL Lac Objects (Wardle, Moore and Angel,
1984), lie horizontally to the right of the XBLs in
Figure 1. The relative positions of these two
classes in Figure 1 is consistent with both
exhibiting almost exclusively unbeamed radio and
optical emission but with the XBLs having x-ray
emission boosted by a factor of ~200 (relativistic
Γ s~5).

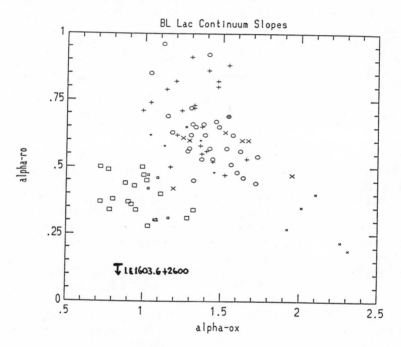

Fig. 1   The overall energy distributions for
several classes of active extragalactic object.
Alpha-ox and alpha-ro are the two point optical
to x-ray and radio to optical spectral indices
(see Stocke et al, 1985). The boxes are x-ray
selected BL Lacs (large boxes $f_x > 10^{-12}$
ergs/s/cm^2; small boxes $f_x < 10^{-12}$; hand-drawn box
is previous "radio quiet" candidate of Chanan,
et al, 1982), the ellipses are radio selected
BL Lacs, the plus signs are x-ray selected AGN,
the crosses are radio galaxies (large crosses
are FR type IIs, small crosses are FR type Is).
A 3 sigma radio upper limit for the only radio
quiet BL Lac **candidate** is also shown.

THE NUMBER COUNTS AND COSMOLOGICAL EVOLUTION OF BL LACS

The Log $N(\geq S)$ vs. Log S plot for XBLs is shown in Figure 2 utilizing data from both the EMSS and from the all-sky survey of HEAO-1 A-2. The EMSS point at S=8 X 10^{-13} is the complete sample of 16 XBLs discussed above. The EMSS point in parentheses to the left of that is the incomplete sample of 7 objects with fluxes below 8 X 10^{-13}; 145 sources remain unidentified out of a total of 675 in the EMSS north of -20°. Figure 2 shows that even at 8 X 10^{-13} the XBL number counts differ significantly from those of the QSOs in our sample (4 sigma level) and appear to be flattening dramatically within the flux range of the EMSS.

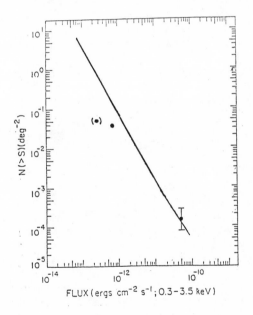

Fig. 2   The Number Count Diagram for X-ray Selected BL Lacs. The point at 5x10^{-11} erg/s/cm^2 is the complete sample of 4 objects from the HEAO-1 A-2 all-sky survey; the 8 X 10^{-13} point is the complete sample of 16 BL Lacs from the bright subsample of the EMSS; and the point in parentheses includes the above 16 plus the incomplete sample of 7 BL Lacs and BL Lac candidates with f_x < 8x10^{-13} from the entire EMSS. The solid line indicates the "Euclidean" slope.

Proposed reasons for this flattening shown in Figure 2 are:

(1). A large "K-correction" due to XBLs having very steep (i.e. "soft") x-ray spectra. This suggestion is plausible due to the steep spectra observed for some of the brightest BL Lacs like PKS 2155-304 (Urry, Mushotzky and Holt, 1986; Giommi et al, 1987) but for the spectral slopes observed ($\alpha = -1$ to -2) the K-correction is small for $Z \leq 2$ and is mitigated in some of the objects due to the presence of a harder component above a few keV. Moreover, the IPC spectral indices for the XBLs found in the EMSS have a distribution of x-ray slopes indistinguishable from that of the EMSS QSOs.

(2). "Beaming" distorts the luminosity function and number counts of the XBLs (Urry and Shafer, 1984; Cavaliere, Giallongo and Vagnetti, 1986). Since relativistic beaming is thought to be important to some extent in all wavelength bands for BL Lacs, this effect must be considered even though it may not be dominant.

(3). "Negative" cosmological evolution, meaning that there are fewer BL Lacs at higher redshift and/or that they were less luminous in the past (Stocke and Perrenod, 1981).

These three reasons cannot be distinguished using the number counts alone. And yet further tests as to what causes the unusual distribution of x-ray fluxes among XBLs require information on the distance to these sources, a very difficult quantity to obtain given their "lineless" nature. Here the x-ray selected samples like the EMSS have a considerable advantage over radio samples because the XBLs have a higher starlight fraction than do the RBLs. In practice this means that 5 of the 16 (8 out of 21 if the marginal XBLs are included) have spectroscopic redshifts determined without using the "tricks" necessary previously (e.g. observing through an annulus; Oke and Gunn, 1974). In addition, 6 of the XBLs without spectroscopic redshifts have been resolved on R-band CCD frames taken with the 24" telescope of the F.L. Whipple Observatory on Mt. Hopkins and so placing an upper limit on their redshift of $Z=0.4$. This leaves 3 not optically resolved and 2 not well observed. Therefore, although extremely preliminary (primarily due to the absence of solid distance information), a first measurement of the evolutionary status of BL Lacs using the V/V_{max} test is possible.

To carry out this test using the 16 EMSS BL
Lacs, the following assumptions were made:
1. Distances were assumed using spectroscopic
redshifts where available (5 objects with Z=0.2-0.6
range); failing that a redshift of 0.3 was assumed
for all objects resolved with the 24" (6 objects
resolved at near the 24" CCD limit). All other
objects (5) were assumed to have Z=0.5.
2. Although an IPC spectral index distribution for
the EMSS BL Lacs as a class is available (see
Maccacaro et al, 1988 for the technique used), here
a slope of -1.0 was assumed for all sources. This
value is not too different from the actual observed
distribution.
3. V/Vmax values were calculated using both the x-
ray (8 X 10^{-13}) and radio (1 mJy) limits. Only 3
objects were radio limited.
4. Due to the variation in solid angle surveyed at
different limiting fluxes, the Avni and Bahcall
(1980) Ve/Va method was employed.
    Under these assumptions the complete sample of
16 EMSS XBLs yields the <V/Vmax> values shown in
Table II. The lack of solid distance information for
the majority of the sample is the biggest
uncertainty in concluding that these **preliminary**
values indicate decreasing numbers and/or luminosities for BL Lacs in the past. We are currently
engaged in obtaining higher SNR spectra and deeper
direct imaging to address this crucial uncertainty.

TABLE II   <V/Vmax> Values for the EMSS XBLs
            (q_o=0.5 assumed)

Sample Description	<V/Vmax>
16 source complete sample	0.37 ± 0.07
16 sources plus 4 which could be extended in x-rays	0.36 ± 0.06
16 sources with 5 unresolved optically assigned Z=2.0	0.42 ± 0.06
22 X-ray Selected AGNs from same EMSS subsample which are radio detected	0.65 ± 0.06

    If this low <V/Vmax> value is confirmed, the
most straightforward conclusion is that BL Lacs are
decreasing in number and/or luminosity beyond Z~0.5.

The members of the EMSS team thank the organizers of this conference for giving us the opportunity to present these preliminary results from our
on-going survey. Besides the support acknowledged in
the previous article, J.T.S. specifically thanks the
MMT night assistants and staff for their excellent
support of this work. J.T.S. also acknowledges the
support of NASA grant NAG8-658 and of a "start-up"
grant from the University of Colorado.

REFERENCES

Avni, Y. and Bahcall, J. 1980, Ap. J., **235**, 694.
Burbidge, G., Crowne, A., and Smith, H. 1977, Ap. J.
    Suppl., **33**, 113.
Cavaliere, A., Giallongo, E., and Vagnetti, F. 1986,
    Astr. Ap., **156**, 33.
Chanan, G., Margon, B., Helfand, D., Downes, R., and
    Chance, D. 1982, Ap. J. (Letters), **261**, L31.
Doxsey, R. et al, 1983, Ap. J. (Letters), **264**, L43.
Giommi, P., et al, 1987, Ap. J., **322**, 662.
Halpern, J., Impey, C., Bothun, G., Tapia, S.,
    Skillman, E., Wilson, A., and Meurs, E. 1986,
    Ap. J., **302**, 711.
Hewitt, A. and Burbidge, G. 1987, Ap. J. Suppl., **63**,
    1.
Kuhr, H. 1986, private communication.
Kuhr, H., Witzel, A., Pauliny-Toth, I., and Nauber,
    U. 1981, Astr. Ap. Suppl., **45**, 367.
Maccacaro, T., Gioia, I., Maccagni, D., and Stocke,
    J. 1984, Ap. J. (Letters), **284**, L23.
Maccacaro, T., Gioia, I., Wolter, A., Zamorani, G.
    and Stocke, J. 1988, Ap. J., in press.
Oke, J. and Gunn, J. 1974, Ap. J. (Letters), **189**,
    L5.
Piccinotti, G., Mushotzky, R., Boldt, E., Holt, S.,
    Marshall, F., Serlemitsos, P., and Shafer, R.
    1982, Ap. J., **253**, 485.
Schmidt, M. and Green, R. 1983, Ap. J., **269**, 352.
Stocke, J. and Perrenod, S. 1981, Ap. J., **245**, 375.
Stocke, J., Liebert, J., Schmidt, G., Gioia, I.,
    Maccacaro, T., Schild, R., Maccagni, D., and
    Arp, H. 1985, Ap. J., **298**, 619.
Urry, M. and Shafer, R. 1984, Ap. J., **280**, 569.
Urry, M., Mushotsky, R., and Holt, S. 1986, Ap. J.,
    305, 369.
Wardle. J., Moore, R., and Angel, J. 1984, Ap. J.,
    279, 93.
Wegner, G., Yachovich, F., Green, R., Liebert, J.,
    and Stocke, J. 1985, PASP, **97**, 575.

DISCUSSION

Miller : I was intrigued by this turnover in the log N - log S relation. Is your sample in any sense volume limited?

Stocke : It is a flux-limited sample.

Miller : I couldn't understand how you could deal with computing a luminosity function in order to explain that turnover.

Stocke : Basically, beaming flattens the observed luminosity function relative to the intrinsic one. To amplify, if beaming is important then the brightest sources in this sample should be high luminosity objects at high redshift, redshifts comparable to quasars. But, we don't see that in our sample. In fact, the brightest X-ray sources are the ones which we either have resolved or have determined that the redshifts are not larger than a few tenths. My prejudice is that we are seeing evolution and that there are fewer BL Lac's beyond $z = 0.2$.

Weymann : Could you comment on the environment of BL Lac's?

Stocke : We have 4 cases where we see BL Lac's embedded in extended X-ray emission. There was one previous case where a BL Lac was seen centered in a cluster of galaxies with a redshift of a few tenths. At low redshifts (all we have), BL Lac's appear to be in richer environments than AGN's.

Peacock : You mentioned earlier in your talk the possibility that the Fanaroff-Riley class was 1, that radio galaxies are the parents of BL Lac's. Doesn't this provide another possible explanation for why you don't have any high redshift ones in the sample? Any BL Lac that was going to lie in your sample at high redshift would have a radio luminosity such that it was no longer FR 1 but FR II. In other words, I'm suggesting you modify your V over V max test to allow whether the object falls above or below the FR limit. Would that make a difference?

Stocke : We haven't done that, but if I understand what you're suggesting, at higher luminosities and lower redshifts, what we call BL Lac's should have emission lines and look like quasars.

Peacock : Yes, and be called highly polarized quasars rather than BL Lac's.

Stocke : Well, in this particular connection, I would offer an alternate to Richard's suggestion yesterday, that the quasars seen at redshift near 0.6 are not obscured AGN's but would be called BL Lac's at lower redshift.

Cowley : I'd like to comment on the apparent low discovery rate by optical surveys. I don't think it's real; it's our failing to record negative observations. I know that in our survey, and I think Ann Savage said something similar, about 5% of the objects turn out to be featureless at our S/N. We don't know if they are DC's, some other kind of blue star, or what. They may be BL Lac's, but since you can get 5 good quasars in the time it takes to get a good spectrum of one of the faint, featureless objects, we don't follow up on them.

Stocke : I realize that these objects seem boring to some of you who love contrasty features in optical spectra, but I remind you of my cartoon at the beginning showing that one BL Lac is now worth 100 quasars!

Cowley : But, we don't even know - they may be white dwarfs.

Stocke : That's why the radio limit is so important, at least until we find an example of a highly polarized, continuous spectrum object with no radio emission.

Cowley : We'll send you all the positions and finding charts.

Schmidt : We are highlighting the reason why X-ray information is very interesting.

Chaffee : I wanted to understand why in the cases where you have redshifts, you feel that the calcium is associated with the BL Lac rather than coming from a foreground object.

Stocke : Ostriker has suggested that perhaps BL Lac objects are lensed and that what we're seeing is a foreground galaxy. My only response is that it would then be very unusual, for in every case so far, the putative foreground object looks like a giant elliptical. There are no spirals. There is no offset from the center of the object.

Hartwick : Can you estimate the space density of BL Lac's?

Stocke : Yes. Not from this work yet, but I recall other estimates that their space densities are comparable to those of bright Seyfert galaxies, which makes it unlikely that they are the parent population. But the FR I's are much higher in space density and therefore could be the parents.

MODELS OF THE QUASAR POPULATION:
II. THE EFFECTS OF DUST OBSCURATION

Julia Heisler
Steward Observatory, University of Arizona, Tucson, Arizona,
85721

Jeremiah P. Ostriker
Princeton University Observatory, Princeton University, Princeton, New Jersey, 08544

ABSTRACT A model for dust obscuration is proposed to account for the apparent lack of high redshift quasars which are assumed to exist, but to be obscured by dust in intervening galaxies. The model successfully reproduces the apparent magnitude and redshift distributions of observed quasar samples at bright, intermediate, and faint magnitudes. Because spotty absorption tends to remove objects from the sample rather than redden them, observable quasars with redshifts less than 3.5 are reddened by only small amounts in conformity with observations. Since X-rays penetrate the dust, the predicted 2-10 Kev X-ray flux of quasars with an absolute blue magnitude brighter than -21.5 is larger than it would otherwise be and equals 82% of the observed background, with roughly one half of the observed background arising from dust obscured quasars that appear to have an absolute magnitude fainter than -21.5. The predictions in the radio are consistent with the observations of QSOs in the Parkes radio survey. If the dust model is correct and radio quasars exist at high redshift there should be many "blank" fields in fainter radio surveys that detect very low luminosity objects.

I. INTRODUCTION

The sharp decline in the quasar population beyond a redshift of about 2 leads us to consider the possibility that our knowledge of the distant universe has been distorted by dust in intervening galaxies. The typical *observed* quasar has, we found, suffered only one third to one half as much extinction as the mean extinction to a given redshift and so may be relatively unreddened. The net result is a picket fence obscuration where a larger and larger fraction of the quasar population is simply removed from a magnitude limited sample as z increases.

In §IIa we discuss the dust properties of intervening galaxies, and in §IIb we describe the quasar properties adopted for the models. Optical results are given in §III for the comoving number density, and mean color excess and optical depth. The model cannot be easily tested using optical results since it has been designed to fit the extant optical surveys. But our predictions for other wavelength bands such as X-ray and radio are given in §IV and provide a means of testing the overall picture. Section V presents conclusions.

II. QUASAR MODELS

We will carry out all of our calculations for a Friedmann cosmology with $\Omega = 1$, $(q_0 = 0.5)$ and a Hubble constant of $H_0 = 100$ km/s/Mpc. We adopt this fiducial model to explore the effects of dust obscuration and reduce the number of free parameters by fixing (H_0, q_0).

a) Dust Properties
The optical depth encountered by light emitted at a redshift z is the sum of the optical depths due to galaxies along the line of sight at redshifts $z_{gal} < z$. For light observed in the $(z = 0)$ blue frequency band we can define τ_B to be the total optical depth encountered by the emitted light. (In this paper the optical depth will always be referred to the blue band, and we will generally drop the subscript B for convenience.)

In this work we assume that galaxies have "soft" edges; for comoving observers the optical depth through the disk decreases exponentially with the distance from the galaxy's center. Furthermore, we allow the disks of galaxies to be tilted with respect to the line of sight. We adopt a frequency distribution for the grain opacity that is based on observations of the Galaxy's UV interstellar extinction curve. We have used Seaton's (1979) fitted analytical expressions for the optical depth in the vicinity of the 2200Å bump observed in our Galaxy.

To determine the distribution function for the total optical depth two other constants must first be specified. These constants are, $\Delta\tau_0(z)$, the optical depth in the blue frequency at the center of a local galaxy, and the constant τ_g, which determines the number of disks along a line of sight. In this work we have set $\Delta\tau_0(z) = 0.5$ and $\tau_g = 0.2$ This value for τ_g gives a covering fraction for the sky that is consistent with the fraction of quasars detected by Wolfe et al. (1986) that have damped Lyα absorption.

b) Quasar Properties
A fiducial quasar with which other quasars at a given epoch may be compared with is assumed to have a luminosity $L(\nu, z) = L_\star(\nu, z) = A_0(\nu/\nu_B)^\alpha \exp(kt)$, with $A_0 = 5.2 \times 10^{29}$ergs s^{-1} Hz^{-1}, corresponding to an absolute magnitude $M_\star = -22.55 - 1.086kt$; k is a free parameter, ν_B is the blue frequency, and $t = 1 - (1 + z)^{-1.5}$ is the dimensionless

light travel time written as a fraction of the age (look back time) of an $\Omega = 1$ universe. The spectral index, α, is set equal to -0.5.

We shall describe the comoving quasar number density by a luminosity function that combines luminosity evolution with luminosity dependent density evolution, but corrected for the effects of dust. The total comoving number density of quasars per unit "indicative" absolute magnitude is given by an integral over the distribution function for τ and is fully specified by three free parameters, n_0, j, and k. The parameter n_0 describes the local quasar number density, j controls the luminosity dependent density evolution and k the luminosity evolution. All absolute magnitudes quoted are indicative rather than actual in the sense that these are the absolute magnitudes that would be inferred from observations of flux and redshift by an observer who was not aware of the dust obscuration. (When we refer to the true, unobscured absolute magnitude we use the subscript "tr"). The value of the A_0 was chosen to fit the curvature of the local luminosity function.

A model is determined once n_0, j and k are specified. We determine these parameters by fitting the observed differential apparent magnitude distribution at the bright end using the Bright Quasar Survey (Schmidt and Green 1983) and at the faint end using the Koo-Kron survey (Kron 1985 and Koo, Kron and Cudworth 1986). Model A is our best fit to the observations, including dust; Model D is our best model *without* dust. Model AN represents Model A without dust; hence it describes the *true* quasar population.

III. OPTICAL RESULTS

a) Redshift Distribution

Figure 1 shows as a function of redshift the observed comoving number density per cubic Gpc of quasars brighter than $M_{in} = -25$;

$$n_{25}(z) \equiv \int_{-\infty}^{-25} n(M_{in})dM_{in}.$$ The left hand vertical axis gives the logarithm of the comoving number density in intervals of $\log(e) = 0.43$; Models A, D, and AN are shown. The right hand vertical axis gives the optical depth in intervals of 0.43, and the average optical depth to a redshift of z is graphed. The figure reveals that in the presence of dust, $n_{25}(z)$ is decreased by about $\exp(a(z)\bar{\tau})$, where a is a slowly varying function of z; for $z = 2$, $a = 1.3$ and for $z = 4$, $a = 1.2$. Large redshift quasars exist, as shown by the curve AN, buth they cannot be observed. Convolving the true increase in density with redshift shown in AN with the obscuration $\bar{\tau}(z)$ produces a broad flat maximum for $1.5 < z < 3.5$ and a peak in the apparent distribution A at redshift $z = 2.2$.

Redshift distributions for the interval $19 < B < 22$, are shown in Figure 2 for Models A and D. The models are each compared with the observed distribution of the Koo-Kron survey. Only the dust model accounts for the observed decline in redshift.

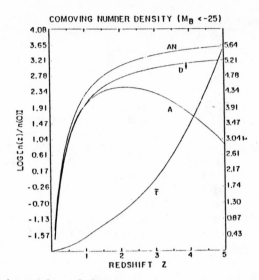

Fig. 1 The logarithm of the comoving number density of quasars brighter than $M_{in} = -25$ is plotted as a function of redshift for Models A, D, and AN. Plotted with respect to the right hand axis is the average optical depth to redshift z. (The left hand axis should read $\log[n_{25}(z)]$.)

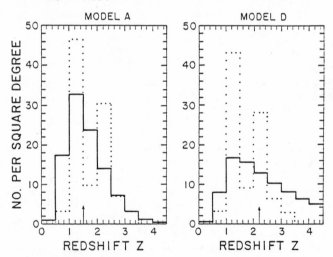

Fig. 2 Redshift histogram for faint quasars ($19 \leq B \leq 22$). The results of the Koo-Kron survey are shown in a dotted histogram.

b) Mean Color Excess and Optical Depth

A potential problem with the suggestion that quasars are obscured by dust is that the dust may excessively redden large redshift quasars and prevent their detection in surveys that identify candidates by color. In this work the observed color of a quasar is assumed to be the sum of the intrinsic color of the source and the dust reddening. (The K correction to the color for a power law source, $L \sim \nu^{\alpha}$, is zero.)

In Figure 3a we have plotted the mean $B - V$ color excess, $\langle E_{B-V} \rangle$, as a function of redshift for quasars brighter than $B = 22$. The $1\,\sigma$ deviation from the mean is indicated by dots. The curve lies consistently below the average reddening, $\overline{E}$, that would be observed if dust were distributed uniformly. For example, for $z = 1$, $\overline{E} = 0.10$, but $\langle E_{B-V} \rangle = 0.070$ for a sample brighter than $B = 22$; for $z = 2.5$ $\overline{E} = 0.14$ and $\langle E_{B-V} \rangle = 0.022$. The mean color excess is less because of the constraint of a magnitude limit; heavily reddened quasars are likely to be fainter than the magnitude limit. Reddening out to a redshift of 3 is typically quite small; it does not, in the mean, exceed $E_{B-V} = 0.1$ with the variance broadening the range to $-0.1 \lesssim E_{B-V} \lesssim 0.2$ for $z < 3$.

In Figure 3b we have plotted the mean optical depth to detected quasars, $\langle \tau \rangle$ as a function of redshift and for $B < 22$. The dashed line gives the average optical depth to a random point, $\overline{\tau}(z)$. For $z \gtrsim 2$ the average optical depth is at least a factor of two greater than the mean optical depth to a detected quasar. If the magnitude limit is reduced to 19, the average optical depth becomes a factor of $2.5 - 3$ greater than that to a detected quasar for $z \gtrsim 2$. This correction, due to observational selection, must be noted if observed samples are used to determine dust obscuration. The effect of selection is obvious but it has not been allowed for in all published studies of this problem.

IV. PREDICTIONS IN OTHER WAVEBANDS

In the previous section we argued that high redshift quasars tend not to be detected in optical surveys due to dust obscuration. We have set up a luminosity function that involves an integral over a dummy variable τ that we physically interpreted as representing dust. However, it is possible that this indicative luminosity function describes the *true* quasar distribution and that the integral over τ introduced in the equation for the luminosity function does not have a physical interpretation. Using optical data alone, there exists two logically indistinguishable possibilities, namely, that either a) high redshift quasars exist, but are obscured by dust (which we refer to as Model A), or b) high redshift quasars do not exist, but they are described by a luminosity function identical to the indicative (or apparent) luminosity function of Model A. We shall refer to this second possibility as Model A$'$. In this section we will make predictions for both of these models, since it is only in the other wavelength bands that we can discriminate between the two.

a) X Rays

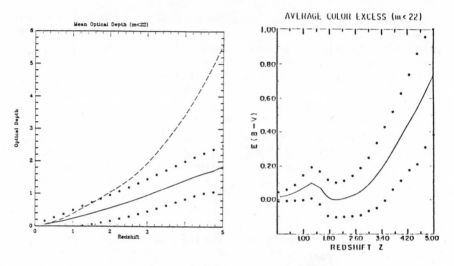

Fig. 3a. Mean reddening of quasars brighter than $B = 22$ is plotted as a function of redshift. The solid curve gives the mean value of E_{B-V} and the dots give the 1σ deviation. The 2200Å bump in the frequency distribution of the opacity is clearly evident. Note that reddening is not expected to be significant for quasars with $z < 3$.

Fig. 3b. Mean optical depth of quasars brighter than $B = 22$ as a function of redshift; true shown dashed and apparent shown as a solid line. The 1σ deviation from the mean is indicated by dots. Note that optically selected quasars are less obscured than the true average obscuration by typically a factor of two for this magnitude limit.

We compute the expected contribution by quasars to the 2-10 Kev soft X-ray background. We adopt Anderson and Margon's (1987) specific results for the average dependence of the X-ray to optical ratio (i.e. α_{ox}) upon optical luminosity, which are based upon the Avni and Tananbaum (1986) finding that this ratio decreases with increasing optical luminosity and that no more than a few percent of optically selected quasars can be X-ray quiet. Furthermore, we impose a true absolute magnitude limit equal to what Anderson and Margon use, $M_{tr} \leq -21.5$ (blue band, $H_0 = 100$ km/s/Mpc), and a lower limit of -30.

Counting all quasars of any apparent magnitude we find that 82 percent of the 2-10 Kev background measured by Marshall et al. (1980) is due to quasars for Model A, 27% for Model A'. About one-half of the X-ray flux from Model A is generated by quasars that have indicative magnitudes $M_{in} \geq -21.5$, (but $M_{tr} \leq -21.5$). Only 40% of the flux from Model A can be attributed to quasars with $B < 22$ and $M_{in} \leq -21.5$. Thus, an observer unaware of the presence of dust would attribute about 42% of the X-ray background to $(B > 22, M_{in} > -21.5)$

objects, that due to their faintness would probably not be classified as quasars. For the X-ray properties we have assumed, the dust model contribute a large fraction and luminosity evolution of intrinsically faint quasars and Seyferts could severely limit the X-ray contribution from apparently faint quasars in a dust model. However, as noted above, a large fraction of the background in Model A is due to quasars that have $M_{tr} \leq -21.5$ but appear much fainter with $M_{in} > -21.5$.

The total number of quasars contributing to the background is 1000 quasars/deg^2 for Model A, for the absolute magnitude cutoffs given above. Since AXAF should be able to see a quasar such as 3C273 to redshifts beyond 5, we would expect it to detect quasars that are optically faint (i.e. $M_{in} > -21.5$), and many of these quasars should appear as blank fields. Without further study it is not possible to say if the number of sources contributing to the background is sufficient to account for the "smoothness" of the X-ray background in the Einstein long exposure measurements (Hamilton and Helfand 1987).

b) Radio

Since radio photons penetrate dust, we might, in principle, expect to see high redshift radio sources that are invisible in the optical due to dust obscuration, assuming such radio sources exist. How many "blank fields" should be present in radio surveys is an important question to consider.

There are two difficulties to consider. The first is a general lack of correlation between radio and optical luminosities (e.g. Marshall 1987). Instead, a wide range of radio luminosities can be observed among quasars with similar optical luminosities. To deal with this problem, we adopt Marshall's (1987) analytical fit of the guillotine function, $g(\alpha_{or})$, which describes the probability that a quasar has a given value of $\alpha_{or} \equiv 0.1854 \log(L_r/L_{opt,tr})$. ($L_r$ and $L_{opt,tr}$ are the true monochromatic luminosities at 5 Ghz and 2500Å, respectively, in the rest frame of the quasar.) (Furthermore, in all of these calculations we use a magnitude limit $M_{B,tr} \leq -22.5$, as suggested by Peacock *et al.* (1986).) Second, it is not known whether the fraction, f, of quasars that are radio loud is a constant. In particular, if f decreases sharply with z, then blank fields will not occur. In this case, radio quasars exist only at low redshifts where dust obscuration is small; hence, one would expect all radio sources to have optical counterparts. However, if f is relatively constant, then high redshift radio quasars do exist, some may be heavily obscured and their existence or nonexistence could provide a test of our model. For the purpose of determining whether dust could be noticed in already existing radio data, such as the Parkes survey, we will assume f is constant.

We will examine two subsamples of the Parkes survey. The first is described by Wills and Lynds (1978, hereafter WL). This sample consists of 69 quasars brighter than $S = 0.35$ Jy at 2.7 Ghz and selected from an area of 0.619 sr; sixty sources (87%) have $V \leq 19.5$ (corresponding to $B \lesssim 19.8$). Fifty-four sources have measured redshifts and $B < 19.8$; of these nine have $z > 2$ and none have $z > 3$. The second subsample is described by Downes *et al.* (1986, hereafter DPSC). This sample consists of 28 steep spectrum and 24 flat spectrum quasars brighter than 0.1 Jy. Out of an additional 14 flat spectrum sources with blank fields or uncertain optical identifications, 7 are suspected to be quasars due to their variability. Thus, out of 31 probable flat spectrum quasars, 7 (23%) appear to be fainter than $B \sim 21$; however, this percentage probably represents an upper limit. Twenty-nine steep and flat spectrum sources have $B < 21$ and measured redshifts; of this group, 3 have $z > 2$ (10%) and none have $z > 3$.

The predicted fraction of quasars with $B > 19.8$ and brighter than 0.35 Jy is 30% for Model A and 3% for A'; the observed fraction in the WL sample is 13%. The fraction for $B > 21$ (and $S > 0.1$ Jy) for Model A is also in closer agreement to the observed fraction of 23% for flat spectrum sources in DPSC.

For $B < 19.8$ both Models A and A' predict similar redshift distributions beyond $z = 3$. For an area the size surveyed by WL both models predict of order one quasar above $z = 3$ as compared to the none observed. A similar conclusion holds with respect to the DPSC sample $(B \lesssim 21)$. Furthermore, Model A predicts that among quasars with $B < 19.8$, 15% should have $z > 2$. This compares well with the observed percentage of 17% (WL). However, for quasars with $B < 21$ the predicted $z > 2$ fraction is 25% for A as opposed to 13% for A'; the observed (DPSC) fraction is 10%. Model A's estimate is somewhat high, but the discrepancy may not be significant because of the small number statistics: out of 29 flat and steep spectrum quasars with measured redshifts, only 3 have $z > 2$. This number is within 1.6σ of Model A's estimate using Poisson statistics.

Fainter radio limits are needed to effectively distinguish between Models A and A'. At a limit of 0.1 mJy, about 30% of all radio quasars should have optical counterparts dimmer than $B = 22$ if dust exists. However, if there is no dust and Model A' is correct *no quasars should have $B > 22$ in an 0.1 mJy sample.* To test this prediction one must obtain such a faint sample and attempt to identify as many sources as possible. Two surveys (Windhorst *et al* 1985 and Windhorst *et al.* 1986) appear to weakly support Model A by identifying quasars dimmer than $B = 22$; however, no radio quasars are dimmer than $B = 24$. If Model A is correct, it is odd that none were found beyond $B = 24$. High redshift $(z > 3)$ quasars, if they exist, should be heavily obscured and have optical magnitudes $B \sim 28$ for absolute magnitudes $M_{B,tr} \sim -22.5$. (Those quasars that happen not to lie along the path of an intervening galaxy must have $B < 24$ $M_{B,tr} < -22.5$.) A lack of quasars or suspected quasars with $B > 24$ might only be explained if there a separate, true redshift cutoff in the radio exists at $z \sim 2.5$, thus effectively making the dust question unanswerable with radio data.

V. CONCLUSION

This paper examines one solution to a persistent problem with previously adopted quasar models, namely, the overestimate of the expected number of high redshift quasars. We propose that large numbers of high redshift quasars do indeed exist, but they are not observed because their light is obscured by dust in intervening galaxies.

We have constructed a model (A) to fit the observed apparent magnitude distribution and then demonstrated that dust imposes an apparent redshift cutoff (Figures 1 and 2). In the presence of dust, it was found that the observed comoving number density of objects brighter than $M_{in} = -25$ is decreased by about $\exp(a(z)\bar{\tau}(z))$, where $a(z) \sim 1.2$ and $\bar{\tau}(z)$ is the average optical depth along a random line of sight. Furthermore, a sample brighter than $B \sim 22$ will have an apparent redshift cutoff at $z \sim 3$.

The mean color excess due to dust as a function of redshift for a magnitude limited sample was also computed. We found that the mean reddening was very small; $\langle E_{B-V} \rangle \lesssim 0.1$ for $z < 3$. Apparently, due to selection, the observed quasars are not excessively reddened by the dust, and it is likely they will still be detected in surveys that select candidates by color. For $z > 1.5$, the mean reddening of a magnitude limited sample tends to be smaller than the average reddening along the line of sight by more than a factor of two.

Finally, we considered how the model affects observations in the X-ray and radio regimes. We computed the expected contribution of quasars to the soft X-ray background. We found that quasars with $-30 \leq M_{B,tr} \leq -21.5$ could contribute 82% of the observed background in the 2-10 Kev band, with one half of the flux coming from dust obscured quasars. In the radio we expect about 30% of radio loud sources with radio fluxes above ~ 0.1Jy to have optical counterparts fainter than $B = 21$, and for most of these sources to be located at low redshift ($z \lesssim 2.5$). At mJy flux limits it should be possible to discriminate between the dust and no dust models by determining the fraction of blank fields that are quasars, assuming there is no separate redshift cutoff in the radio. All radio quiet quasars should be identifiable with objects brighter than $B = 22$ if dust does not exist.

ACKNOWLEDGMENTS

We thank R. Kron for providing us with his data on faint quasars, as well as a long list of people with whom we had very helpful discussions. These were M. Fall, J. Gunn, C. Hogan, C. McKee, P. Osmer, J. Peacock, M. Schmidt, J. Stocke, E. Turner, S. White, and R. Windhorst. We also express our gratitude to E. Wright, who outlined for us in detail the method used to compute the optical depth distribution, and to P. Joss for providing material support. This work was supported by NSF Grants AST-8307654, AST-8352062, and AST86-15845, and NASA grant NAGW-765.

REFERENCES

Anderson, S. and Margon, B. 1987. *Astrophys. J.* **314**, 111.

Avni, Y. and Tananbaum H. 1986. *Astrophys. J.* **305**, 83.

Downes, A.J.B., Peacock, J.A., Savage, A. and Carrie, R. 1986. *Mon. Not. Roy. Astron. Soc.* **218**, 31.

Hamilton, T.T and Helfand, D.J. 1987. *Astrophys. J.* **318**, 93.

Koo, D.C., Kron, R.G. and Cudworth, K.M. 1986. *Pub. A.S.P.* **98**, 285.

Kron, R.G. 1985. Private Communication.

Marshall, H.L. 1987. *Astrophys. J.* **316**, 84.

Marshall, H.L. *et al.* 1980. *Astrophys. J.* **235**, 4.

Peacock, J.A., Miller, L. and Longair M.S. 1986. *Mon. Not. Roy. Astron. Soc.* **218**, 265.

Schmidt, M. and Green, R.F. 1983. *Astrophys. J.* **269**, 352.

Seaton, M.J. 1979. *Mon. Not. Roy. Astron. Soc.* **187**, 73.

Wills, D. and Lynds, C.R. 1978. *Astrophys. J. Suppl.* **36**, 317.

Windhorst, R.A., Miley, G.K., Owen, F.N., Kron, R.G. and Koo, D.C. 1985. *Astrophys. J.* **289**, 494.

Windhorst, R.A., Dressler, A. and Koo, D.C. 1986, in *IAU Symposium 124, Observational Cosmology*, eds. G. Burbidge and L.Z. Fang (Dordrecht: Reidel).

Wolfe, A.M., Turnshek, D.A., Smith, H.E., and Cohen, R.D. 1986. *Astrophys. J. Suppl.* **61**, 249.

DISCUSSION

Peterson : When you have a clumpy dust distribution, there's always a certain fraction of your objects that are not obscured at all. Can you give us an idea of what this fraction would be?

Heisler : For the exponential disks we are assuming, essentially any quasar at high redshift will be at least slightly obscured.

Marshall : Have you tested the models against other data sets that have come out in V magnitudes? How was the optical depth of the galaxies determined? Were they integrated over an evolutionary galaxy luminosity function?

Heisler : If I understand your question correctly, no, the distribution property is defined by specifying how dusty each disk is, how many of them there are, and by assuming a certain frequency dependence for the optical depth.

Marshall : I was in need of the numbers.

Heisler : The numbers of objects. We took a number which gave us the correct redshift distribution and then checked to see if it was at least reasonable in terms of the covering fraction for all galaxies.

Peacock : As you know, I think the redshift cutoff in the radio distribution is real, but I agree that the radio surveys can test the dust hypothesis. We do know that there are radio-selected quasars with redshifts up to virtually 4, which are presumably unaffected by dust.

Heisler : I know, but you only have a few, and I don't think there's very much you can say from that.

Boyle : A number of years ago the Durham group produced tentative evidence for dust in clusters of galaxies based on the observed anticorrelation on the sky between UVX objects and low redshift galaxies. We have since extended this work based on our UVX spectroscopic survey, and we confirm that it is present in all 7 fields, although at a reduced level. It requires about 0.2 mag absorption in the B band per cluster. These results are in press in the MNRAS. You might like to incorporate these parameters in your models.

Surdej : If the dust absorption has an important effect on the observed properties of quasars, wouldn't you expect to see a correlation between dust and the the number of absorption lines, for example?

Heisler : We haven't looked for it yet.

Chaffee : Now that we know of a half dozen redshift 4 quasars, it would seem we are building up an adequate sample to look for such effects.

ACTIVITY PATTERNS AND EVOLUTION OF AGNs

A. Cavaliere
Astrofisica, Dip. Fisica II Università di Roma, Italy

E. Giallongo
Osservatorio Astronomico di Capodimonte, Italy

P. Padovani
Università di Padova, Italy; STScI, Baltimore, USA

F. Vagnetti
Astrofisica, Dip. Fisica II Università di Roma, Italy.

ABSTRACT   Scenarios for the QSO population evolution are discussed in the context of three basic patterns for the activity of individual AGNs: long-lived, short-lived and recurrent activity.

FACTS AND PROBLEMS

Evidence has been growing over the past 20 years (since Schmidt 1968) to show that the QSO population "evolves": the (comoving) optical luminosity function $N(L, z)dL$ depends very effectively on z, and in fact – looking back into our past light cone – it increases at a given L by a large factor, up to $\sim 10^4$ out to $z \sim 2.2$. This means

$$\frac{\partial N}{\partial z} > 0 , \qquad (1)$$

or $\partial N/\partial t < 0$ along the cosmological arrow of time, on a fast scale $\tau_p \sim 1$ Gyr.

The evidence rested for a while on integral tests like the number counts rising steeply up to magnitudes $B \sim 19$, or the $V/V_{max} > 0.5$. Now it is increasingly based upon direct resolution of $N(L, z)$: Boyle et al. 1987, see also these Proceedings, using multi-objects spectroscopy resolve $\Delta z \sim 0.5$ out to $z \sim 2.2$ and up to $B \sim 21$. The data are consistent with a "luminosity evolution" (Mathez 1976, Braccesi et al. 1980): between $z \sim 2 - 0.5$ the *number* of QSOs is statistically conserved at $N_a \sim$ a few % of the bright galaxies, while their *average luminosity* $\langle L(z) \rangle$ decreased by some 10^{-2}.

Actually, Cavaliere, Giallongo and Vagnetti 1985, Weedman 1986, Boyle et al. 1987 and others, pointed out that the data (to within their uncertainties) even suggest a close conservation of the *shape* of the distribution per magnitude interval $\sim N(L, z)L$: this looks like almost bodily shifted to lower L to join onto the local data of Cheng et al. 1985, *as if* each distant QSO dimmed down almost uniformly to become a Seyfert $1 - 1.5$ in our local environment. The time scale required for such a dimming ($\dot{L} < 0$) is $\tau_L \equiv L/|\dot{L}| \sim 3$ Gyr, corresponding to a scale $\tau_p = \tau_L/(\gamma - 1) \sim 1$ Gyr for

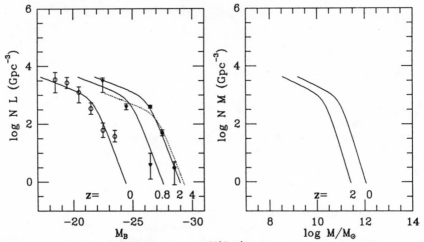

Fig. 1a. A model for the luminosity function $N(L, z)$. For $z < 2$, luminosity evolution of type C, uniform with $\tau_L = 3$ Gyr. From $z = 10$ to 2, the QSOs are born with $\tau_S = 1.5$ Gyr. Note that a flat l.f. sticks out at high z when $\tau_L/(\gamma - 1) \lesssim \tau_S$ as luminosity increase offsets number decrease. Data from Cheng et al. 1985, Boyle et al. 1987. $H_o = 50$ km/s Mpc, $\Omega_o = 0.2$.
Fig. 1b. Corresponding distributions of nuclear masses in AGNs, "bolometric correction" $\kappa = 10$.

a population with the observed $N(L)$ in the form of a steep power-law at the bright end, $N(L) \propto L^{-\gamma}$ with $\gamma \sim 3.5$ (cf. fig. 1).

But 3 Gyr is an odd time scale for individual behaviours, being $\ll t_o$ and yet $\gg$ the natural time scale associated with an accreting black hole, the most appealing prime mover to energize a QSO on account of stability and efficiency $\eta \sim 10^{-1}$ (cf. Rees 1984). In fact, the time scale $\eta t_E = \eta c \sigma_T / 4\pi G m_p \sim 4 \ 10^{-2}$ Gyr governs the simplest accretion flow: self-limited at the Eddington luminosity, from an unlimiting supply, to yield $L \propto \exp(t/\eta t_E)$.

To force in an effective dimming scale ~ 3 Gyr it takes hence a "Great Coordinator", and obvious candidates are: the gradual exhaustion in splendid isolation of a large mass stockpile by a formed hole, or the petering out of social interactions of the host galaxies with companions that recurrently replenished the mass supply, or tuned initial conditions ordering from inside the sequence of black holes formation. All these have been appointed to synchronize the activity of the *individual* objects, producing three major scenarios for the *population* evolution.

From the other end, all scenarios must face a beginning, that early epoch when the first QSOs populated the $L - z$ plane: $\partial N/\partial t > 0$, i.e., a decline toward high z:

$$\frac{\partial N}{\partial z} < 0 \,. \tag{2}$$

In fact, a paucity of high-z objects was long inferred from preliminary data by Osmer (1982). It was recently confirmed or discussed in the light of new results from surveys based on multicolor selections and from searches based on slitless spectroscopy, reviewed or presented in these Proceedings.

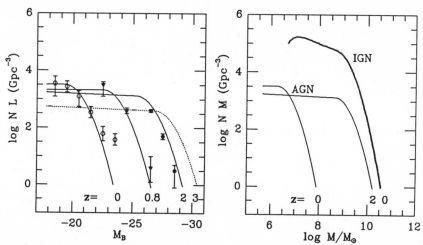

Fig. 2a. Luminosity functions, model of type E. The objects are born at low L down to the present, on a scale τ_S changing from 1 to 17 Gyr for $z > 2$, < 2, respectively. Once born, they brighten up with $L \propto \exp(t/\eta t_E)$, but beyond $L_b = 10^{43}(1+z)^5$ erg/s they turn off statistically (emulating an average dimming) at a rate $1/\tau_{S-} = 25 \ln L/L_b$ Gyr^{-1} (cf. Cavaliere et al. 1983). The quenching phase is assumed short so as to contribute negligibly to N.

Fig. 2b. Corresponding distributions of nuclear masses for Active and Inactive Nuclei, $\kappa = 10$.

These data show that high-z QSOs do exist, with spectra similar to their closer counterparts, and often with large intrinsic power. We have heard at this Meeting some agreement on an apparent decline by $1/5 - 1/10$ between $z \sim 2.5 - 3.5$, not unlike the outcome of Véron's 1986 compilation of slitless searches. But the statistical weights and the selection biases are not yet fully assessed, so the detailed structure of the decline will remain for some time under scrutiny.

Problems here arising: is some curtain lowered at $z \gtrsim 2.5$ to obscure our view, or are we reaching the physical edge of the QSO universe to witness their real turning on? How come a QSO rise so sharp?

A most intriguing feature of the QSOs is how do they manage – certainly for $z \lesssim 2$ and perhaps beyond – to evolve on scales $< H^{-1}(z)$, faster than the Universe as a whole.

AMBIGUITIES AND CONSTRAINTS

The diversity of the evolutionary scenarios may seem surprising, but in fact this is permitted by the structure of the continuity equation

$$\frac{\partial N}{\partial t} = S(L,t) - \frac{\partial(\dot{L}N)}{\partial L} \tag{3}$$

(Cavaliere, Morrison and Wood 1971) governing the population evolution along the cosmological arrow of time: $\partial N/\partial t$ may be driven by a *combination* of death or birth rates for the objects (term $S < $ or > 0), with average dimming or brightening (term $\partial(\dot{L}N)/\partial L$ with $\dot{L} < $ or > 0).

Less ambiguous is the early phase of rise $\partial N/\partial t > 0$, due to birth and/or brightening to observability which superpose to the extent of differing only nominally.

As to the phase of fall, $\partial N/\partial t < 0$ for $z \lesssim 2.2$, the literature existing (Cavaliere et al. 1983, Blandford 1986, Cavaliere and Vagnetti 1987, also Schmidt 1987) shows that *diverse* combinations of the 1st and 2nd term on r.h.s. describing reasonable activity patterns can produce remarkably *close* behaviours of $N(L, t)$, consistent with the data. Relevant examples are shown in figs. 1 and 2.

But additional constraints to the individual activity patterns may be set from nuclear mass determinations: these are increasingly being attempted on the basis of the relationship $M \propto \alpha v^2 d/G$ ($\alpha \simeq 1$) both in *Active* and in *Inactive* Galactic Nuclei with complementary uncertainties (see Woltjer 1959, Kormendy 1987, Cavaliere and Padovani in preparation). The constraints to be expected under the *accretion* assumption for the evolutionary scenarios have been worked out in some detail by Cavaliere Padovani and Vagnetti 1987 (CPV), comparing two simple relationships for the properly normalized form of the specific luminosity L/M, i.e., for the Eddington ratio L/L_E of the bolometric to the Eddington luminosity $L_E = Mc^2/t_E = 1.3 \ 10^{46}M_8$ erg/s.

In the last reference we expanded somewhat on these relationships to convey a feeling for their robustness confirmed by detailed numerical work, but their basic content is intuitive: First, given that the average L decreases with t increasing while $M \propto \int dt L(t)$ is to build up (mostly at the beginning), the longer the activity lasts in each nucleus the smaller will be L/M locally. Second, the shorter is the activity, the more numerous will be the generations of AGNs required to reproduce the statistically conserved number of objects: easily all galaxies reasonably bright are involved as hosts.

Formally, a first relationship holds for any object deriving its power from accretion:

$$\frac{L}{L_E} = \frac{\eta t_E L}{\int dt L(t)} \ , \tag{4}$$

which is easily computed from a given model.

To link in with the statistical approach to follow, note that eq. 4 has also a statistical meaning (to within factors $1/2 - 2$, not worse than the data uncertainties) when computed for the representative objects near the "break" of the luminosity functions.

Finally, the fully statistical approach of Soltan 1982 and of Phinney 1983 may be taken up (CPV) to relate the average Eddington ratio with the observed density of light (dominated by $B \simeq 20\pm1$, at $z \simeq 1.5$) or with ρ_S, the mass density of AGNs associated to the former under the accretion assumption. The other quantity intervening is the total number N_a/δ of the nuclei ever been active ($\delta \leq 1$ being the duty cycle or the ratio on/on+off), and the connecting relationship reads

$$\langle \frac{L}{L_E} \rangle = \frac{t_E \langle L \rangle N_a}{c^2 \rho_S \delta} \ . \tag{5}$$

This is independent of the value of the bolometric corrections that appear both in L and in ρ_S, and applies to local values as well as to $z \neq 0$ as long as the objects involved contributed substantially to ρ_S. With this proviso, eliminating L/L_E from eqs. 4 and 5 yields an estimate for the number N_a/δ of host galaxies required by a given model.

Numerically, a local value $\langle L/L_E \rangle \sim 10^{-4}\eta_{-1}/\delta$ obtains from the following data: the luminosity function of the Seyfert 1 – 1.5 galaxies (Cheng et al. 1985), to derive

$\langle L \rangle \sim 2 \ 10^{44}\kappa_1$ erg/s and $N_a \sim 6 \ 10^3$ Gpc^{-3}; the optically-selected QSO counts, as reviewed e.g. by Koo et al. (1986), yielding $\rho_S \sim 7 \ 10^{13} \ \kappa_1/\eta_{-1}M_\odot$ Gpc^{-3}, with account duly taken of the big bump redshifted into the B band. We use a "bolometric correction" $\kappa \equiv 10 \ \kappa_1$ over $L_{2500} \ \nu_{2500}$. Note consistency of eqs. 4 and 5 for $\delta = 1$, the case of continuous activity.

BASIC ACTIVITY PATTERNS

Including continuous activity, three basic activity patterns span the domain of interest:

C (for *continuous*). Starvation caused by a given, limited mass stockpile characterizes the extreme case of literal luminosity evolution, with each object undergoing a long, *continuous* dimming after a short formation phase at $z > 2$. From eq. 4 local values $2 \ 10^{-4}\eta_{-1} \lesssim L/L_E \lesssim 5 \ 10^{-3}\eta_{-1}/T_1$ obtain. The lower bound refers to continuity down to the present. The upper limit is provided by a population evolution extremely differential in L, i.e. one with no change at the faint end since the turning-on epoch (look-back time $T \equiv T_1 \ 10$ Gyr). Evolutions non-uniform in t are described by a time scale $\tau_L \propto t^n$ to yield $L \propto \exp(t/\tau_{L_o})$ for $n = 0$ and $L \propto (1+z)^\beta$ for $n = 1$: the latter case with $\beta \sim 3.5$ (cf. Boyle et al. these Proceedings), yields L/L_E different from the above numbers only by a factor of $\sim 1/2$. A strong trend for L/L_E increasing with z is expected, stronger than the increase of $L(z)$ alone.

E (for *event*). At the other extreme, activity may represent a *single, short event* in the host galaxy lifetime, i.e., the very formation phase of the black hole. If accretion proceeds at the fiducial – but not guaranteed – rate $\dot{M} \sim \eta^{-1}\dot{M}_E$ for a duration inversely correlated with epoch in many successive generations of different Active Nuclei, an evolutionary scenario obtains (Blandford 1986, Cavaliere and Vagnetti 1987 and fig. 2) that emulates a luminosity evolution yet implying at all z ratios $L/L_E \sim 1$: suitable initial conditions are required, like a tuned production rate of black holes or a tuned sequence of latency times after a nearly coeval formation. Note: not necessarily $L/L_E = 1$, see Lightman, Zdziarsky and Rees 1987; some indirect trend of L/L_E with z may still be possible if $\dot{M}(z)$ is finely tuned to the distribution of initial hole masses.

R (for *recurrent*). An intermediate class is provided by intermittent activity consisting of episodes lasting $< \tau_L$ and *recurring* after longer times on average, for a cumulative effective duration $t_e \equiv t_{e,-2} \ 10^{-2}$ Gyr: this behaviour may be driven by accretion rekindled by decreasing events of interactions of the host with companion galaxies (Norman and Silk 1983, Gaskell 1985; for recent evidence cf. Smith et al. 1986, Yee and Green 1987). Here $L/L_E \sim 5 \ 10^{-2}\eta_{-1}/t_{e,-2}$. The corresponding evolutionary scenario can emulate case C on average as shown by ongoing numerical work based on interaction rates $\langle N_g\sigma v \rangle \propto t^{-n}, n \gtrsim 2$. A clear trend of L/L_E with z obtains anyway, essentially following the increase of $L(z)$.

Fig. 3 in its lower part illustrates the three kinds of mass distribution expected in currently *Active* Nuclei from the above activity patterns. In its higher part, it visualizes also the corresponding mass distributions predicted for currently *Inactive* Nuclei.

As to the number of host galaxies, in case C $(\delta = 1)$ we recover $N_a \sim 6 \ 10^3$ Gpc^{-3}. For patterns of class R, eq. 5 yields for the total numbers of nuclei entertaining activity $N_a/\delta \sim 3 \ 10^6$ Gpc^{-3}. With pattern E such extrapolations to local and weak emitters do not hold when the corresponding objects never contributed materially to ρ_S; here eq. 5 gives instead $N_a/\delta \sim 7 \ 10^5$ Gpc^{-3} already for $z \gtrsim 1.5$, the range contributing $\sim \rho_S/2$. These numbers are given for $H_o = 50$ km/s Mpc and scale $\propto H_o$.

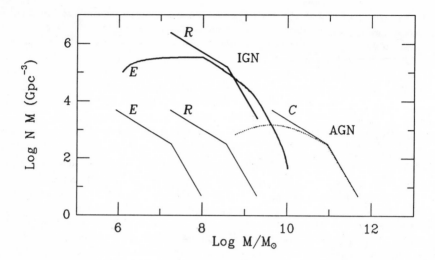

Fig. 3. Schematic mass distributions at $z = 0$ for Active (lower, thin lines) and Inactive (upper, thick lines) Galactic Nuclei as predicted by the three activity patterns discussed in the text: C with a local Eddington ratio $L/L_E \simeq 2 \; 10^{-4}$, R with $L/L_E \sim 5 \; 10^{-2}$ and E with $L/L_E \sim 1$ always. "Bolometric correction" $\kappa = 10$. The dotted line shows a differential variant ($\dot{L} \to 0$ for $L \to 10^{43}$ erg/s) minimizing the masses at the faint end of C, and likewise of R, models.

The first constraint to these patterns is that $\langle L/L_E \rangle > 10^{-2}$ locally will constitute a threshold ruling out *continuous* accretion models, however well simulated by the statistical observations. Recent data reductions or collations – in X-rays by Wandel and Mushotzky (1986) (but see Lawrence et al. 1987), and in the IR-optical-UV by Padovani and Rafanelli (1987) – yield values for $\langle L/L_E \rangle$ tham seem to be settling into the range $10^{-2} \div 10^{-1}$. *If* this will hold out, then *discontinuous* models in the range from the intermediate R to the extreme case E will be selected. Telling signatures of these two alternatives are discussed.

Pattern E – although it tends to blot out the evidence of stability based on recurrence, long held a signature of black hole activity – will be demonstrated by finding: values $L/L_E \sim 1$ with no strong trend for Nuclei *Active* at any z; and masses systematically larger (by as much as 10^2, see fig. 2) in currently *Inactive* Nuclei. The inequality $M_{IGN} > M_{AGN}$ is expected to weaken with z increasing, mirroring the observed increase of the average luminosities.

Patterns of class R are consistent with the morphological evidence (cf. Smith et al. 1986, Yee and Green 1987) of frequent associations of nuclear activity with *distortions* of the galactic bodies: presumable signs of social interactions of the host with companion galaxies in groups. They imply that holes more massive than $10^6 \div 10^7 M_\odot$ will be found in all galaxies reasonably bright, with $L_B > 0.1 L_*$. As said, a clear increase of L/L_E with z is predicted, from values $\sim 5 \; 10^{-2}$ locally up to ~ 1 out to $z \sim 2$.

Both patterns of types R and E requiring so many hosts are constrained from the other end by the flat shape of the galaxy luminosity function (cf. Phinney 1983). For example, when $\langle L/L_E \rangle \gtrsim 5 \; 10^{-2}$ holds, eq. 5 predicts that even galaxies with $L_B \lesssim 0.1 \; L_*$ be involved. Observations such as those of Smith et al. (1986) that tend to associate active nuclei with *bright* galaxies are potentially telling, provided it is assessed whether the body's enhanced luminosities are associated with larger masses or instead with starburst activity.

Patterns R are easily consistent with relatively small galaxies bright by star bursting, as they envisage nuclear fueling from the same interactions conducive to first trigger extensive rejuvenation of the star populations. A similar connection for the radiogalaxies is strongly indicated by growing evidence, cf. Windhorst et al. 1985, Heckman et al. 1986. In addition, our poster paper (these Proceedings) considers random variances plus systematic trends introduced by emission lines and by photometry to show that the effective QSO timescale may be renormalized up to ~ 4 Gyr: this becomes consistent with the timescale for the evolution of starburst galaxies found necessary by Danese et al. 1987 to explain the counts of mJy radiosources and the IRAS counts.

Pattern E, instead, is less easily related to interactions, whilst it involves at least as many galaxies ever gone through an active phase: in fact, after eq. 5 some 10^2 generations are required down to $z \sim 1.5$, with an addition $> 2 \ 10^2$ from there to now. We are exploring the possibility that a nuclear super-cluster of stars (as envisaged by Duncan and Shapiro 1983 and Scoville and Norman 1987) may be catalyzed by interactions to switch from slow evolution to fast self-destruction yielding a large outburst.

HIGH z

The interpretation of the high z decline that is closer to a scenario of short-lived objects and is concerned with the requirement of limiting the AGN contributions to the X-Ray Background to under 50% (De Zotti et al. 1982), envisages the *amplitude* of the luminosity functions declining for $z \gtrsim 2$, while the average L still increases (see fig. 1). The rise time indicated for the apparent turn-on of the bulk of the QSOs – i.e. $\tau_S \sim 1$ Gyr, intriguingly shorter than $H^{-1}(z \sim 2.5)$ – stimulates questions concerning connections with the process of galaxy "formation". One envisages a high probability of deep collapses and rapid accretions during the era of protogalaxy contractions and settlings, when sufficiently massive gravitational wells first took shape in a highly non-steady environment.

Qualitatively, any such QSO-protogalaxy connection must be rooted at $z \gtrsim 5$: this is becoming obvious for the QSOs now being found at $z > 4$, but it ought to hold also for those at $z \sim 2 - 3$ *if* the bulk of their decline is really so sharp.

Quantitatively, the source $S(t)$ may be connected (Cavaliere and Vagnetti 1987) with the rate of galaxy formation dN_g/dt as visualized by:

$$S(t) \propto \frac{dN_g}{dt}(t' = t - t_q) . \tag{6}$$

The time lag t_q represents the time to trigger and to build up a massive black hole. The effective rise time τ_S is set by the balance of steepening and smoothing effects.

An average $\langle t_q \rangle \sim 1$ Gyr tends to steepen the QSO turn-on as the formations of the parent objects are mostly confined to considerably earlier cosmic times when all time scales were naturally shorter. But the statistical distribution of t_q around $\langle t_q \rangle$ tends to smooth out the rise: its dispersion $\sigma = \langle (t - \langle t_q \rangle)^2 \rangle^{1/2}$ reflects variances of the initial conditions and alternative routes to collapse (from prompt dynamical infall, to Eddington-limited accretion lasting $\sim \eta t_E \ln M_{fin}/M_{in}$, to slow dissipation of angular momentum $\mathbf{j}$ or of random kinetic energy $\langle v^2 \rangle$ before the runaway terminal collapse, see Rees 1984).

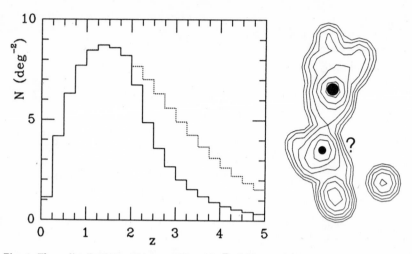

Fig. 4. The z-distributions of high z QSOs with $B \leq 21$ as predicted by two models: Continuous line, same as in fig. 1. Dotted line, evolution like in fig. 1 but extrapolated to high z with no source. Note some decline still present in the latter case, due to the observed flux "sliding down" the assumed $N(L)$ with z increasing.

Fig. 5. To visualize AGNs in primeval galaxies; the model primeval galaxy is adapted from fig. 3 of Baron and White 1987, with their permission.

Another, intrinsic smoothing effect arises from the distribution of the times taken by a galactic M_g to recollapse, i.e., to become non-linearly bound. In the framework of the hierarchical clustering scenario, at each epoch a typical mass $M_c(t)$ recollapsed, but a decreasing number of larger M_g also recollapsed being associated with larger initial perturbations, while a number of smaller M_g still survived the reshuffling into larger units: this shapes the galaxy mass distribution $N_g(M_g, t)$ intervening in eq. 6.

Cavaliere and Szalay 1986 used for N_g a model simple yet representative of hierarchical clustering, and included in S the above three steepening or smoothing effects. They found interestingly short rise times τ_S when $\langle t_q \rangle \sim 1$ Gyr ($\Omega_o = 1$), implying most triggering events to occur at $z_g \gtrsim 5$. In addition, the dispersion σ had to be limited to $\langle t_q \rangle / 2$, suggesting near critical accretion for most objects with a forerunning minority of prompt collapses. Two additional, observationally relevant effects from the shape of the luminosity function are illustrated in figs. 1, 4 and noted in their captions.

But if one accepts values $z \gtrsim 5$ for the onset of an extensive hole formation, then in keeping with the thrust of the hierarchical clustering scenario (Silk and Wyse 1986, Baron and White 1987) embodied in $N(M_g, t)$ one expects (CPV) the formation process to begin within $10^9 - 10^{10} M_\odot$ clouds or clumps that are gradually coalescing – with much dissipation – to build up a protogalaxy: the host masses involved will scale down from the values $\sim 10^{12} M_\odot$ at $z \sim 2$ following in a critical universe $M_g \propto t^{2/3a} \propto (1+z)^{-6/(n+3)}$ ($\Omega_o = 1$), where $n \simeq -2$ is suggested by the cold dark matter scenario (cf. Blumenthal et al. 1984). At high z the limiting factor for a sizeable hole was the small mass of the clumps then prevailing: $z \lesssim M_g^{-(n+3)/6} - 1$ yields a firm upper limit $z \lesssim 7$ ($\Omega_o = 1$) for the bright objects formed promptly in clumps with total mass

$\sim 5 \ 10^9 M_\odot$.

DISCUSSION

As for alternative explanations of the apparent decline, J. Heisler in these Proceedings discusses obscuration by absorbtion in dusty intervening galaxies and counters the expectations of a decline smeared out in z and associated with much precursory reddening. On the other hand, obscuration hardly can explain the similar decline of the radio-loud quasars both in the radio and in the optical band (cf. Peacock 1985, also these Proceedings). Finally, a canonical emission of many objects beyond $z \sim 2$, even though optically obscured, would still overproduce the XRB (cf. Cavaliere and Vagnetti 1987).

So for $z \gtrsim 2$ one is led back to consider fast turning on of QSOs, largely originated in a pre-galactic, hectic era and in protogalactic clumps probably too weak and too faint then to be observed by their star light only. Whence one is tempted into two speculative expectations (see fig. 5): multiple active nuclei in some protogalaxies, implying spectral signatures (cf. Gaskell 1983) of binary objects among high-z QSOs; a few isolated objects after all, stragglers from their original protogalaxies.

As for $z < 2$, fast evolution is likely associated with steeply decreasing interactions. Two final comments concern the reality of this evolution. First, our poster paper shows that noise and trends added to photometry by the emission lines and by the sky limit may tend to rise somewhat the effective evolutionary scale and its uncertainties, but cannot seriously jeopardize the evidence of an evolution much faster than H_o^{-1}. Second, just as that scale becomes consistent with the scale for starburst activity, interest is added to the view that one might dispense with the lore of population evolutions after all, and instead blame all apparent excesses of bright sources at $z > 0$ simply to cosmology, in the form either of high values of q_o (with $\Lambda \neq 0$ to resettle the cosmological timescales, cf. Wampler 1987), or of the minimum in the flux-redshift relationship implied by chronometric cosmology (cf. Segal and Nicoll 1986). However, Cavaliere and Colafrancesco (1988 preprint) note one counterexample offering some anchoring if preliminary evidence from the results of the Medium Sensitivity Survey in X-rays (Gioia et al. 1987): the number counts for Clusters of Galaxies are so flatter than the AGN's that one hardly envisages the difference (a factor ~ 3.5 at the faint end) explained out only in terms of incompleteness for faint, extended and cool clusters or groups. If the extension of the MSS and the softer survey expected from ROSAT will confirm the difference, the notion will be nailed down that population evolutions – however complicated by uncertain activity patterns, by shape of the luminosity functions and by cosmology – still bear some fundamental imprints of the different astrophysics of the sources.

REFERENCES

Baron, E., and White, S.D.M. 1987, *Ap. J.* **322**, 585.
Blandford, R. D. 1986, in *Quasars*, Proc. I.A.U. Symp. Nr. **119**, Bangalore, p. 295.
Blumenthal, G., Faber, S., Primack, J. and Rees, M.J., 1984, *Nature* **311**, 517.
Boyle, B. J., et al. 1987, *M.N.R.A.S.* **227**, 717.
Braccesi, A., Zitelli, V., Bonoli, F., and Formiggini, L. 1980, *Astr. Ap.* **85**, 80.
Cavaliere, A., Giallongo, E., Messina, A., and Vagnetti, F. 1983, *Ap. J.* **269**, 57.
Cavaliere, A., Giallongo, E., and Vagnetti, F. 1985, *Ap. J.* **296**, 402.
Cavaliere, A., Morrison, P. and Wood, K. 1971, *Ap. J.* **170**, 223.

Cavaliere, A. and Szalay, A.S. 1986, *Ap. J.* **311**, 589.

Cavaliere, A., and Vagnetti, F. 1987, Workshop *Supermassive Black Holes*, Fairfax, Virginia, to appear in the Proceedings.

Cavaliere, A., Padovani, P., and Vagnetti, F. 1987, Proc. Erice Workshop *Toward Understanding Galaxies at Large Redshift*, in press (CPV).

Cheng, F. Z., Danese, L., De Zotti, G., and Franceschini, A. 1985, *M.N.R.A.S.* **212**, 857.

Danese, L., De Zotti, G., Franceschini. A., and Toffolatti, L. 1987, *Ap.J. (Letters)* **318**, L15.

De Zotti, G., Boldt, E. A., Cavaliere, A., Danese, L., Franceschini, A., Marshall, F. E., Swank, J. H., and Szymkowiak, A. E. 1982, *Ap. J.* **253**, 47.

Duncan, M. J., and Shapiro, S.L. 1983, *Ap. J.* **268**, 565.

Gaskell, C.M. 1983, in *Quasars and Gravitational Lenses*, Proc. 24th Liège Int. Ap. Colloquium, p. 473. Institut d'Astrophysique, Universitè de Liège.

Gaskell, C.M. 1985, *Nature* **315**, 386.

Gioia, I.M., et al. 1987, preprint.

Heckman, T.M., et al. 1986, *Ap. J.* **311**, 526.

Koo, D.C., Kron, R.G., and Cudworth, K.M. 1986, *P.A.S.P.* **98**, 285.

Kormendy, J. 1987, Workshop *Supermassive Black Holes*, Fairfax, Virginia, to appear in the Proceedings.

Lawrence, A., Watson, M. G., Pounds, K. A., and Elvis, M. 1987, *Nature* **325**, 694.

Lightman, A.P., Zdziarsky, A.A., and Rees, M.J. 1987, *Ap. J. (Letters)* **315**, L113.

Mathez, G. 1976, *Astr. Ap.* **53**, 15.

Norman, C., and Silk, J. 1983, *Ap. J.* **266**, 502.

Osmer, P.S. 1982, *Ap. J.* **253**, 28.

Padovani, P., and Rafanelli P. 1987, *Astr. Ap.*, submitted.

Peacock, J.A. 1985, *M.N.R.A.S.* **217**, 601.

Phinney, E. S. 1983, Ph. D. thesis, Univ. Cambridge, England.

Rees, M. J. 1984, *Ann. Rev. Astr. Ap.* **22**, 471.

Schmidt, M. 1968, *Ap. J.* **151**, 393.

Schmidt, M. 1987. Talk at 13th Texas Symposium on Relativistic Astrophysics

Scoville, N., and Norman, C. 1987, preprint.

Segal, I.E., and Nicoll, J.F. 1986, *Ap. J.* **300**, 224.

Silk, S.J., and Wyse W.R.F.G. 1986, in *Structure and Evolution of Active Galactic Nuclei*, p.173, Giuricin, Mardirossian, Mezzetti and Ramella eds., Reidel, Dordrecht.

Smith, E. P., Heckman, T. M., Bothun, G. D., Romanishin, W., and Balick, B. 1986, *Ap. J.* **306**, 64.

Soltan, A. 1982, *M.N.R.A.S.* **200**, 115.

Véron, P. 1986, *Astr. Ap.* **170**, 37.

Wampler, E. J. 1987, *Astr. Ap.* **178**, 1.

Wandel, A., and Mushotzky, R. F. 1986, *Ap. J. (Letters)* **306**, L61.

Weedman, D.W. 1986, in *Structure and Evolution of Active Galactic Nuclei*, p.215, Giuricin, Mardirossian, Mezzetti and Ramella eds., Reidel, Dordrecht.

Windhorst, R.A., Miley, G.K., Owen F.N., Kron, R.G., and Koo, D.C. 1985, *Ap. J.* **289**, 494.

Woltjer, L. 1959, *Ap. J.* **130**, 38.

Yee, H.K.C. and Green, R.F. 1987, preprint.

DISCUSSION

Marshall : Several years ago I pointed out that the chronometric cosmology could not fit the observed number count - redshift distributions when extended to 20th magnitude. Extreme conditions are needed to fit the chronometric cosmology in particular.

Cavaliere : I agree with that.

Marshall : I also noticed a few years ago that the Eddington timescale is the natural timescale for the turning on of quasars in an interval which is much shorter than the local expansion timescale. Therefore, it makes it easy to turn quasars at a z of, say, 3, even if you start galaxy formation at a z of 6.

Cavaliere : Well, we must take into account the total timescale. If you start with stellar masses, then it can be quite long.

Marshall : It doesn't take much of an initial mass to get you only a few e-folding times, so it's not too hard to bring the time down to one half or even one third of a gigayear, which is what is needed.

Green : What is now the pacing timescale for formation? Is it the black hole formation, or is it enough stellar processing to produce abundances close to the high values that are seen?

Cavaliere : There are three processes in competition. The first is the formation of the black hole itself, the second is the inevitable spread of the times for the black hole formation, and the third is the inevitable spread of the collapse to the nonlinear regime.

Boyle : There appears to be very good agreement in the luminosity functions in the regions where faint quasars and Seyferts overlap. However, I noticed in your luminosity functions for the recurrent activity model, the low z function was about a factor of a hundred lower in space density. How do you reconcile this?

Cavaliere : That is the point you made the other day during your talk. There may be rather strong density evolution at the low luminosity end of the luminosity function. Unfortunately that makes it much more taxing to fine tune the parameters needed to obtain a model which will fit your data.

LUMINOSITY DEPENDENT EFFECTS IN QSO SPECTRA

J. A. BALDWIN
Astronomy Department, The Ohio State University
174 W. 18th Ave., Columbus, Ohio 43210

ABSTRACT   We summarize the results of careful
spectrophotometric observations of two complete samples
(Baldwin, Wampler and Gaskell 1988), which give further
information about luminosity-dependent ionization effects.

INTRODUCTION

Previous work (Wu, Boggess and Gull 1983; Mushotzky and
Ferland 1984; Kinney et al. 1985; Kinney et al. 1986) has
shown that for very inhomogeneous samples of QSOs and Seyfert
galaxies in addition to the well-known correlation between
emission-line equivalent widths and the continuum luminosity,
there are luminosity-dependent effects governing emission line
intensity ratios. It was suggested that these are due to a
systematic _decrease_ in the ionization parameter with
increasing luminosity, coupled with a systematic dependence of
the covering factor on luminosity. This was deduced from the
finding that both the Lα and CIII] λ1909 emission lines get
stronger relative to CIV as luminosity increases.
    However, it is dangerous to try to investigate the
systematic range of any property of a sample of objects if the
sample is selected partly on the basis of that property in
some unsystematic and unquantified way. Unfortunately, the
great majority of studies of QSO emission-line properties are
based on observations of samples of objects where the
detection of strong emission lines on low signal:noise survey
spectra or objective prism plates was an extremely important
selection criterion. These samples may or may not be biased in
a damaging way. It depends on whether or not the (generally
unknown) lower limit on equivalent width or other emission-
line parameters affecting the survey procedure actually
excluded objects in a systematic fashion. This can only be
checked by knowing what was missed.

COMPLETE SAMPLES

The underlying aim of the investigation described here is to investigate in more detail the luminosity dependence of the ionization level and velocity structure in QSO emission-line regions, using well-defined samples in which the results of interest are unlikely to be influenced by selesction effects. What is needed are samples in which the selection criteria depend as little as possible on emission line strengths. Suitable radio samples are in fact fairly easy to define. Wampler et al. (1984), in their investigation of the correlation between emission line equivalent width and continuum luminosity, used a subset of the PKS±4° quasar sample in which *all* objects lying in the small (6 arcsec) radio error boxes and brighter than photographic magnitude 19 were observed spectroscopically. This subset was further defined by eliminating certain right ascension zones and restricting the study to objects with high frequency radio spectra flatter than $\alpha = 0.5$. The resulting sample included 45 QSOs with broad emission lines, 5 BL Lac objects, and one mysterious object with one and possibly a second unidentified narrow emission line. The flat-spectrum radio sample also included 7 sources for which no optically detected objects other than stars lay within the radio error boxes, and 9 sources which were in positional coincidence with optical objects below our magnitude limit. We are confident that we have found and observed all of the objects identified with the radio sources down to the optical magnitude limit except for the possible effect of variability taking objects above or below the optical plate limit during the 25 year gap between taking the discovery plates (the Palomar Sky Survey) and our spectroscopic measurements. Despite the latter caveat, this is by far the best defined QSO sample we know of for which reasonably good optical data are available, and is in fact well suited to looking for general trends in the emission line properties. The only trouble is that it is not representative of all radio-loud QSOs, but only of ones with flat radio spectra.

It was reasonably straight forward to define and observe the PKS±4° QSO sample because a very large fraction of radio sources are in fact QSOs. Selecting radio-*quiet* QSOs independent of their emission line properties is far more difficult because the vast majority of radio quiet objects are not QSOs. Here we avoid objective prism surveys, because to date they have had a strong and usually ill-defined dependence on recognizing emission lines in QSO spectra. The optically selected sample which is defined least through the presence of strong emission lines, and one for which a complete set of spectroscopic data of reasonably good quality exists, is the Bright Quasar Survey (BQS) of Schmidt and Green (1983). These

QSOs were discovered on the basis of showing an ultraviolet excess (UVX) on Schmidt plates covering a large area of sky but extending down only to a very bright limiting magnitude ($m_{pg} \leq 16$). Besides the obvious bias against objects with red continuum spectra, this technique also selects against various combinations of emission-line strengths in a redshift dependent manner as the different emission line pass through one or the other of the filter passbands used in the survey (see Wampler and Ponz 1985). However, it is not clear how a more complete sample could be obtained unless one observed basically every object in some patch of sky. An additional strong point in favor of the BQS is that Wampler and Ponz (1985) have already obtained reasonably high quality spectra of essentially all of its QSOs as part of another study. We have therefore used it as our optically selected sample.

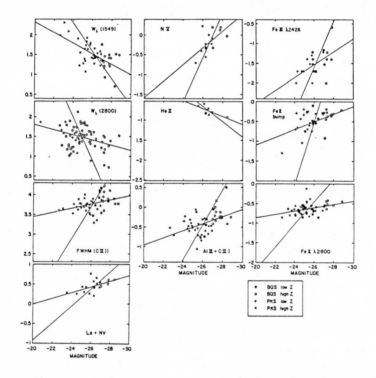

Fig. 1   Variables showing correlation (>2σ significance) with continuum luminosity. Wλ refers to equivalent width, FWHM to full width at half maximum intensity. Other variables are intensity ratios of designated emission lines relative to CIV. From Baldwin, Wampler and Gaskell (1988).

RESULTS

Baldwin, Wampler and Gaskell (1988) have combined these two samples, which are the best-defined and virtually only completely observed samples selected on, respectively, radio and optical criteria, to study systematic trends in emission line properties. We have probably confirmed the earlier results in our much more systematically chosen complete samples, as can be seen from Figure 1 in which intensity ratios and other emission-line parameters are plotted against continuum luminosity. There is very large (real) scatter in the correlations, to the extent that it is hard to believe that some of them are significant except that standard statistical tests say this is so. We assume that this extra scatter is because the continuum luminosity is not the only underlying variable. However, there is no doubt that not only the equivalent widths but also the intensity ratios of all of the strong uv emission lines (Lα, CIV, CIII] and MgII) depend on luminosity. Additional statistical tests show that they do not depend on redshift. The general behavior continues to be very consistent with the idea that the ionization parameter decreases with increasing luminosity. Thus, there definitely seems to be an interlinking of the continuum luminosity, covering factor and ionization level which shows a steady progression over a luminosity range of 5 orders of magnitude. An acceptable QSO model should predict this effect.

FURTHER WORK

A basic problem in tying together these two samples immediately becomes obvious as soon as we look at a plot of absolute continuum magnitude against redshift, as in Figure 2. There is virtually no overlap in luminosity between the two samples at any given redshift. This is because the BQS sample picks off the very most luminous QSOs by searching over a wide area of sky, while the radio-selected sample picks up quasars in a much more heavily populated part of the luminosity function by going much deeper in a smaller area of sky. This luminosity separation makes it difficult to know if differences between the two samples are due to luminosity effects or to their different radio properties. The BQS does include some radio-loud objects, but not many and not ones chosen in the same way as the PKS$\pm4°$ QSOs.

What is needed before a proper study can be made of the dependence of emission-line properties on other variables such as luminosity or radio power is to observe some additional carefully selected QSO samples which will provide better

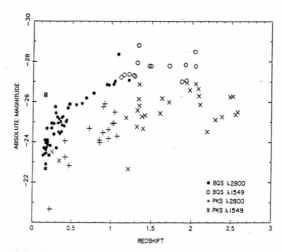

Fig. 2   Absolute continuum magnitudes (for q_0 = 1) vs.
redshift for the BQS and PKS±4° samples. Magnitudes are
either measured directly at rest wavelength 1549Å, or at
rest wavelength 2800Å and then extrapolated to 1549Å. From
Baldwin, Wampler and Gaskell (1988).

overlap in their optical and radio continuum luminosities. The
obvious additions to make to the PKS±4# and BQS samples would
be to observe small but complete samples of a) luminous, high-
redshift radio-selected QSOs; b) fainter uv-excess optically
selected QSOs; and c) QSOs with steep radio spectra. Filling
in these samples would not in fact take a great deal of
telescope time, and can partly be done from results already in
the literature, but would be a tremendous step towards
removing the possibility that selection effects control our
picture of systematic trends in QSO emission-line properties.

REFERENCES

Baldwin, J.A., Wampler, E.J. and Gaskell, C.M. 1988, Ap.J.
    (submitted).
Kinney, A.L., Huggins, P.J., Bregman, J.N. and Glassgold,
    A.E., 1985, Ap.J. 291, 128.
Kinney, A.L., Huggins, P.J., Glassgold, A.E. and Bregman,
    J.N., 1987, Ap.J. 314, 145.
Mushotzky, R. and Ferland, G.A., 1984, Ap.J. 278, 558.
Schmidt, M. and Green, R.F., 1983, Ap.J. 269, 352.
Wampler, E.J., Gaskell, C.M., Burke, W.L. and Baldwin, J.A.,
    1984, Ap.J. 276, 403.
Wampler, E.J. and Ponz, D., 1985, Ap.J. 298, 448.
Wu, C., Boggess, A. and Gull, T.R. 1983, Ap.J. 266, 28.

A JOINT DEEP X-RAY AND UV-EXCESS SURVEY IN PAVO: QUASARS AT Z ~ 1 CONTRIBUTING TO THE EXTRAGALACTIC X-RAY FLUX

R. E. GRIFFITHS
Space Telescope Science Institute

I. R. TUOHY and R. J. V. BRISSENDEN
Mt. Stromlo and Siding Spring Observatories
Australian National University

M. WARD
University of Washington

S. S. MURRAY and R. BURG
Center for Astrophysics

ABSTRACT   The nature of discrete x-ray sources contributing to the all-sky x-ray background has been investigated, by performing multi-object spectroscopy of the counterparts to deep survey x-ray sources. X-ray selected quasars, with a surface density of ~ 32 per square degree, account directly for ~ 30% of the all-sky x-ray background in the energy range of the Einstein Observatory. Quasars which were not detected in x-rays may account for a further ~ 10% of the x-ray background. No examples have yet been found of any class of sources ,other than quasars, which might make a comparable or substantial contribution to the extragalactic x-ray background.

INTRODUCTION

The only direct method of finding out the nature of the discrete sources contributing to the all-sky x-ray background is to resolve as many sources as possible in an x-ray deep survey, and to obtain optical spectroscopy of the corresponding optical counterparts.

The early results of the Einstein Observatory deep surveys in Draco and Eridanus were reported by Giacconi *et al.* (1979), who established that discrete sources contribute about a quarter of the extragalactic x-ray background in the energy range 1-3 kev, where such sources were detected at least five standard deviations above background. Giacconi *et al.* assumed that the spectrum of the x-ray background in the 1 to 3 kev range was a smooth extrapolation of the 3-40 kev background spectrum measured by Marshall *et al.* (1980) with HEAO A2. These discrete source detections were made partially with the Imaging Proportional Counter on the Einstein Observatory, with resulting x-ray astrometric errors of 60 arc seconds. The large 'error circles' and insufficient time for full optical spectroscopy of the possible candidates meant that the nature of most of the discrete sources remained unknown. Giacconi *et al.* identified only four of their 43 sources with quasars. In contrast, the Pavo deep survey was performed predominantly with the micro-channel plate High Resolution Imager (HRI), with source positional errors of 10 arc seconds or less. The initial optical spectroscopy of sources in the 0.44 square degree Pavo deep survey resulted in identifications with four quasars out of seventeen sources, of which only one redshift was definite, with a further seven objects having the broad-band colours of quasars but without the confirming spectroscopy because of the faintness of the optical counterparts (Griffiths *et al.* 1983, Paper I).

Large numbers of quasars have been identified as x-ray emitters as the result of Einstein Observatory and other satellite experiments which have examined opticaally and radio-selected samples. In addition, large numbers of x-ray selected quasars have been found in the Medium Sensitivity Survey performed with the Einstein Observatory (Gioia *et al.* 1987). Nevertheless, the sky surface densities of the objects found in these surveys are such that they do not contribute more than a few percent of the extragalactic x-ray background.

The purpose of this paper is to present a progress report on new optical spectroscopy of sources reported in Paper I, together with results on some of the objects found by re-analysis of the x-ray data. The Pavo field was also studied by a variety of methods in order to search for optically selected qso's or other emission line objects which were not detected in the deep x-ray survey. These observations have been used to estimate the contributions to the all-sky background of those quasars and emission-line objects which were below the x-ray detection threshold.

II. X-RAY OBSERVATIONS

The x-ray observations of the Pavo deep-survey field have been described in Paper I, in which seventeen x-ray sources were reported, of which all but one were detected with the HRI, and were therefore positioned to accuracies of 10 arc seconds or better. Only one source was found to be extended (source #5), which was associated with an elliptical galaxy at a redshift of 0.13, probably in a small group of galaxies. The other sixteen sources showed no evidence for extension, and were therefore considered to be likely to have compact optical counterparts. Further analysis of the x-ray data has confirmed the existence of all of the original sources, and added a further eight detections with the HRI, together with a further seven detections with the IPC. The latter were largely at the edge of the IPC field where there was no HRI coverage. Identification of these new sources is the subject of continuing optical spectroscopy. The purpose of the present paper is to report further on the identifications of the original seventeen sources, together with a few of the new sources for which optical spectroscopy has already been carried out.

The significance of source detection in the HRI was lowered for the purpose of finding fainter sources than had hitherto been found. The detection significance is four standard deviations above background in some cases, so that flux determinations are not well known. Although some sources have fluxes of $\sim 7 \times 10^{-15}$ ergs cm^{-2} s^{-1} ($\sim 1 - 3$ kev) the sample is not complete at this level. No attempt will be made here to define a complete sample for x-ray log(N) versus log(S) purposes (q.v. Giacconi et $al.$ 1979) but rather to find and identify optical counterparts to the faintest sources detected in the Deep Survey. If there is a class of discrete objects other than quasars contributing significantly to the all-sky background, then examples of such a class might be found amongst the faintest Deep Survey sources.

III. THE X-RAY SOURCE COUNTERPARTS— OPTICAL SPECTROSCOPY

Optical candidates for the x-ray selected sources were observed at the 3.9 metre Anglo-Australian Telescope on September 6 and 7 1986. The fibre-optical coupled aperture plate, FOCAP (Ellis et $al.$ 1984; Gray and Sharples, 1985) was used to feed the RGO spectrograph (Robinson, 1985) and Image Photon Counting System (IPCS, see Lucey and Taylor, 1983) for the 3300 to 7300 Angstrom range. The output from another set of FOCAP fibers was fed to the Faint Object Red Spectrograph (FORS) for the 5500 to 9800 Angstrom range.

Fourteen of the x-ray selected objects were found to be quasars, of which only four were previously known, and for which the redshift distribution is shown in Figure 1, compared with the redshift distribution of the x-ray selected quasars from the Einstein Observatory Medium Sensitivity Survey (MSS), taken from Gioia et $al.$ (1987). The average redshift for the MSS objects is 0.4, compared with an average redshift of 0.9 for the Pavo Deep Survey quasars.

The present work shows that the deep survey selects more distant

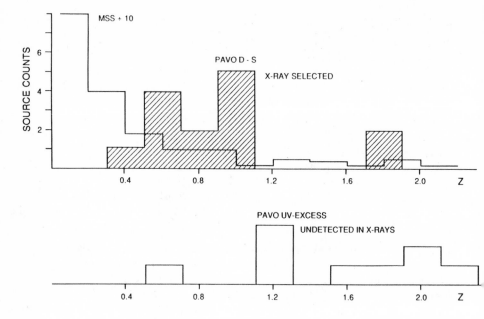

Figure 1. Redshift distribution of x-ray selected quasars in the Pavo deep survey, compared with that for the Medium-Sensitivity Survey (Gioia *et al.* 1987). The redshift distribution is also shown for the UV-excess quasars which were not detected in the deep x-ray survey.

objects, rather than closer objects at the faint end of the x-ray luminosity function for Seyferts or quasars. This is illustrated in figure 2, which compares the luminosity of the Pavo x-ray selected quasars with those in the MSS. The Pavo quasar luminosities fall within the envelope of the MSS qso's. At an average redshift of 0.9, the observed energy range of 0.5 to 3 kev corresponds to an emitted x-ray energy range of 1 to 6 kev, where x-ray absorption within the source is less important (c.f Elvis and Lawrence, 1985).

The nominal dividing line between 'Seyfert galaxies' and quasars has been drawn at $M_b = -23$ by Schmidt and Green (1987). The deep-survey x-ray selected objects fall largely on the quasar side of such a division. This is in contrast to the predictions of Schmidt and Green (1986) and those of Khembavi and Fabian (1982), who predicted that the majority of the x-ray background is made up of low luminosity active galactic nuclei (LLAGN). In particular Schmidt and Green predicted that the contribution of LLAGN to the x-ray background is more than twice that of quasars, in direct contradiction with the results reported here.

The x-ray selected quasars occupy a different part of the $L_{opt} - z$ diagram from objects found in grism surveys or objects in the bright quasar survey of Schmidt and Green, for example. In particular, the x-ray deep survey quasars have typical values of log (l_{opt}) between 29.5 and 30.5, at a

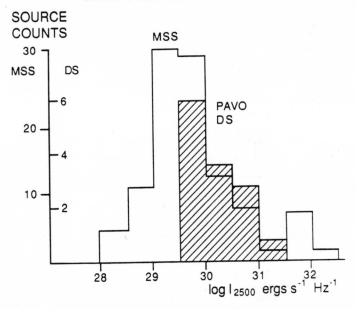

Figure 2. Luminosities of the x-ray selected quasars in the Pavo deep survey compared with the Medium Sensitivity Survey objects (Gioia *et al.* 1987).

median redshift of 0.84. For the same log (l_{opt}), the bright quasar surveys find objects at much lower redshifts, typically 0.1. At similar redshift to the x-ray selected sample, optically selected quasars typically have greater log (l_{opt}), between 30 and 32 (e.g. Avni and Tananbaum,1986).

The x-ray selected quasars had a mean $\langle \alpha_{ox} \rangle$ of 1.4, with a standard deviation of 0.13. α_{ox} shows a dependence on L_{opt} (fig. 3), as found for the optically selected samples (Avni and Tananbaum, 1986). A best fit regression analysis for this dependence shows that

$$\alpha_{ox} = 1.38 + 0.15(l_{opt} - 30.0)$$

The data suggest a steeper dependence of α_{ox} on l_{opt} than that found for optically selected objects (Avni and Tananbaum, 1986, Kriss and Canizares, 1985), but the statistical significance of this difference is small, and is the subject of continuing work. The residuals to this best fit are consistent with a symmetric Gaussian distribution with $\sigma = 0.1$. This σ on the α_{ox} residuals is significantly smaller than the value of 0.2 typically found for optically selected quasars (Avni and Tananbaum, 1986), and explains the discrepancy between x-ray source number counts and the counts predicted from optically selected samples (this possibility was discussed by Franceschini *et al.* 1987). There is marginal evidence that the sigma (α_{ox} residuals) found here is less than that for the Medium Survey quasars, viz. 0.14. The Pavo object luminosities are similar, but may be slightly higher to those of the MSS (i.e. $\langle M_v \rangle = -25.0$, c.f. -23.7 for the MSS).

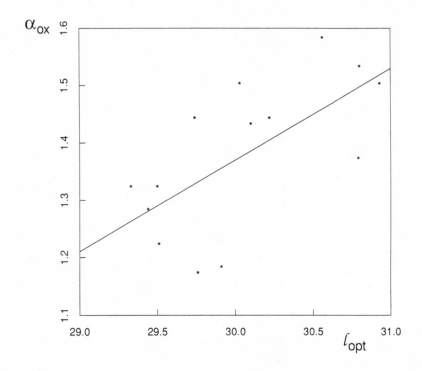

Figure 3. α_{ox} plotted against lopt for the x-ray selected quasars in the Pavo field. The line is the best fit by regression.

Of the 14 x-ray selected quasars, at least eight have (U-J) < −0.5, while at least four are not 'uv excess' objects under this criterion. A further six uv-excess quasars were not detected in x-rays, along with three which were found by prism or blue-excess selection. The incompleteness of the UV-excess method of quasar selection was therefore found to be about 40% for this sample, exceeding even the 30% estimate of Boyle *et al.* (1987) on the incompleteness of uv-excess surveys for quasars in the redshift range 0.5 to 0.9, using the same colour criterion. This result would also imply that x-ray selection is capable of finding most of the remaining quasars in a magnitude-limited survey, given sufficient x-ray sensitivity.

IV. SELECTION OF UV-EXCESS QUASARS

Plates of the Pavo field were taken at the United Kingdom Schmidt Telescope (UKST) in 1983, in the U, J, R and I passbands. These plates were scanned and digitised using the SERC Automatic Plate Measuring facility (APM) at the Institute for Astronomy in Cambridge (Kibblewhite *et al.* 1983). Apparent magnitudes for all objects with stellar appearance were calculated by standard techniques (Bunclark and Irwin 1983).

UV-excess objects were selected based on the criterion U-J < −0.5. Of the objects selected in this way, it was found that 14 of the total of 35 had already been selected by virtue of their x-ray emission, but there was no overlap between uv-excess selection and selection by prism or J-R colours.

V. SPECTROSCOPY OF UV-EXCESS QUASARS

The uv-excess quasars which were undetected in the deep x-ray survey were observed spectroscopically at the AAT in the same manner to that described above for the x-ray selected objects.

The redshifts of the uv-excess quasars undetected in x-rays (UVNX objects) are compared in figure 1 with the x-ray selected objects. The redshift distribution of the UVNX objects is marginally different from that for the x-ray selected objects, with an median z of 0.84 for the x-ray selected quasars and a median z of 1.2 for the UVNX quasars. These UVNX objects have, on average, the same Jmag as the x-ray selected sample $\langle J \rangle = 20.2$, but a higher average optical luminosity $\langle M_v \rangle = -26.1$ compared with the $\langle M_v \rangle = -25$ of the x-ray selected quasars. The dependence of L_x on L_{opt} for optically selected qso's, viz. $L_x \sim (L_{opt})^{0.84}$ (Avni and Tananbaum, 1986) is in the sense that qso's with higher optical luminosity and greater redshifts have relatively lower x-ray luminosities.

VI. CONTRIBUTION OF OBSERVED X-RAY SELECTED QUASARS TO THE X-RAY BACKGROUND

The HRI had no energy resolution over its operating range of 0.1 to 4.5 kev, which was defined at the low energy end by the aluminum filter, and at high energy by the response of microchannel plates to x-rays, folded with the telescope mirror response function. An effective energy can be defined by the known instrument response function and the spectrum of incoming radiation. In order to convert source counts from the HRI into a fraction of the all-sky background, the assumption of spectral shape can reasonably be made in at least two ways:

1) from measurements of the specral shape of the background made by other x-ray satellite experiments. Whereas the spectrum of the all-sky x-ray background has been relatively well determined in the energy range 3 to 40 kev (Marshall et al. 1980), the HEAO A-2 experiment indicated a turn-up in the spectrum below 3 kev, in agreement with the specrum determined by the Low Energy Detectors on HEAO-A2 (Garmire and Nousek 1980). Extrapolation of the Marshall et al. spectrum is therefore inappropriate, and an energy power law index of ∼ 1 is possibly indicated; or

2) to assume an average source spectrum which is the same as that of the qso's found in the MSS, or, better still, the average of qso's detected with the IPC in the Deep Surveys. This average spectrum for the MSS qso's has a hardness ratio (ratio of counts in the ~ 0.5 to 1.5 kev channel to counts in the ~ 1.5 to 4.5 kev channel) of 0.9, consistent with a photon power law of index ~ 2.0 (Maccacaio $et\ al.$ 1987). The Pavo qso's, with an average redshift of 0.9, are observed in the emitted energy range from ~ 0.2 to ~ 9 kev ,so that the low energy excess found by Wilkes and Elvis (1987) in the lowest IPC energy channel ($<$ 0.28 kev) is generally red-shifted into the lowest part of the detected energy range of the HRI. This may account, in fact, for the success of the HRI in detecting a large number of quasars in the deep survey fields.

Flux determinations for the individual x-ray sources reported here are generally subject to large errors. The individual sources have therefore been summmed to find their total flux, for comparison with the total predicted flux from the HEAO-A2 LED extragalactic spectrum in the 0.1 to 4.5 kev range. For an energy spectral index of 1.0, the observed flux corresponds to ~ 30% of the extragalactic x-ray background.

VII. CONTRIBUTION OF QUASARS TO THE X-RAY BACKGROUND

The UVNX quasars probably make a further contribution to the x-ray background, in addition to that of the x-ray selected objects. The overall distribution and limits on aox have been used to estimate that the UVNX quasars may contribute a further ~ 10% of the x-ray background, on the same assumptions as above.

VIII. SUMMARY AND CONCLUSIONS

The discrete sources detected in the Einstein Observatory's deep survey in Pavo are predominantly quasars of redshift ~ 1, with an average absolute visual magnitude M_v ~ −25 (L_{opt} ~ 10^{30} ergs s^{-1} at 2500Å). There is a distinct absence of low-luminosity AGN amongst the discrete sources.

It has been shown by Leiter and Boldt (1983), and Giacconi and Zamorani (1987) that the remaining contribution to the all-sky background must originate in discrete sources with Comptonized spectra. No evidence has yet been found for any unusual features in the optical spectra of the deep survey objects. The spectra of the x-ray selected quasars discussed here are likely to have power law indices of about 1, by analogy with the Medium-survey objects discussed by Maccacaro $et\ al.$ (1987). This may explain a turn-up in the background spectrum below 3 kev, but it is significantly different from the overall x-ray background spectrum above 3 kev. The present work does not directly address the 3-40 kev background, of which the origin is largely unknown.

REFERENCES

Avni, Y. and Tananbaum, H., 1986, *Ap. J.*, **305**, 83.

Boyle, B. J., Fong, R., Shanks, T. and Peterson, B. A., 1987, *M.N.R.A.S.*, **227**, 717.

Bunclark, P. S., and Irwin, M. J., 1983, in Proc. Symp. "Statistical Methods in Astronomy", ESA SP-201, p. 195.

Ellis, R. S., Gray, P. M., Carter, D., and Godwin, J., 1984, *M.N.R.A.S.*, **206**, 285.

Elvis, M. and Lawrence, A., 1986, in "Astrophysics of Active Galaxies and Quasars", ed. J. Miller, p. 289.

Franceschini, A., Gioia, I. M., and Maccacaro, T., 1987, *Ap. J.*, **301**, 124.

Garmire, G. and Nousek, J., 1980, private communication.

Giacconi, R., *et al.* 1979, *Ap. J.*, **234**, L1.

Giacconi, R. and Zamorani, G., 1987, *Ap. J.*, **313**, 20.

Gioia, I. M., Maccacaro, T., Schild, R. E., Wolter, A., Stocke, J. T., Morris, S. L., and Danziger, I. J., 1987, in Proc. IAU Symposium 121 on "Observational Evidence of Activity in Galaxies", eds. Khachikian, E. Ye., Fricke, K. J., and Melnick, J., p. 329.

Gray, P. and Sharples, R., 1985, AAO Fiber System Users Guide and Technical Manual.

Griffiths, R. E., *et al.* 1983, *Ap. J.*, **269**, 375.

Khembavi, A. K. and Fabian, A. C., 1982, *M.N.R.A.S.*, **198**, 921.

Kibblewhite, E. J., Bridgeland, M. T., Bunclark, P. S., and Irwin, M. J., 1983, in Proc. Conf. on "Astronomical Microdensitometry", NASA, CP-2317, ed. D. A. Klinglesmith, p. 277.

Kriss, G. A. and Canizares, C. R., 1985, *Ap. J.*, **297**, 177.

Leiter, D. and Boldt, E., 1982, *Ap. J.*, **260**, 1.

Lucey, J. and Taylor, K., 1983, IPCS User's Manual, AAO UM-10.

Maccacaro, T., Gioia, I. M., Wolter, A., Zamorani, G., and Stocke, J. T., 1988, *Ap. J.*, in press.

Marshall, F. E. *et al.* 1980, *Ap. J.*, **235**, 4.

Robinson, R. D., 1985, "The RGO Spectrograph", AAO UM-2.

Schmidt, M. and Green, R. F., 1986, *Ap. J.*, **305**, 68.

Wilkes, B. J., and Elvis, M., 1987, *Ap. J.*, **323**, 243.

ACKNOWLEDGEMENTS

We would like to thank the Australian Time Assignment Committee for awarding us time at the Anglo-Australian Telescope, and also the staff at the AAT, especially Ray Sharples for observing assistance. We also thank the staff at the UK Schmidt, and Paul Hewett for help with the APM scans.

CURRENT QSO SURVEYS: A COMPENDIUM

CRAIG B. FOLTZ
Multiple Mirror Telescope Observatory, University of
Arizona, Tucson, Arizona 85721

PATRICK S. OSMER
National Optical Astronomy Observatories, P.O. Box 26732,
Tucson, Arizona 85726-6732

ABSTRACT We present a summary of ongoing and recently-
completed QSO surveys compiled from information provided
by participants in the NOAO Workshop on Optical Surveys
for Quasars. The detail of the material varies from
survey to survey; in some cases explicit information on
selection techniques, magnitude calibration, etc. is
provided, in others, only the barest description. This
reflects the detail in the material provided to us by the
principal investigators. In most cases the information is
presented with only minimal editing, although it was necessary to condense some of the longer contributions.

I. OPTICAL SURVEYS FOR QSOS

A. Principal Investigators: D. Koo and R. Kron in collaboration
with: S. Majewski, J. Munn, M. Bershady, J. Shields, Yin Zhan,
D. Trevese, and G. Pittella.
1. Multi-band color search in the photographic U J F N system:

U: IIIaJ + UG5	λ 3650	
J: IIIaJ + GG385	λ 4650	
F: 127-02 + GG495	λ 6100	
N: IV-N + RG8	λ 8000	

Selection is for stellar image structure and colors unlike
ordinary stars. Magnitude limit is J=22.5 or F=21.5. The
following fields have been surveyed with Mayall prime focus
plates (each 0.3 square degrees):

SA57	$13^h 05^m$	$+29^o$	
SA68	00 16	+16	
SA28	08 44	+45	
Hercules	17 20	+50	

Status: SA57, SA68 surveyed, but spectroscopic checks are
so far only extensive for SA57 (Koo and Kron, 1988 Feb. 1

Ap.J.). Analysis of other fields including selection by F
magnitude in J-F, F-N diagram and further spectroscopic checks
are planned.

    2. Multi-epoch variability search. Analysis is underway
for an excellent series of Mayall IIIaJ prime focus plates
dating to 1974.

B. US Survey -- Principal Investigator: P. Usher

Selection via Tonantzintla 3 color (u,b,v) images from Palomar
1.2 meter camera. Each 14-inch plate covers 44 square degrees.
Ocular selection of UV-excess objects was followed by iris-
photometric calibration of colors using known subdwarfs as
standards. Extended sources were intentionally selected
against.

Field Centers		SA	Completeness Limit (B mag.)
$02^h\ 53^m$	$+00.3^\circ$	94	18.5:
03 14	+15.2	71	17.8:
08 43	+45.8	28	18.7:
09 42	+44.6	29	18.5
11 34	+29.5	55	18.5:
13 06	+29.6	57	18.5:

C. Hamburg Quasar Survey -- Principal Investigators: D. Engels, D. Groote, H.-J. Hagen and D. Reimers

Object selection is from objective prism plates taken with the
former 80 cm Hamburg Schmidt Telescope on Calar Alto, Spain. A
1.7° prism gives a dispersion of 1390 Å/mm at Hγ. IIIa-J
plates are supplemented occasionally by IIIa-F plates. Fields
are located north of 0° declination and outside of the galactic
plane ($|b| > 20^\circ$).

    Plates are processed in Hamburg by scanning with a PDS
1010G microdensitometer, with background subtraction accomplished during scanning. Spectra are searched for emission
lines and blue continua. Bright quasar candidates (B < 17.5)
will be provided for all fields observed and complete samples
of QSOs with a magnitude limit around 18 and z~ 3 will result
in selected areas.

D. APM Survey -- Principal Investigators: S. Anderson, F. Chaffee, C. Foltz, P. Hewett, G. MacAlpine, D. Turnshek, and R. Weymann

Object selection and magnitude calibration is accomplished from
direct UK Schmidt Telescope plates. Non-stellar objects are
discarded. QSO candidates are chosen from APM scans of UK
Schmidt IIIa-J objective prism plates. Candidates are selected
for (1) emission lines, (2) UV excess, (3) continuum breaks,
and (4) non-stellar absorption features. Selection criteria
are sensitive to QSOs in the redshift range 0.3 ~ z ~ 3.1.

Each plate covers 33 square degrees. Objects are selected in the magnitude range 16.0 ~ m_J ~ 18.5. The aim is to discover about 1000 bright QSOs, requiring coverage of about 600 square degrees. All candidates are observed at about 5-10 Å spectral resolution with a S/N greater than about 12. Spectroscopy and magnitude calibration are carried out on the MMT, the DuPont 2.5-m telescope, and the McGraw Hill telescopes. Completeness will eventually be calculated as a function of magnitude and QSO spectral characteristics via numerical simulation.

The first installment of the survey presenting 192 QSOs in a 102 square degree field has recently been published (Foltz et al. 1987, A.J. **93**, 1423). Status of other plates is given below:

Plate Designation	Plate Center R.A.		Dec.		No. of QSOs	Status
UJ 6543 P	00^h	00^m	$+00^o$	00'	34	ID Spectroscopy complete.
UJ 11474 P	00	20	+00	00	35	Initial object selection complete. ID Spectroscopy underway.
SGP	00	53	-28	00	61	ID spectroscopy of high-probability objects complete over 36 deg.^{2}.
UJ 11757 P	11	40	+00	00	0	Object selection underway.
UJ 10732 P	12	16	+11	12	33	Virgo field. Spectra and magnitudes published (A.J. **93**, 1423, 1987).
UJ 10742 P	12	16	+15	27	42	As UJ 10732 P.
UJ 10738 P	12	34	+11	12	53	As UJ 10732 P.
UJ 10749 P	12	34	+15	27	64	As UJ 10732 P.
UJ 5853 P	12	40	+00	00	29	ID spectroscopy complete.
F 287	21	28	-45	00	43	ID spectroscopy of high-probability objects complete over 28 deg.^{2}.
MTF	22	03	-18	55	20	ID spectroscopy of high-probability objects complete over 18 deg.^{2}.
891	22	40	+00	00	30	ID spectroscopy complete.

E. The Case Low-Dispersion Northern Sky Survey-- Principal Investigators: P. Pesch and N. Sanduleak

Observational Parameters:

    Telescope: Burrell Schmidt Telescope

    Field: 5.1^o x 5.1^o

    Prism: 1.8^o apex angle, Schott UK50, yielding a dispersion of 1300 Å/mm at Hγ.

    Wavelength Coverage: λλ3300-5350

    Plates: IIIa-J, 75 minute exposure.

    Limiting Magnitude: B~18.0 (stellar-like continuum), B~19.0 (strong emission-line objects).

Areal Coverage: Ultimately expect to cover approximately 8000 square degrees lying above +30° north galactic latitude and north of the celestial equator. Completion of the plate acquisition as of November 1987 was: 221 (97%) of 227 fields covered in the declination range from +90° to +29° and 31 (18%) of 174 fields in the range from +29° to 0°.

Plates are scanned independently by both observers using 15X magnification under a binocular microscope. Spectral categories selected are described in the first two installments of the survey (see below). Finding lists provide approximate magnitudes, coordinates good to +3 arcsec in declination and +6 arcsec in right ascension, and finding charts where necessary.

On average 2-3 QSOs per 25 square degree field are detected on the basis of emission-line features. A majority prove to be QSOs with $1.7 < z < 3.3$ in which the strongest spectral feature is generally Ly α. Among the objects listed with stellar, blue featureless continua, about one-third have been found by higher-resolution follow-up spectroscopy to be QSOs (generally with $z < 1.7$) in which the emission features were too weak ($W_\lambda < 50$ A) to be detected on the objective prism spectra. In addition, some very blue low-z QSOs have been found to be included in the category of suspected blue stars where they were placed as a result of their very blue continua ($U-B < -1.0$). A few high-z QSOs have been found among the emission-line galaxy candidates where a feature interpreted as [O III] λλ4959, 5007 turned out to be Ly α.

When all follow-up observations are made, it appears that the surface density is about 0.5 QSOs per square degree to a limiting magnitude of B~18.0.

Survey Results:
Pesch, P. and Sanduleak, N. 1983, Ap.J. Suppl. 51, 171.
Sanduleak, N. and Pesch, P. 1984, Ap.J. Suppl. 55, 517.
Pesch, P. and Sanduleak, N. 1986, Ap.J. Suppl. 60, 543.
Pesch, P. and Sanduleak, N. 1987, Ap.J. Suppl. in press.
Follow-up Studies:
Zotov, N. 1985, Ap.J. 295, 94.
Wagner et al. 1988, Ap.J., submitted.
Tifft et al. 1986, Ap.J. 310, 75.

F. Principal Investigators: E. Gosset, J. Surdej, and J.P. Swings

Object selection from dual exposure (U and B) 5 X 5 degree Schmidt plates from Palomar and ESO. Exposures are balanced for similar brightness for U-B around -0.4. Plates are scanned twice by two different investigators. The limiting magnitude of the survey is about 19.75 for the Palomar plates and somewhat fainter for those from ESO. Grism plates are also obtained for selected areas in the fields.

Plates are centered on NGC 450, NGC 520 + surroundings,

and at 3^h -40°. The maximum redshift of QSOs in the survey is about 2.2.

G. The Montreal-Cambridge Survey of Southern Subluminous Blue Stars -- Principal Investigators: S. Demers, M.J. Irwin, G. Fontaine, F. Wesemael, and R. Lamontagne

Objects are selected from double exposure (U and B) plates. The sky area covered will be approximately 8600 square degrees south of -5° declination in the range -30° < b < -90°. The limiting magnitude is B = 16.5. This survey is the southern counterpart of the PG Survey, reaching 0.5 mag. fainter. Plates are obtained at CTIO with the Curtis Schmidt Telescope and are scanned by the APM.

The acquisition of the photographic material is more than 50% complete. Follow-up spectroscopy is in its early stages.

H. Principal Investigator: L. Miller

Objects are selected from broad-band UBVRI plates obtained with the UK Schmidt Telescope. The plates are measured on COSMOS. Limiting magnitude is R = 18.5. The region surveyed is a 325 square degree contiguous zone in the range 12^h 30^m to 14^h 50^m, -7.5^o to +2.5^o. Galactic latitude is about 55 degrees.

I. The CFHT/MMT Blue Grens Survey -- Principal Investigators: D. Crampton, A.P. Cowley, and F.D.A. Hartwick

Earlier work using objects selected from CFHT blue grens plates and identification spectroscopy with the MMT is being continued in two new fields. Ocular searches of the plates can detect QSOs with z < 3.4 and m < 21.5, although the samples are believed to be complete to m < 20.5.

The 1950 coordinates of the field centers are:
Field A 13^h 39^m +27° 18'   6.5 square degrees
Field B   16   35   +39   30   8.5 square degrees

J. The Calan-Tololo Survey -- Principal Investigator: J. Maza

The survey covers 3400 square degrees in the southern sky, using unwidened objective-prism plates taken at CTIO with the Curtis Schmidt Telescope and the thin UV prism. Plates are taken on unfiltered IIIa-J plates and cover the range from 3300 to 5300 Å. Quasars are detected in the redshift range 1.8 < z < 3.3 and 15 < m_{pg} < 18.5.

The approximately 200 plates in 160 fields are still being searched. Some 300 candidates have been discovered, 80 of which have been confirmed spectroscopically. The survey should yield more than 1000 high-probability candidates.

K. Automated Survey of CFHT Grens Plates -- Principal Investigator: E. Borra

Object selection is by automated computer analysis of slitless spectra. The spectra cover the range from 3500 to 5300 Å at a dispersion of about 1000 Å/mm (~70 Å resolution). Limiting magnitude is B ~ 20.5 to 21.5 (complete to B ~20 to 21). The total area surveyed is 44 square degree (complete over 31 square degrees). Coordinates of the survey fields follow:

Field	R.A.			Dec.			l^{II}	b^{II}
A1	11	24	26	+67	30	07	134.86	47.82
A2	11	30	51	+70	02	53	132.42	45.81
A3	11	36	06	+69	59	54	131.86	46.04
B1	13	09	26	-13	36	03	310.55	48.79
B2	13	13	29	-15	40	37	311.59	46.60
B3	13	07	45	-16	16	22	309.51	46.18
C1	16	05	49	22	22	15	37.98	45.35
C1.5	16	06	17	22	39	22	38.41	45.33
C2	16	11	28	22	07	24	38.18	44.03
C3	16	10	32	23	58	11	40.58	44.76
D1	21	24	03	-11	53	42	40.36	-39.70
D1.5	21	25	08	-11	51	36	40.51	-39.94
D2	21	24	59	-13	22	37	38.69	-40.55
D3	21	22	26	-12	43	52	39.11	-39.71
D4	21	25	00	-11	00	00	41.49	-39.53
E1	23	54	46	19	00	37	105.68	-41.85
E2	23	55	00	18	00	00	105.38	-42.84
E3	23	55	00	17	00	00	104.98	-43.88
F1	03	59	47	-13	58	38	206.23	-43.72
F2	03	59	52	-14	58	49	207.52	-44.11
F3	03	59	50	-15	58	48	208.80	-44.52
G1	08	26	46	44	30	28	176.10	35.72
G2	08	26	43	43	30	31	177.32	35.62
H1	09	26	47	20	00	27	210.63	43.43
H2	09	26	47	19	00	26	211.92	43.09
H3	09	26	48	18	00	17	213.20	42.75
H4	09	26	49	17	00	23	214.47	42.39
H5	09	26	43	16	00	33	215.70	41.99
H6	09	26	43	15	00	17	216.94	41.60
NGP1	13	07	05	30	57	02	76.75	84.63
NGP2	13	04	21	29	43	29	69.51	85.85
NGP3	13	05	43	28	15	04	48.25	86.18
SGP1	01	00	00	-18	00	00	138.8	-80.35
SGP2	01	00	00	-19	00	00	140.48	-81.31
SGP3	01	00	00	-20	00	00	142.58	-82.27
S3	17	06	31	16	35	54	37.22	29.95
S6	15	03	22	27	29	33	41.25	60.27
ASM2	04	29	47	19	00	30	178.33	-19.27
ASM3	04	04	49	12	00	47	179.97	-28.37
ASM6	02	24	48	-07	58	32	177.01	-60.21
ASM7	01	59	40	-13	59	14	177.72	-68.75
SA68	00	14	53	15	36	48	111.07	-46.29
UKST	00	50	00	-28	00	00	285.27	-89.28
2670	23	52	32	-10	35	31	81.82	-68.65

L. Principal Investigators: B. Marano, G. Zamorani, and V. Zitelli
Survey material is a set of direct and grism plates in a 0.69 square degree field centered at $3^h 13.8^m -55°25'$. The direct plates were obtained in three photometric bands (J, U, and F; essentially identical to those used by Koo and Kron -- see A., above). Non-stellar objects were intentionally selected against. The ability to discriminate between stellar and extended images at faint magnitudes limited the survey to objects brighter than J = 22.0.

Candidate selection was carried out using three techniques: (1) color selection, (2) grism selection for objects showing emission features, and (3) variability analysis of plates separated by a one year temporal baseline.

Spectroscopic follow-up has been carried out for all high-priority objects brighter than J = 20.9, yielding a sample of 23 confirmed QSOs. At fainter magnitudes some spectroscopic confirmation has been carried out. A paper describing the survey will appear in M.N.R.A.S. (1988, in press).

M. CTIO 4m Grism Survey at $12^h 10^m$, $-11°$ -- Principal Investigators: P.S. Osmer and P.C. Hewett
Sky Coverage: A strip of sky extending $12°$ in right ascension and $0.5°$ in declination. Field center is $12^h 10^m$, $-11°$. Area is 6.1 square degrees.

Plate Material: CTIO 4m grism plates. Baked IIIa-J emulsion, spectral coverage 3300-5400 Å.

Search Techniques: Plates were first searched visually, and 75 candidates were confirmed spectroscopically as quasars and emission-line galaxies. The plates were then scanned with the APM, and the confirmed quasars were used to establish criteria for the selection of additional candidates.

Calibrations: Direct CCD observations with the CTIO 1.5m telescope and direct UK Schmidt plates have been used to establish a magnitude scale for the candidates.

Properties of the Sample: Further spectroscopic observations have produced a sample of 127 quasars meeting the APM selection criteria. The quasars cover the redshift range 0.3 < z < 3.4. Quasars as faint as 21.5 mag. are detected, but the completeness limit depends on both magnitude and line strength. The sample is first being used for a study of clustering (see paper by Osmer and Hewett, this volume). It will also be useful for investigating the space density for 1.5 < z < 3.4 and for absorption-line studies.

N. APM Multi-Color Quasar Survey -- Principal Investigators: S. Warren, P. Hewett, P. Osmer and M. Irwin
Survey Material: UK Schmidt Telescope plates, two in each of U, B_J, V, R, and I bands in each of three fields: $00^h 53^m$, $-28°$

20'; 21 20, -40 00; and 20 40 -35 00, totalling 100 square
degrees. Limitations on spectroscopic follow-up may limit
final survey to a sub-area of 50 square degres. CCD and photoelectric photometry is available in the fields for most of the
colors.

Search Techniques: Plate material is digitized using the
APM at Cambridge. Object selection is performed in a 4 or 5
dimensional space defined by the available color combinations.
The aim is to include any objects not coincident with the well-
defined locus of stars evident in the data. Various subsets
are obtained depending on the nature of the quasars being
sought.

Principal Aims: (1) To obtain a sample of non-UVX quasars
over a large field to magnitude limits m_R = 20 and m_B = 21 to
allow direct comparison of the quasar population at redshifts 2
< z < 3 with results of UVX surveys for redshifts z < 2. (2)
To detect a large sample of very high redshift quasars (z > 3)
in order to place limits on the quasar luminosity function at
very high redshift.

Current Status: All the scanning and object selection is
largely complete. A total of about 160 spectra have been
obtained, principally in the South Galactic Pole field. Most
of these spectra relate to the very high redshift project,
where a sample of 25 quasars with z > 3 now exists in the 30
square degrees of the SGP field. An effort has been made on
acquiring a large number of quasars in the lower redshift
range. Further allocations of telescope time for spectroscopic
follow-up have been made in Summer 1988.

O. Palomar CCD Grism Survey for Large Redshift Quasars -- Principal Investigators: M. Schmidt, D.P. Schneider, and J.E. Gunn
Principal Aim: To study the evolution of the quasar luminosity
function for redshifts above 2.2, on the basis of well-defined
samples. This will form a quantitative basis for evaluation of
whether or not the "redshift cutoff" is caused by dust. If it
is not, the results will document the early formation of
quasars and galaxies.

Survey Material: Two-color (F555W and F785LP) and grism
surveys have been obtained with the 200-inch telescope in
transit mode with a 4-CCD camera in the following strips 8.5
arcmin wide in declination:
   (a) 20^h 50^m to 4^h 54^m at +1° 34' (equinox 1985.8)
   (b)   7   08 to 10   30 at +56 16 (1986.0)
   (c) 11   41 to 16   44 at +37 45 (1986.2)
   (d) 21   07 to 4   35 at +2   30 (1987.9)
The grism observations are taken under photometric conditions and calibrated. The spectroscopic resolution of the
grism spectra is between 80 and 150 Å, depending on seeing.

Search Technique: Object selection is based on detection
of line emission, though a computer algorithm involving a mini-

mum signal-to-noise ratio of detection against background and a minimum observed equivalent width. Follow-up slit spectroscopy is obtained for all emission-line candidates.

Status: Slit spectroscopy of all 180 emission-line candidates for sample (a) has been obtained. The results are discussed in our contribution elsewhere in this volume. Slit spectra for one-third of the emission-line candidates in samples (b) and (c) are in hand. Further transit surveys will be obtained until around 50 to 100 quasars with redshift > 3 have been detected.

P. Four-Shooter Multicolor Survey for Faint QSOs -- Principal Investigators: Scott Anderson and P. Schechter in collaboration with D. Koo, R. Windhorst, and S. White

Selection Technique: Objection selection is based on stellar point spread function and location in multicolor space away from the locus of normal stars. Direct CCD images are taken with the Four-Shooter on the Palomar 5 m telescope in pseudo-Johnson B, Gunn g, r, and i.

Area Surveyed:
(a) 0.2 square degrees for m < 22.5 in SA 57 ($13^h 06^m$, $+29^o$); field overlaps that surveyed by Koo and Kron et al.
(b) 0.3 square degrees for m < 22.5 in Draco ($17^h 09^m$, $+71^o$)
Planned Extensions of Survey: 0.25 square degrees for m < 24.5 in two fields:
(c) SA 57
(d) Lynx.3A ($08^h 45^m$, $+45^o$)

Q. AAT Survey -- Principal Investigators: B. Boyle, B. Fong, T. Shanks, and B. Peterson

Object selection: Objects are selected from COSMOS machine measurements of UK Schmidt U and J plates in seven high galactic latitude fields. Objects meeting the following selection criteria were included as candidates: stellar images, 16.9 < B < 20.9, U-B < -0.35.

Spectroscopy: Object identification was carried out on the Anglo-Australian Telescope using the FOCAP fiber feed to the RGO Spectrograph and IPCS detector. The field of view of 0.35 square degrees allowed 50-60 candidates to be observed per field. The spectrograph gave 145 Å/mm for 10Å resolution.

Results: 32 FOCAP fields surveyed
1400 objects observed
420 QSOs (396 with z < 2.2)
56 narrow emission-line galaxies
Remainder are Galactic stars (431 white dwarfs)

Field Centers:

Center	No. of FOCAP Fields	Area	Limiting Magnitude
$0^h 53^m -28^o$	7	2.45	B < 20.9
3 44 -45	4	1.40	B < 20.9

Center			No. of FOCAP Fields	Area	Limiting Magnitude
10	40	0	2	0.70	B < 20.9
12	40	0	5	1.75	B < 20.9
13	40	0	1	0.35	B < 20.4
15	16	+2	5	1.75	B < 20.9
22	04	-20	8	2.80	B < 20.65

R. AQD Surveys -- Principal Investigator: R.G. Clowes
The following table list the UK Schmidt fields involved in AQD
surveys:

294	00 26 00	-40 00 00	ROE/ESO
295	00 52 00	-40 00 00	ROE/ESO
296	01 18 10	-40 00 00	ROE/ESO
297	01 44 00	-40 00 00	ROE/ESO
351	00 48 00	-35 00 00	ROE/ESO
352	01 12 00	-35 00 00	ROE/ESO
411	00 46 00	-30 00 00	ROE/ESO
119	05 04 00	-60 00 00	ROSAT
120	05 42 00	-60 00 00	ROSAT
416	02 41 00	-30 00 00	
899	01 20 00	+05 00 00	
927	10 40 00	+05 00 00	
N1	02 53 00	+00 20 00	
SGP	00 53 00	-28 03 00	

The ROE/ESO survey and the survey for ROSAT are described by
Clowes, Iovino, and Shaver (1987, M.N.R.A.S. 227, 921) and
Beuermann and Clowes (1987, in Observational and Analytical
Methods Relating to Large-scale Structures in the Universe,
Workshop of the Astronomisches Institut der Universitaet
Muenster, in press), respectively; the other fields are for
various special-purpose, single-plate surveys. A new, major
survey of approximately 10 fields is planned.

S. The Edinburgh-Cape Blue Object Survey -- Principal Investigators: R.S. Stobie, D.H. Morgan, D. Kilkenny, and D.
O'Donoghue
A major survey to discover blue stellar objects in the
southern hemisphere is in progress to complement the Palomar-
Green survey in the north. The aim is to obtain a complete
catalog of blue objects brighter than B ~ 17 for fields with
galactic latitude $|b| > 30°$. U and B pairs of photographic
plates taken with the UK 1.2 m Schmidt Telescope are scanned
with the COSMOS measuring machine. The blue stellar objects
are selected on the basis of U-B color yielding typically 50
candidates per field. So far 50 fields have been processed and

it is clear from fields with a photometric calibration that blue objects are being detected to fainter than B = 18. Follow-up spectroscopy and photometry is being obtained to classify the candidates. Results based on spectroscopy of ten fields give the following percentage classification: 36% hot subdwarfs, 7% DA white dwarfs, 11% A and B types or horizontal branch stars, 14% peculiar objects (including some obvious cataclysmic variables), and 32% F and G type subdwarfs.

T. The CTI Quasar Survey -- Principal Investigators: J.T. McGraw in collaboration with the CTI Group

The goal of this survey is to discover quasars in a region of the sky utilizing as many detection criteria as possible. Each criterion will be applied independently of the others in an attempt to measure selection effects, as well as to search for correlations among physical properties.

The CTI is a fixed telescope pointed in the meridian at a declination of +28 degrees. It utilizes the "time delay and integrate" (TDI) readout technique on two RCA CCDs in the focal plane. These are aligned E-W such that a color (U,B,R, or I) is observed first, followed by an integration in V. The effective integration time is one minute resulting in a nightly limiting magnitude of V=21 (S/N=3). The survey vital statistics are given below. The CTI programs underway include:

1. Selection of quasar candidates by color. Because the CTI observes time-averaged colors of objects, it is possible to select candidates either by cluster analysis or by arbitrary color-color projections of the color four-space. We shall co-add data to achieve photometric precision approaching 0.01 mag. at V=20. Selection of candidates and spectroscopic follow-up are now underway.

2. Selection of quasar candidates by variability. Variability on timescales of days to years may well be the most unbiased technique for the selection of quasar candidates. The CTI is amass a data base of V light curves with one sidereal day sampling which will be searched for stochastic variability indicating a quasar candidate. Again, data will be co-added to provide weekly and monthly mean magnitudes with correspondinly lower uncertainties.

3. Radio observations of the CTI strip. We have used the VLA to observe snapshots of the CTI strip over approximately two-thirds of its extent (excluding the galactic plane). Observations were made in C array at 20 cm to a flux limit of about 5 mJy. Point-like sources in the radio survey are compared to corresponding images in the CTI data and spectra of the quasar candidates are acquired. This portion of the CTI survey has thus far yielded 15 new radio-loud quasars in the redshift interval 0.7 < z < 3.3.

Survey Vital Statistics:
- 1.8 m, f/2.2 automated telescope.

- Two RCA CCDs - 320x512 pixels, 30 microns in size.
- Field scale of 52 arcsec/mm - each pixel subtends 1.55 arcsec.
- Declination of strip is 28 degrees.
- Width of surveyed strip is 8.26 arcmin.
- Total area surveyed is 43.7 square degrees.
- Time history recorded in V bandpass, time-averaged UBVRI colors obtained.

Detectors -
- CCDs have no cosmetic blemishes.
- "V" CCD has 40 electron readout noise, 10.78 electrons/adu.
- Nightly limiting magnitude V = 21.
- "Color" CCD has 72 electron readout noise, 9.18 electrons /ADU.

Control and Acquisition Computer System -
- 16-bit minicomputer.
- Realtime program requires minimal human response.
- Data stored on magnetic disk during acquisition.
- Data written to magnetic tape for transport to Tucson.

U. PALOMAR BRIGHT QUASAR SURVEY -- Principal Investigators: M. Schmidt and R.F. Green

Survey Material: Palomar 18-inch Schmidt telescope films, with 266 U and B double images on IIa-O emulsion. Sky coverage is 10,714 sq. deg. in absolute galactic latitudes above 30 degrees and declinations above -12 degrees. Limiting B magnitudes range from 15.5 to 16.7, with an effective limiting magnitude for the survey of 16.16.

Search Technique: The films were digitized with the scanning microdensitometer at the Image Processing Laboratory of the Jet Propulsion Laboratory. Magnitudes and colors were standardized to the Johnson system with respect to some 700 photoelectrically measured stars in the range 11th to 17th magnitude, with the instrumental system tied together by a model of star counts as a function of galactic latitude and longitude. Candidates for spectroscopic follow-up were selected to have U-B < -0.44 mag.

Results: Spectroscopic identification was made for a complete sample of 1715 ultraviolet-excess stellar objects. Of these, the Bright Quasar Survey consists of 114 objects, of which 92 are quasars with M_B < -23. The rather large measuring uncertainty leads to systematic effects with respect to the color and magnitude limits, which can be taken into accout explicitly when analyzing the statistical results of the survey.

References:
Schmidt, M. and Green, R.F. 1983, Ap.J., **269**, 352.
Green, R.F., Schmidt, M., and Liebert, J. 1986, Ap.J. Suppl., **61**, 305.

II. SURVEYS AT OTHER WAVELENGTHS

A. Parkes Flat-Spectrum Survey -- Principal Investigators: A. Savage, D. Jauncey, G. White, B. Peterson, W. Peters, J. Condon, and S. Gulkis

A sample of flat-spectrum radio sources drawn from the Parkes 2.7 GHz Survey and complete to 0.5 Jy is being investigated. The sample covers all right ascensions and declinations from $+10^o$ to -45°, but excluding the galactic plane (b < 10°).

The sample contains some 403 sources. Accurate radio positions (< 2 arcsec in either coordinate) have been measured for all of the sources. Optical identifications based on positional coincidence alone can be made to the 22.5 mag. limit of the SERC/UKST IIIa-J sky survey. CCD 'R' frames are being obtained on the AAT for those sources which show no counterpart on the IIIa-J plate material. This is a long-term project, underway for about 15 years and making use of a wide variety of telescopes, both optical and radio.

B. The PKS 4° Flat Radio Spectrum Sample -- Principal Investigators: J.A. Baldwin, E.J. Wampler, and C.M. Gaskell

This is a subset of the PKS $+4^o$ quasar sample in which all objects lying in the small (6 arcsec) radio error boxes and brighter than photographic magnitude 19 were observed spectroscopically. This subset was further defined by eliminating certain right ascension zones and restricting the study to objects with high frequency radio spectra flatter than $\alpha = 0.5$. The resulting sample included 45 QSOs with broad emission lines, 5 BL Lac objects, and one mysterious object with one and possibly a second unidentified narrow emission line. The flat-spectrum radio sample also included 7 sources for which no optically detected objects other than stars lay within the radio error boxes, and 9 sources which were in positional coincidence with optical objects below our magnitude limit. We are confident that we have found and observed all of the objects identified with the radio sources down to the optical magnitude limit except for the possible effect of variability taking objects above or below the optical plate limit during the 25 year gap between taking the discovery plates (the Palomar Sky Survey) and our spectroscopic measurements.

C. The Einstein Extended Medium Sensitivity Survey (EMSS) -- Principal Investigators: I.M. Gioia, T. Maccacaro, R. Schild, A. Wolter (CfA), J.T. Stocke (Colorado), and S. Morris (MWLCO)

This is an x-ray survey of the high (b > 20°) galactic latitude sky. The survey is conducted by searching x-ray images of the sky obtained with the Imaging Proportional Counter (IPC) on board the Einstein Observatory for serendipitous sources. These images were obtained during the period November 1978 - April 1981. The exposure time of the images ranges from 800 to

40,000 sec. This converts into a minimum detectable flux (for a four sigma detection) in the range 1×10^{-12} to 6×10^{-14} ergs cm^{-2} s^{-1} on the 0.3 - 3.5 keV band. 1435 images have been analyzed over a total of 780 square degrees yielding 836 sources. Optical identification of the sources is in progress. Spectroscopic follow-up is carried out for the candidate objects in the IPC error circle (typically 45 arcsec radius). At the time of this writing, 605 sources have been positively identified.

Active galactic nuclei (quasars and Seyfert galaxies) constitute the largest subclass of objects. 280 AGN and 24 BL Lac objects have been found during the identification process. Upon completion of the optical identification program we expect in excess of 400 x-ray selected AGNs and about 30 x-ray selected BL Lacs. B, V, and R photometry, as well as a VLA map at 5 GHz is available for all of the AGNs and BL Lacs with declinations north of -20 ^{o}

III. SURVEYS FOR GRAVITATIONALLY-LENSED QSOS

A. Principal Investigators: B. Burke, E. Turner, J. Gunn, J. Hewitt, C. Lawrence, M. Schmidt, and D. Schneider
Objects are selected by passage through the following set of filters:
  1. VLA snapshots (typically 150 sec. integrations) are obtained for 4200 sources with 6 cm fluxes ~ 50 mJy and declinations in the range from 0° to 20°.
  2. Objects with two or more unresolved (with 0.4 arcsec beam) components within 60 arcsec are identified for optical follow-up.
  3. Objects with non-colinear structure, component separations < 10 arcsec, more than two unresolved components, and galactic latitude > 20° are given higher priority for optical follow-up.
  4. Direct CCD imaging in R band to 24th magnitude is carried out under seeing conditions sufficient to resolve optical counterparts (if they exist) of the two most widely separated unresolved radio components.
  5. Objects showing two or more optical components and/or evidence for a plausible lensing object near the radio source are identified for spectroscopic follow-up.
  6. Spectra of sufficient quality to test the lensing hypothesis are obtained for the optical components.
  Extrapolation of current results suggest that 10 to 30 lens systems will be found along with a somewhat larger number of binary radio sources and other peculiar systems.

B. Principal Investigators: R. Webster, P. Hewett, and M. Irwin
Objects are selected from UKST J band plates in high galactic

latitude fields which are scanned by the APM. Images are para-
metrized and image classification is carried out and a magnitude-limited sample is obtained. Significantly elliptical
images (typically ~6000 per plate) are flagged as possible
lenses.

Further refinement of the list of possible lenses is
accomplished via analysis of their spectra on UKST objective
prism plates digitized with the APM. Selection of quasar candidates is carried out in a way such that an image with a
spectrum not consistent with a normal star is included as a
candidate. It is estimated that 50 + 5 percent of the candidates are quasars.

237,000 images with magnitudes $15 < m_B < 20$ surveyed over
130 square degrees. Of these, 2500 quasar candidates were
selected, from which a list of 73 candidate lenses was
compiled. This latter list may contain one or two new gravitational lenses.

An important aspect of the survey is the ability to quantify the probability of finding lenses for a wide range of lens
properties. This allows any models of the frequency and properties of lensed quasars to be directly related to the observations.

The survey and its initial results are described in
Webster et al. 1988, A.J. **95**, 19.

C. Principal Investigators: J. Surdej, J.P. Swings, and P.
Magain, in collaboration with U. Borgeest, T.J.-L. Couvoisier,
R. Kayser, K.I. Kellermann, H. Kuehr, and S. Refsdal
In order to better understand the effects of gravitational
lensing on the observed quasar luminosity function, on the
source counts of extragalactic objects, etc., we have initiated
a high angular resolution direct imaging survey of an optically-selected sample of highly luminous quasars (HLQs; $m_V <$
18.5, $M_{abs} < -29$, Dec. $< +20°$). The observations are being
carried out with the ESO 2.2 m telescope for CCD imaging
through wide- and/or narrow-band filters, and the VLA at 6 cm
in A configuration. More than 100 HLQs have already been
observed: several (~ 5) very good lens candidates have been
identified.